Antifungal Compounds Discovery

Antifungal Compounds Discovery

Natural and Synthetic Approaches

Marina Soković
Konstantinos Liaras

ELSEVIER

Elsevier
Radarweg 29, PO Box 211, 1000 AE Amsterdam, Netherlands
The Boulevard, Langford Lane, Kidlington, Oxford OX5 1GB, United Kingdom
50 Hampshire Street, 5th Floor, Cambridge, MA 02139, United States

Notices
Knowledge and best practice in this field are constantly changing. As new research and experience broaden our understanding, changes in research methods, professional practices, or medical treatment may become necessary.

Practitioners and researchers must always rely on their own experience and knowledge in evaluating and using any information, methods, compounds, or experiments described herein. In using such information or methods they should be mindful of their own safety and the safety of others, including parties for whom they have a professional responsibility.

To the fullest extent of the law, neither the Publisher nor the authors, contributors, or editors, assume any liability for any injury and/or damage to persons or property as a matter of products liability, negligence or otherwise, or from any use or operation of any methods, products, instructions, or ideas contained in the material herein.

Library of Congress Cataloging-in-Publication Data
A catalog record for this book is available from the Library of Congress

British Library Cataloguing-in-Publication Data
A catalogue record for this book is available from the British Library

ISBN: 978-0-12-815824-1

For information on all Elsevier publications
visit our website at https://www.elsevier.com/books-and-journals

Publisher: Susan Dennis
Acquisitions Editor: Emily M. McCloskey
Editorial Project Manager: Mona Zahir
Production Project Manager: Debasish Ghosh
Cover Designer: Miles Hitchen

Typeset by SPi Global, India

Dedications

I dedicate the book to the person who taught me to love books, to my father, and to my children to maintain and keep this habit.

Dr. Marina Soković

I dedicate this book to my mother for her unconditional love and continuous support in everything I do.

Dr. Konstantinos Liaras

Contents

Acknowledgments

Dr. Marina Soković
I am very thankful to my husband Miroslav and my children Jana and Luka for their love, understanding, patience, and continuous support to complete this book. I am grateful to them for the time they spent without my attention, when I was dedicated to writing this book.

I am extremely grateful to my parents for their love, caring, and sacrifices for educating and preparing me for my future and to complete all my ideas successfully, including this. I am extending my thanks to my brother for supporting me all the time.

Also, I express my thanks to my mother and sister in law for their support and valuable help with children.

I would like to say thanks to my friends and research colleagues, Dr. Jasmina Glamočlija, Dr. Ana Ćirić, Dr. Jovana Petrović, Dr. Marija Ivanov, Dejan Stojković, Marina Kostić for their constant help and for understanding me all the time. Their work and results are also presented in this book.

Dr. Konstantinos Liaras
I am very grateful to my family for their love and patience during this long period that I was focused in writing this book.

I would also like to thank my colleagues in the pharmacy for their understanding and extra work when I was concentrated in completing this book.

I would also like to express my gratitude to all my lab colleagues from Aristotle University of Thessaloniki for our fruitful collaboration in our common research projects and papers, some of which were also reviewed in this book.

Common acknowledgements of the authors
We would both like to thank Prof. Athina Geronikaki who introduced us years ago and worked with us in many research projects the results of which are included in this book.

We are also grateful to the architect and artist Sophia Spirliadou for her substantial help in designing the artistic figures of the book.

Chapter 1

Introduction

Marina Soković[a] and Konstantinos Liaras[b]
[a]*Mycological Laboratory, Institute for Biological Research "Siniša Stanković", University of Belgrade, National Institute of Republic of Serbia, Belgrade, Serbia,* [b]*Department of Pharmaceutical Chemistry, School of Pharmacy, Aristotle University of Thessaloniki, Thessaloniki, Greece*

Chapter outline

Infectious diseases in humans have markedly increased during the past 10 years, especially in immunocompromised patients. Consequently, as high as 10% of hospital-acquired systemic infections are caused by fungi (Soković et al., 2006). Fungal infections could be provoked by primary and opportunistic pathogens. Primary pathogens are capable to provoke infection in the healthy population, while opportunistic ones could be responsible for infection in immunosuppressed organisms.

The research efforts concerning antimicrobials were focused mainly to bacterial infections and antibacterial resistance, delaying that way the progress in testing fungal diseases and resistance to antifungal agents. First of all, fungal diseases were for many years neglected and fungi as pathogens were not recognized and accepted as important (Wey et al., 1988). A large increase in fungal infections was noticed in late 1960s, when antibiotic treatments were developed in high extent and fungi started to represent a global health problem and threat (Vandeputte et al., 2012). The increase of fungal diseases is associated with changes in medical practice such as uncontrolled usage of antibiotics (especially broad-spectrum ones), frequent use of therapies that suppress immune system, utilization of indwelling intravenous devices, and spread of chronic immunosuppressive viral infections such as AIDS (Ghannoum and Rice, 1999).

Fungal infections still remain a therapeutic problem despite the availability of a number of treatments and can become an important issue in the care of patients with compromised immune system (Nychas, 1995). This could be a consequence of infectious diseases, chemotherapy, transplantations, and treatment of autoimmune diseases such as rheumatoid arthritis, AIDS, Crohn's disease, etc., or might be a result of aging, but also as consequence of uncontrolled use

Antifungal Compounds Discovery. https://doi.org/10.1016/B978-0-12-815824-1.00001-9

of antimycotics. Increased numbers of immunocompromised persons lead to a rise in the number of people that have succumbed to fungal diseases. The frequency of fungal diseases is higher in low socioeconomic status communities: bad living conditions lead more often to skin-to-skin contact and close proximity to animals, while hygiene may be suboptimal (Havlickova et al., 2008).

Diseases provoked by fungi are often more difficult to treat than bacterial infections. Either topical or oral treatments are long term and with limited success in controlling the fungus, or completely ineffective. The majority of infections are chronic, tending to recurrence of the disease. One of the difficulties in antifungal therapy is that existing antifungal drugs usually cause harmful side effects due to similar cellular structure and functions between eukaryotic human and fungal cells.

Fungi are in general being controlled by drugs and the use of antimycotics is limited due to their undesirable effects, including teratogenicity, carcinogenicity, acute but also chronic toxicity and extended degradation period with consequences affecting environmental pollution (Shin and Lim, 2004).

It is obvious that some barriers are present in the effective reduction of mycoses: difficult elimination of fungi due to the fact that they are widespread in the nature; difficult diagnosis because fungi have various clinical manifestations that are host dependent; difficult to establish proper therapy due to restricted access to a number of available drugs; successful treatments are available only for a limited number of fungi (Marković et al., 2016).

Two main populations are in risk to acquire invasive infections by fungi. In the first group are included individuals who are susceptible to infection because of their geographic location and their infections can be referred to as endemic mycoses. In the second group are included individuals who possess increased host susceptibility, developing in that way opportunistic infections (immunocompromised, severely ill, malnourished patients, etc.) (Fridkin and Jarvis, 1996).

There is also some discordance and limitation in the interpretation of results obtained from in vivo and in vitro data; infections provoked by susceptible strains respond to appropriate therapy in ~90% of cases, whereas infections due to resistant strains respond in ~60% of cases (Rex and Pfaller, 2002).

Determination of minimal inhibitory concentration (MIC) in vitro that could inhibit fungal growth is also limited, because MIC levels do not always represent the most optimal measurement of resistance. Knowing that for the majority of fungal species, or at least for those which are important in medicine, there is a determined MIC for a specific studied drug, it is necessary to estimate its average MIC. Consequently, one of the common ways to define a strain as resistant is by observing a MIC higher than the average for the corresponding drug. There are suggestions and considerations that determination of minimum effective concentration of some drug should be done at hyphal cells, not only at spores and that might be a more appropriate end point than MIC obtained by testing the susceptibility of filamentous fungi to echinocandins (Kurtz et al., 1994).

In any case, it is widely accepted that testing antifungal activity means determination of MIC and minimal fungicidal concentration (MFC) against conidia rather than sensitivity of clinically more important hyphal structures (Kanafani and Perfect, 2008).

The interpretation and comparison of the results obtained from different research groups are often very difficult due to differences in methodology, fungal strains, control drugs, etc. This has to be adapted at first for in vitro investigation and afterwards for testing antifungal activity in vivo and in clinical trials. Once reproducible methodologies and standards are established, it will be much easier to transfer knowledge from labs to patients.

One of the acknowledged ways to define clinical resistance is by evaluating the persistence of progression of a microbial infection despite administering the appropriate therapy. Especially for fungal infections, characterizing clinical response to therapy as successful, depends not only on the susceptibility of the pathogenic fungus, but also on the immune system condition of the host, patient compliance, drug pharmacokinetics, and absence of a persistent or protected focus of infection (f.i. abscess or catheter).

One of the greatest scientific achievements of 20th century is the development of new, effective, and mass-produced antimicrobial agents. Antibiotics and antimycotics comprised western medicine's primary defense against infectious diseases for more than 60 years. After a long "antibiotics era," when millions of lives were saved, our chemical shield has become increasingly leaky. World Health Organization published that "the world is on the brink of losing these miracle cures (antibiotics)" and that "in the absence of urgent corrective and protective actions, the world is heading towards a post-antibiotic era, in which many common infections will no longer have a cure and, once again, kill unabated" (Liljeqvist et al., 2012).

Antimycotic agents are in most of the cases the choice of preference in order to treat this kind of infections, being alongside with antibiotics the most widely used therapeutics around the globe. Despite the successful application of conventional antifungals for many years, their uncontrolled and injudicious use, as mentioned, led to the emergence of numerous resistant fungal species.

It should also be noted that in many countries, the administration of topical antifungal drugs is relatively uncontrolled. The use of dermatological products that contain antifungal drugs, alone or in combination with corticosteroids, has become very familiar to patients, since they consider them as totally safe, giving extreme dimensions to the problem.

Another factor that should be taken into account is the economic status of patients. There are many cases in which due to financial problems, combined with difficult access to public health institutions, patients turn to the dangerous "solution" of self-treatment. This phenomenon has the result of overuse of certain Over-The-Counter (OTC) antifungals with dosage/duration treatment schemes, which are self-defined from the personal experience of each patient. False self-diagnoses and wrong treatment schemes can prove to be even worse than no treatment at all.

We should also take into account some kinds of human behavior that can also potentially promote the emergence of resistance. An excellent example is the case of noncompliance of patients with the appropriate treatments. Noncompliance takes place when patients forget to take their medication, prematurely forfeit treatment when they start to feel better, or are not able to cover the whole cost of their therapeutic scheme.

In a number of countries, the issues of noncompliance and self-medication become even greater due to the fact that a series of the available antimicrobial drugs are not produced under the proper standards, they are counterfeit, or have expired in terms of efficacy (Knobler et al., 2003).

Another key factor that contributes to the resistance issue is the administration of series of antimicrobials in animals that are raised for food (poultry, pigs, cows, etc.). The only thing that has not been clarified yet is the contribution of that kind of agricultural in the emergence and spread of resistance pathogens that are dangerous for humans. While in some studies it was stated that this problem is under control by different programs, in others it was proposed that the level of risk is much higher than we suppose. It was mentioned that by using antimicrobial compounds in animal care, either for growth promotion or for therapeutic reasons, there is a possibility to induce the emergence of drug-resistant strains that can ultimately be transmitted to humans via food products (Knobler et al., 2003).

It is clear that the benefits that derive from the use of antimicrobials in animals reflect on more factors than just the animal health, taking into account the economic benefits for the producers and healthier and safer food products for the public. Despite the previously mentioned benefits, many scientists suggest the reduction of current use of antimicrobials due to serious risks, connected to resistance, for both animal and human health. Others propose that such restrictions in the use of antimicrobials in animals could turn out to be economically unfavorable for producers and ultimately, customers. Nevertheless, the vast majority of scientists agree that the injudicious use of antibiotics and/or antifungals should be taken into serious consideration and potentially be restricted when alternative solutions are available and recommended, or when conventional drugs are found to be undoubtedly ineffective.

Having all these facts in mind, human and veterinary doctors are entitled to promote the judicious and safe use of antimicrobial drugs. In addition, through health systems and higher education we should be in position to ensure the sufficient training of human and animal health professionals concerning the rational and considerate use of antimicrobials.

Despite the already-known facts about the way hospitals could minimize the spread of infectious diseases, intensive studies are still in need to give us better perspective on the matter. In addition, the development of fast and effective diagnostic methods for the identification of an infection, the specific microorganism and its susceptibility to marketed drugs remains a constant priority.

Precision in diagnostic tests is crucial in our efforts to alternate the current strategy of extensive empiric administration of antimicrobial agents in infected patients. Moreover, development of materials that can be used with medical devices (e.g., catheters) and are designed to be resistant to colonization by pathogenic strains is still drawing the attention of scientists and companies around the globe, while the ongoing production of innovative antimicrobials also remains a high priority (Knobler et al., 2003).

The progress in techniques and methods in the field of molecular biology led to the deduction that genes, which are responsible for resistance, can be also horizontally transferred among species in higher extent than it was previously thought by scientists. Consequently, this fact leads to a pessimistic scenario for future generations, when most of the known antimicrobial agents could possibly prove to be ineffective, bringing us back to a preantimicrobials era where common bacteria and fungi will again become life threatening.

Regarding the transfer of resistance to pathogens that can infect humans, the number of pathogenic species that can serve as recipients is of great importance. Having that in mind, it is implied that the human microbiome might act in this process, as intermediary reservoir for the emergence of resistance (Sommer et al., 2010).

However, this role of human microbiome has not been studied extensively by scientists so far. Moreover, environmental strains could possibly carry resistance genes into the human microbiome, where they could be subsequently transferred to human-associated species, including pathogens. Nevertheless, for the majority of this kind of microorganisms, the interaction periods with microorganisms of human microbiome can be short. It should be mentioned that animals can also act as intermediates for resistant species and offer an ideal ground for transferring resistance to human pathogens (Allen et al., 2010).

All mentioned facts could potentially lead to **antimicrobial resistance (AMR)** of fungal species that makes the problem of fungal infections even worse and more difficult to control and treat.

Antimicrobial resistance occurs when microorganisms (such as bacteria, **fungi**, viruses, and parasites) change at their exposure to antimicrobial drugs (such as antibiotics, **antifungals**, antivirals, antimalarials, and anthelmintics). Microorganisms that develop antimicrobial resistance are sometimes referred to as "superbugs" (WHO, 2017)

There are two main types of resistance that can be observed in vitro: primary (intrinsic) and secondary (acquired) resistance. When a fungus is resistant to a particular drug before exposure, then it is described as acquiring primary resistance. On the other hand, secondary resistance is obtained when there is previous exposure to an antifungal compound. Additionally, the main factors that could be responsible for the presence of a resistant fungal strain are: intrinsic resistance of endogenous strains, replacement with more resistant species or strains, genetic modifications that lead to endogenous resistant strains, temporary expression of certain genes that have as a result the production of an endogenous strain that is resistant for a certain period of time, cell type alterations, overcrowding and, in general, modifications in the size and variability of population, etc. (White et al., 1998).

This undesirable situation became a constant threat for public health, while the dramatic scale of the issue has alerted health care organizations, governments and pharmaceutical companies.

Economic factors and high costs of antimicrobial agents production make their development unprofitable (Nathan and Goldberg, 2005), while facts that uncontrolled use of antibiotics and antimycotics influence AMR turn us to search for new approaches, which will help us to fight the pathogens that are capable to kill. This idea is not futuristic or even theoretical: we can tap into an ancient arms race between bacteria and fungi and novel natural and synthetic antimicrobials to combat drug-resistant bacteria and fungi. Natural therapy, or the use of new semisynthetic or synthetic compounds based on combined scientific results from different branches (biology, molecular biology, proteomics, chemistry, pharmacy, etc.), represents potentially significant weapons in our ongoing battle against fungal diseases.

Vandeputte et al. (2012) reported that fungi possess mechanisms to counteract the fungistatic or fungicidal effects of different antifungal drugs. There are three basic mechanisms: possibility of fungi to decrease the affinity of the drug for its target, to reduce the accumulation of the drug in fungal cell, and potential to modify metabolism to counterbalance the drug effect.

According to mentioned facts, the need for novel approaches, methodologies, and studying of all of three basic mechanisms of resistance is now greater and more necessary than ever.

Although resistance to antimicrobials is a natural biological phenomenon, the extensive introduction of antibacterial and antifungal agents into clinical practice was followed by the laboratory detection of resistant strains, able to multiply in drug concentrations that are higher than the therapeutic ones. This kind of resistance may either characterize the entire species or be acquired by strains of a normally susceptible species via mutation or transfer of genes. Various mechanisms that promote resistance to particular antimicrobials are encoded by corresponding genes of microorganisms. These mechanisms can also promote resistance to antimicrobials of the same or even in some cases of different classes.

Alongside with the extensive use of antimicrobials worldwide was observed an increase of resistance prevalence to each new drug. Although the prevalence of resistance varies in diverse geographical areas, sooner or later applies also on every antimicrobial agent (WHO, 2001). Consequently, resistance has been observed to be acquired for every antimicrobial agent some time after their introduction to clinical practice. A fine example is the acquisition of resistance by *Staphylococcus aureus* to penicillin in hospitals, only a few years (mid-1940s) after its introduction to therapeutic use. Subsequently, the initial emergence of antimicrobial resistance in hospitals soon became widespread also in community-acquired infections.

To support these facts, it would be interesting to take a glimpse into some information regarding infections that are caused by *Aspergillus* spp. This type of infections is still in the spotlight of numerous medicinal research fields mainly due to the rapid development of diagnostic and therapeutic techniques and methods, having always as ultimate goal the reduction of morbidity and mortality rates. Especially invasive pulmonary aspergillosis (IPA) is a serious disease with increasing incidence and fatality rates. The first described case of IPA is from 1953. The incidence of this disease has increased in the past 10–20 years; in the years between 1978 and 1992 IPA incidence increased from 0.4% to 3.1%, while nowadays from 17% to 60%. Moreover, the mortality of IPA increased till 50% in patients with neutropenia and 90% in recipients of hematopoietic stem-cell transplantation (HSCT) (Kousha et al., 2011). Conventional treatment of aspergillosis is not effective in preventing death in up to two-thirds of the patients. The condition is even more severe for the immunosuppressed patients in the form of invasive and life-threatening aspergillosis. Moreover, aspergillosis is responsible for the presence of a series of risk factors, such as neutropenia and myeloablative-immunosuppressive treatments that are connected to different kinds of transplantations (hematopoietic stem cells, whole organ, primarily lung and heart transplantations). It should be pointed out that the mortality rates for this illness range between 30% and 90%, proving that conventional treatments don't exhibit the desirable results for the patients (Steele et al., 2005).

An important variable in such cases and conditions is the immunity that a host develops after exposure to the fungus or on the other hand the pathogenesis that can occur after transition from saprophytic form of some fungi (e.g., *Candida albicans*).

A side effect of antibiotic therapy is the increased growth of some fungal species, e.g., *Candida* sp., *Aspergillus* sp., and others. *Candida albicans* is a commensal microorganism and is usually present in human microbiota (mouth, urogenital, and gastrointestinal tract). Hormones, stress, weak immune system could affect the levels of *C. albicans*. Antibiotic treatment could dramatically increase the number of *Candida* colonies, while organisms treated with high doses of antibiotics and for long duration are more susceptible to *C. albicans* colonization and infection (Noverr et al., 2004). Rational use of antibiotics is very important for controlling *C. albicans* and other fungal species.

To summarize, in that kind of cases both innate and acquired immunitics play a crucial role in confronting pathogenic fungal strains (Bellocchio et al., 2004).

There are numerous factors, which play important role in the emergence of resistance, including overuse of drugs, infections of inmmunosuppressed patients, and increased use of internal medical devices that provide an ideal environment for pathogens.

In order to avoid this threat it is necessary to work not only on simple biological screening of new antimicrobials, but also to find new approaches and new methodologies taking into account common experience and efforts of scientists from different branches (mycologists, bacteriologists, microbiologists, molecular biologists, chemists, pharmacists, medical doctors).

1.1. Toward novel antifungal agents

In addition, under particular stress conditions, pathogenic microorganisms have the ability to evolve via epigenetic mechanisms and genetic variations. Manifold approaches are in need to control and combat effectively resistance to antimicrobials. Such strategies contain rational use of existing antimicrobials, combination of different therapeutic approaches, and development of innovative antimicrobial agents and methods of treatment. The use of microbial genomics gives incredible opportunities for introduction of molecular diagnostic tools in order to overcome resistance.

Incorporating innovative technologies from diverse fields like chemistry, computer-aided studies, bacterial bioengineering, microbiology, and pharmacology could be the answer to the ominous problem of resistance to antimicrobial agents, lowering the cost, speed of development, and potency of drugs.

Despite the discovery of numerous antimicrobial molecules to date, design of new and efficient drug candidates is an issue that has not yet been fully addressed. To overcome the existing problem, combining skills and funds from academia, pharmaceutical companies, and public health systems would prove to be essential. This synthesis of skills could lead to synergy between innovative technological developments, academic sector, and industry, bringing promising results concerning the hunt for novel antimicrobials.

In the recent years, it has come to recognition that traditional therapeutic targets of antimicrobials, such as protein synthesis and cell wall peptidoglycan biosynthesis, can be subjects to rapid emergence of resistance. This led scientists to begin a search for alternative targets for antimicrobial agents like **quorum sensing**, **protein (enzyme) folding**, and formation of **biofilms**.

Biofilm is described as a strongly adherent assemblage of differentiated microbial cells enclosed in a matrix of polysaccharides (Stoodley et al., 2002). Another way to describe biofilms can be "an assemblage of microbial cells that is irreversibly associated (not removed by gentle rinsing) with a surface and enclosed in a matrix of primarily polysaccharide material" (Donlan, 2002). In the form of biofilms, microorganisms stick together and on a surface, embedded in a polymeric matrix containing polysaccharides, proteins, lipids, and DNA, produced by themselves, which is generally named extracellular polymeric substance (EPS) (Flemming and Wingender, 2002). This matrix is highly hydrated and transport of nutrients to the cells within the biofilm is provided through the water channels. Bacteria and fungi are capable to detach and migrate through the biofilms and hence to colonize other surfaces. They can also remodel biofilm matrix by using specific enzymes.

There are many kinds of surfaces and in general places where pathogens can form biofilms. These can be organic or inorganic and diverse, such as lungs, teeth, intestines, lenses, and water tubes. Microorganisms involved in biofilms cause serious infections to humans and animals (endocarditis, cystic fibrosis, periodontitis, rhinosinusitis, and osteomyelitis) and can be often seen in urinary catheters and medical implants. This is also a big problem in the industry, in every procedure where a water-based process is involved (paper manufacturing, nautical shipping, drinking water facilities, cooling systems, health care, medical devices, and food processing) (Gupta et al., 2016).

On the other hand, there are also many kinds of microorganisms that can form biofilms, while it should be noted that bacteria and fungi can potentially coexist in such films. These pathogenic microorganisms that grow in biofilms can acquire resistance to antimicrobials sooner than their planktonic forms. The reason behind this problem is that a community like a biofilm can develop protective mechanisms such as impedance against diffusion and also a pathogen-friendly environment within the matrix. It should be mentioned that even when the antimicrobial therapy is effective for a part of the colony, the pathogens that survived the treatment can develop and multiply by using the nutrients left behind. Once a biofilm has formed, natural defenses such as mucosal secretions will then have limited success against pathogenic microorganisms. Consequently, bacteria and fungi living within biofilms can survive exposure to even 1000-fold greater concentrations of antimicrobial drugs, which are effective against the same pathogens when are in suspension (Raffa et al., 2005). At least 65% of the infection diseases are connected to microbial communities that spread by forming biofilms (Lewis, 2007).

Beside the biofilm formation, there is a whole set of microorganism behaviors (antibiotic production, bioluminescence, conjugation and virulence, competence, swarming, motility, sporulation, and conjugation) named as **quorum-sensing mechanism (QS)** (Fig. 1.1) (Scutera et al., 2014).

This complicated communication and social behavior could be observed as well in a single bacterial and/or fungal species, but also could be formed among different species (combinations of different bacteria and fungi). It is a response of microorganisms through the secretion of signaling molecules known as **autoinducers**. Autoinducers are recruited by bacteria and/or fungi, followed by increase in their motility, which plays the most important role in spreading the colonies. Motility types depend on each bacterial/fungal species. When the number of cells is sufficient, bacteria or fungi are in stage of aggregation, which leads to the formation of biofilm (Fig. 1.2). The last step is the production of exopolysaccharidic matrix (EPS) in which the bacteria and/or fungi are embedded (Czaczyk and Myszka, 2007).

It has been observed that microorganisms often establish an infection by sensing their own population size, i.e., quorum sensing. They can be inert until their numbers are high enough to overcome the host's defenses. Consequently, when the quorum has been reached, the formation of biofilm and pathogenesis

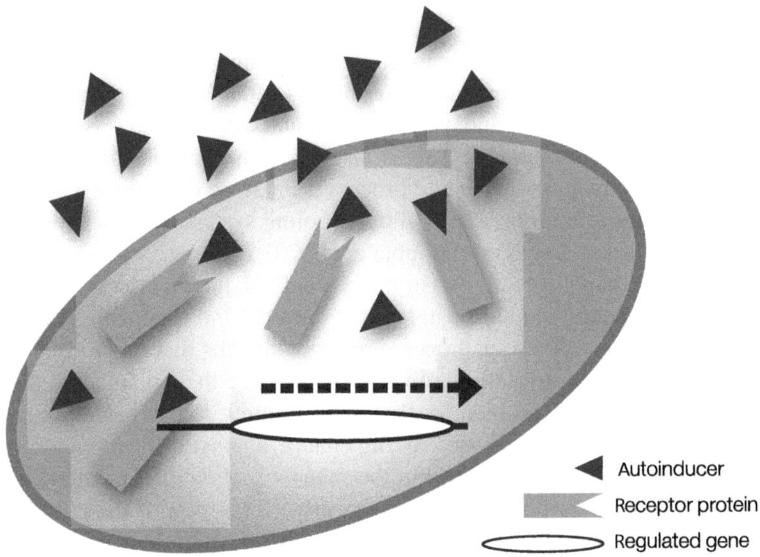

FIG. 1.1 Quorum-sensing mechanism.

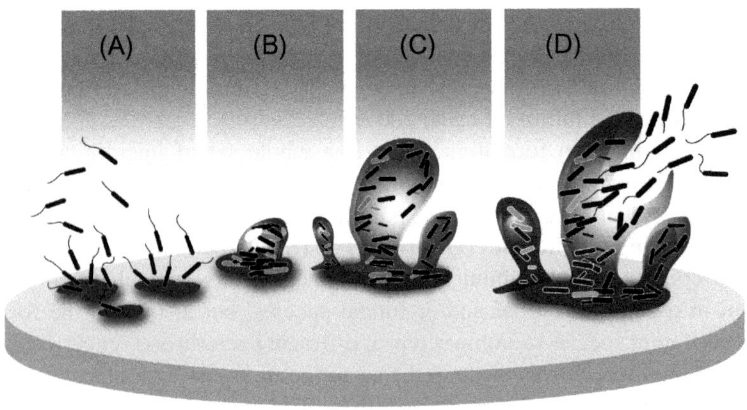

FIG. 1.2 Process of biofilm formation: (A) planktonic cells adhesion; (B) cells proliferation; (C) biofilm maturation; (D) mature biofilm.

are initiated. Potent inhibitors of this process can be an effective way to make pathogens more susceptible to the patient's immune system reaction and prescribed drugs.

Quorum sensing plays an important role in gene regulation connected with pathogenesis mechanisms, including biofilms, elastase, pyocyanin, protease, and swarming motility (Parsek and Greenberg, 2000). Simple and rapid screenings of a broad spectrum of autoinducers involved in QS and biofilm formation process could be extremely helpful in our efforts to comprehend the nature of quorum sensing and design innovative and effective antimicrobials. Knowing that quorum sensing plays an important role in pathogens' virulence, developing potent inhibitors of this process could give us alternatives to existing therapeutic strategies and also reduce the risk for emergence of resistance to antimicrobials.

There are a number of reasons pointing that inhibiting QS could turn out to be an attractive method of therapeutic intervention. Inhibiting QS offers three targets to attack: the signal generation mechanism, the signal molecule, and the signal reception. Obviously, a promising strategy would include screening of compounds that can potentially prevent the synthesis of signal molecules by the regulatory genes. With absence of signal molecules, the pathogens would be unable to sense when quorum is reached, having as a result the QS-controlled genes to remain inactive. If this strategy does not succeed, alternatively the signals can potentially target themselves as subjects of degradation. Additionally, preventing the binding of signal molecules on the targeted protein would result in loss of their ability as transcriptional regulators.

Although it is not completely clear yet that QS inhibitors are the new hope in the ongoing fight against resistant microorganisms, the preliminary results certainly seem to be promising. QS inhibitors are not only responsible for downregulating expression of numerous virulence factors, but also have the ability of rendering biofilms more susceptible to conventional antimicrobials. Consequently, it can be presumed that a combinatory treatment, including both QS inhibition and an antimicrobial agent, may prove to be successful.

Many of the processes that can be observed in the quorum-sensing phenomenon resemble interactions that are well described in pharmacology, such as specific signal molecules binding on and activating receptors, leading to transmission of signals intracellularly and thus activating second messenger responses. Therefore we should point out the similarity of quorum-sensing signaling processes to ligand-receptor interactions. This could suggest that QS inhibitors could be designed by following well-established pharmacological principles, having innovative mechanisms of action and possibly higher potency against resistant microbial species.

Now, more than ever, it is justified that pathogenesis includes processes that are difficult to mimic in cultures due to the fact that they involve particular pathogenic mechanisms using specific enzymes.

Beside research of QS and biofilm processes, targeting these specific enzymes included in virulence and pathogenesis could have more successful results in the treatment of fungal infections. This should lead to development of novel, more selective, and effective enzyme inhibitors (e.g., CYP51 enzyme at *Candida* spp.).

The acquired resistance in fungi (*C. albicans* and *A. fumigatus*) is a consequence of a point mutation in the CYP51A gene that encodes the CYP51 enzyme. This enzyme is responsible for the conversion of lanosterol to ergosterol. Fungal sterol 14α-demethylase (CYP51) is required for ergosterol biosynthesis and is the main target of azole antifungal drugs (Arendrup, 2013).

Azoles inhibitors (selective for the fungal enzyme over the human homologue) are commonly used to treat fungal infections, including infections provoked by *C. albicans* and *A. fumigatus*.

One of the challenges of scientists during the past decades has undoubtedly been controlled drug release. Eventually, formulation of an Active Ingredient (AI) has to be delivered to the desired target. For many years, design of suitable delivery systems in order to increase bioavailability of drugs, minimize their side effects and adverse reactions, while controlling the duration of activity, has drawn the attention of pharmaceutical scientists. Countless innovative delivery systems have been developed using nanoemulsion or microemulsion AI conjugates. Such systems proved to be easy to manufacture, of low cost, versatile in controlling diverse AIs and with a favorable in vivo profile, turning them into the choice of preference for numerous scientists around the globe.

AMR occurs as a natural process, where microorganisms could evolve and become more resistant. This process could be just slowed down by humans, but cannot be stopped completely. Through this process, bacteria and fungi will inevitably develop new ways for obtaining resistance to existing drugs. In order to stop emerging resistance, an aggressive strategy is needed, combining development of new drugs, introduction of modern diagnostic tests, and discovery of new sources, approaches, and methodologies to track and fight evolved strains.

Discovery of new antifungal drugs with desired properties, either from natural or semisynthetic or synthetic products, is a great challenge. For that purpose, combined efforts of scientists are taking place, from basic research on the field of fungi and employment of new sources, to innovative molecular approaches, known as combinatorial biosynthesis.

In this book we are going to review existing antifungal therapies and compounds, the past and current efforts of researchers and their achievements in discovering novel antifungal drugs using combinatorial biosynthesis, but also to highlight new approaches and challenges, novel trends and possibilities.

In particular, extensive data regarding natural and synthetic antifungal agents will be presented in a way to help the reader understand the chemical components or moieties that are responsible for the activity observed. In that way, researchers will have the opportunity to focus on certain directions in their efforts to discover and/or synthesize, in general, compounds that will have higher chances to be effective antifungals. Furthermore, through the discussion, SAR and comparisons that can be found in the chapters of this book, researchers will find useful information for discovering innovative molecules with potentially also unique or alternative modes of action to help overcome the emerging resistance to marketed drugs.

Another important element in "Antifungal Compounds Discovery" is the wise selection of screening methods and techniques that help researchers to determine the biological activity of their lead compounds, extracts, or components of natural products. By reading and using this book as a research tool, significant time and effort could be spared by filtering and understanding which are the best screening methods for each chemical group.

To conclude, knowing in depth the outcome of research efforts from the past and recent years can prove to be crucial for better focused, less time consuming, and eventually more effective research projects in the field of antifungal compounds discovery.

References

Allen, H.K., Donato, J., Wang, H.H., Cloud-Hansen, K.A., Davies, J., Handelsman, J., 2010. Call of the wild: antibiotic resistance genes in natural environments. Nat. Rev. Microbiol. 8, 251–259.

Arendrup, M.C., 2013. Update on antifungal resistance in *Aspergillus* and Candida. In: ESCMID conference. Invasive Fungal Infections. Italy.

Bellocchio, S., Montagnoli, C., Bozza, S., Gaziano, R., Rossi, F., Mambula, S.S., Vecchi, A., Mantovani, A., Levitz, S.M., Romani, L., 2004. The contribution of the toll-like/IL-1 receptor superfamily to innate and adaptive immunity to fungal pathogens *in vivo*. J. Immunol. 172 (5), 3059–3069.

Czaczyk, K., Myszka, K., 2007. Biosynthesis of extracellular polymeric substances (EPS) and its role in microbial biofilm formation. Pol. J. Environ. Stud. 16, 799–806.

Donlan, R.M., 2002. Biofilms: microbial life on surfaces. Emerg. Infect. Dis. 8, 881–888.

Flemming, H.C., Wingender, J., 2002. What biofilms contain—proteins, polysaccharides, etc. Chem. Unserer Zeit 36, 30–42.

Fridkin, S.K., Jarvis, W.R., 1996. Epidemiology of nosocomial fungal infection. Clin. Microbiol. Rev. 9, 499–511.

Ghannoum, M.A., Rice, L.B., 1999. Antifungal agents: mode of action, mechanisms of resistance, and correlation of these mechanisms with bacterial resistance. Clin. Microbiol. Rev. 12 (4), 501 517.

Gupta, P., Sarka, S., Das, B., Bhattacharjee, S., Tribedi, P., 2016. Biofilm, pathogenesis and prevention-a journey to break the wall: a review. Arch. Microbiol. 198 (1), 1–15.

Havlickova, B., Czaika, V.A., Friedrich, M., 2008. Epidemiological trends in skin mycoses world wide. Mycoses 51, 2–13.

Kanafani, Z.A., Perfect, J.R., 2008. Resistance to antifungal agents: mechanisms and clinical impact. Clin. Infect. Dis. 46, 120–128.

Knobler, S.L., Lemon, S.M., Najafi, M., Burroughs, T. (Eds.), 2003. The Resistance Phenomenon in Microbes and Infectious Disease Vectors: Implications for Human Health and Strategies for Containment: Workshop Summary. Factors Contributing to the Emergence of Resistance, vol. 5. National Academies Press, Washington, DC.

Kousha, M., Tadi, R., Soubani, A.O., 2011. Pulmonary aspergillosis: a clinical review. Eur. Respir. Rev. 20 (121), 156–174.

Kurtz, M.B., Heath, I.B., Marrinan, J., Dreikorn, S., Onishi, J., Douglas, C., 1994. Morphological effects of lipopeptides against *Aspergillus fumigatus* correlate with activities against (1,3)-beta-D-glucan synthase. Antimicrob. Agents Chemother. 38, 1480–1489.

Lewis, K., 2007. Persister cells, dormancy and infectious disease. Nat. Rev. Microbiol. 5, 48–56.

Liljeqvist, T., Andresen, D., Zuo, Y., Weston, C., 2012. Antibiotic resistance: moving forward to the past. N.S.W. Public Health Bull. 23 (2), 37.

Marković, T., Nikolić, M., Glamočlija, J., Ćirić, A., Emerald, M., Radanović, D., Zheljazkov, V., Soković, M., 2016. Essential oils for the prevention and treatment of human opportunistic fungal diseases (book chapter). ACS Symp. Ser. 1218, 247–277.

Nathan, C., Goldberg, F., 2005. The profit problem in antibiotic R&D. Nat. Rev. Drug Discov. 4, 887–891.

Noverr, M.C., Noggle, R.M., Toews, G.B., Huffnagle, G.B., 2004. Role of antibiotics and fungal microbiota in driving pulmonary allergic responses. Infect. Immun. 72 (9), 4996–5003.

Nychas, G.J.E., 1995. Natural antimicrobials from plants. In: Gould, G.W. (Ed.), New Methods of Food Preservation. Springer, Boston, MA, pp. 58–89.

Parsek, M.R., Greenberg, E.P., 2000. Acyl-homoserine lactone quorum sensing in gram-negative bacteria: a signalling mechanism involved in associations with higher organisms. PNAS 97 (16), 8789–8793.

Raffa, R.B., Iannuzzo, J.R., Levine, D.R., Saeid, K.K., Schwartz, R.C., Sucic, N.T., Terleckyj, O.D., Young, J.M., 2005. Bacterial communication ("quorum sensing") via ligands and receptors: a novel pharmacologic target for the design of antibiotic drugs. J. Pharmacol. Exp. Ther. 312, 417–423.

Rex, J.H., Pfaller, M.A., 2002. Has antifungal susceptibility testing come of age? Clin. Infect. Dis. 35, 982–989.

Scutera, S., Zucca, M., Dianella, S., 2014. Novel approaches for the design and discovery of quorum-sensing inhibitors. Expert Opin. Drug Discovery 9 (4), 353–366.

Shin, S., Lim, S., 2004. Antifungal effects of herbal essential oils alone and in combination with ketoconazole against *Trichophyton* spp. J. Appl. Microbiol. 97, 1289–1296.

Soković, M.D., Glamočlija, J., Marin, P.D., Brkić, D.D., Vukojević, J., Jovanović, D., Bulajić, N., Kataranovski, D., 2006. Antifungal activity of the essential oil of *Mentha* x *piperita*. Pharm. Biol. 44, 511–515.

Sommer, M.O.A., Church, G.M., Dantas, G., 2010. The human microbiome harbors a diverse reservoir of antibiotic resistance genes. Virulence 1, 299–303.

Steele, C., Rapaka, R.R., Metz, A., Pop, S.M., Williams, D.L., Gordon, S., Kolls, J.K., Brown, G.D., 2005. The beta-glucan receptor dectin-1 recognizes specific morphologies of *Aspergillus fumigatus*. PLoS Pathog. 1 (4), e42.

Stoodley, P., Sauer, K., Davies, D.G., Costerton, J.W., 2002. Biofilms as complex differentiated communities. Annu. Rev. Microbiol. 56, 187–209.

Vandeputte, P., Ferrari, S., Coste, A.T., 2012. Antifungal resistance and new strategies to control fungal infections. Int. J. Microbiol. 2012, 713687. 1–26.

Wey, S., Mori, M., Pfaller, M., Woolsen, R.F., Wenzel, R.P., 1988. Hospital-acquired candidemia—the attributable mortality and excess length of stay. Arch. Intern. Med. 148, 2642–2645.

White, T.C., Marr, K.A., Bowden, R.A., 1998. Clinical, cellular, and molecular factors that contribute to antifungal drug resistance. Clin. Microbiol. Rev. 11 (2), 382–402.

WHO (World Health Organization), 2001. WHO Global Strategy for Containment of Antimicrobial Resistance.

WHO (World Health Organization), 2017. Implementation of the Global Action PLAN on Antimicrobial Resistance. WHO GAP AMR Newsletter No. 32.

Chapter 2

Fungal infections—Background to specific fungal species

Marina Soković[a] and Konstantinos Liaras[b]
[a]*Mycological Laboratory, Institute for Biological Research "Siniša Stanković", University of Belgrade, National Institute of Republic of Serbia, Belgrade, Serbia,* [b]*Department of Pharmaceutical Chemistry, School of Pharmacy, Aristotle University of Thessaloniki, Thessaloniki, Greece*

Chapter outline

Despite the fact that the kingdom of fungi contains various taxa with different kinds of ecological niches, morphologies, and life-cycle strategies, little is known regarding the true biodiversity of fungal species. Based on new data, using new modern techniques of identification, the number of fungal species is around 12 million compared to the 2–3 million species estimated by classic techniques (Wu et al., 2019). Fungi have very important roles in ecosystems, they are decomposers, mutualists, in some cases the role of these organisms in nature is still unknown, but some of them are pathogenic (Schmit and Mueller, 2007; Hyde et al., 2018). Fisher et al. (2012) stated that an incredible number of fungal diseases resulted in some of the most severe die-offs and extinctions ever witnessed in wild species. There are fungal strains, which are parasites for plants, animals, humans, as well as other fungi. Moreover, plant pathogenic fungi can cause serious damage to agriculture and forestry, while other strains

Antifungal Compounds Discovery. https://doi.org/10.1016/B978-0-12-815824-1.00002-0

can cause serious, or even fatal, diseases in humans such as candidiasis, aspergillosis, coccidioidomycosis, mycetomas, cryptococcosis, histoplasmosis, mucormycosis, and paracoccidioidomycosis. Additionally, dermatophytic and keratinophilic fungi can infect skin, eyes, nails, and hair, while several strains and fungal spores can potentially cause allergic reactions. Some of these fungal pathogens have not been described or we had only limited knowledge about them (Farrer et al., 2011).

While the bacterial infections were already a global issue of high importance at the beginning of the 20th century, fungal infections were not considered as equally severe. Only in the late 1960s when production and use of antibiotics rapidly increased, a corresponding rise in fungal infections incidence has been recorded, leading to nowadays, when they represent a serious international health issue. This phenomenon was enhanced by the growing number of immunocompromised patients due to AIDS, cancer, diabetes, organ transplantations, etc.

In general, three are the main groups of fungi that are able to cause pathogenic conditions to humans: **single cells** (yeasts), **filamentous forms**, and **dimorphic fungi**. In the cases where these infections become chronic, this results to prolonged therapies with the additional risks, especially, for the health of immunocompromised patients (Madhani and Fink, 1998).

As parasites in humans and animals, fungi can attack the skin and keratinous tissue (nails, hair) or internal organs (lungs, brain, etc.). In the first case, they are surface parasites, often opportunistic, with localization of the pathological process they cause. **Primary** pathogens are considered the ones that can naturally develop an infection in the healthy populations, while **opportunistic** pathogens (including commensal microorganisms of the healthy population) have the ability to establish pathogenic infections in humans when the conditions allow it (e.g., immunosuppression). The clinical signs and symptoms vary greatly, ranging from small superficial nodules to the hair and skin scarifications, to deep internal diseases that can be fatal. In the second case, the fungi can cause **systemic** or deep mycoses, abscesses in various organs, leukemia, endocarditis, etc. (Kogoj and Čajkovac, 1959; Ryan, 1994).

Fungal pathogens can be divided into two major groups: filamentous fungi and yeasts. The majority of the primary pathogens are filamentous fungi, while most of the opportunistic pathogens are yeasts. Nevertheless, some species of filamentous fungi can be categorized as opportunistic pathogens. At this point should be mentioned that fungal infections can also be classified according to the location of the infected tissue.

2.1 Superficial mycoses

Superficial mycoses, like tinea versicolor, are localized at the external part of skin and hair and they are usually caused by *Malassezia globosa* and *M. furfur*, which are carried by 2%–8% of the healthy population. They can potentially lead to tinea versicolor under conditions that haven't been clarified yet (Magill et al., 2014).

In theory, systemic mycoses can be found in any part of the human body, with many of the formerly considered nonpathogenic species, now acknowledged as opportunistic pathogens that cause severe mycoses, with symptoms ranging from a simple fever to a severe and rapid septic shock. They are common in immunocompromised patients and are frequently connected to an increased mortality rate. Some of the pathogens that are responsible for causing systemic fungal infections are *Candida, Aspergillus, Pneumocystis, Histoplasma, Cryptococcus, Mucor, Rhizopus,* and *Coccidioidomyces* (Kao et al., 1999; Hajjeh et al., 2004; Lockhar et al., 2012).

2.2 Cutaneous and subcutaneous mycoses

Cutaneous and **subcutaneous** mycoses caused by dermatophytes fungi can occur at keratinized structures of the body, having as most common of genera *Trichophyton, Epidermophyton,* and *Microsporum* (Morgan et al., 2005).

2.3 Mucosal infections

Mucosal infections are caused in most of the cases by opportunistic yeasts, with those of the *Candida* genus being the most common ones. It should be noted that vaginal, esophageal, oropharyngal, and urinary tract candidiasis can frequent occur in immunocompromised patients (Vallabhaneni et al., 2015). Moreover, fungal infections of the eye are also considered as mucosal fungal infections, with the difference that are more frequently caused by *Aspergillus* or *Fusarium* genera.

2.4 Opportunistic mycoses

There are several fungal strains, which are able to cause diseases in humans. Although most of these fungi appear to be soil-inhabiting species that live as saprobes, under certain conditions (i.e., when there is an open wound, or there is an immunosuppressed patient) can threaten the health of humans. In such cases, the corresponding infections are caused by opportunistic fungi (i.e., facultative parasites) and not by strains that are specifically responsible for diseases in humans (Hajjeh et al., 2004).

As it was previously mentioned, fungi are considered, in general, as low-risk pathogens for healthy individuals, but they can be life-threatening for patients with compromised immune systems (e.g., AIDS patients, cancer patients, etc.). Especially in the recent years, an essential factor promoting the increase in the incidence of opportunistic infections is the rise in the number of immunocompromised patients who are considered as susceptible hosts. The strains that are usually isolated from this kind of patients are either saprophytic or endogenous.

Opportunistic mycoses in humans that have primary entry routes via the skin and mucous membranes are described by the terms dermatomycosis and candidiasis, respectively.

Dermatomycoses are superficial infections of the skin, hair, and nails tissues in humans, which are caused by dermatomycetes, a group of pathogenic fungi. As it was stated previously, these fungi grow in an area just above where the protein keratin is deposited (Kao et al., 1999) and are able to degrade keratin to use it for their nutritional needs (Alexander et al., 2013). It should be noted that dissemination of this kind of infections is very unlikely due to their localization at keratinized tissues.

Fungal skin infections take place when a fungus contaminates or colonizes the epidermis or hair follicles. In several cases, clinical changes in the skin might not be obvious, while lesions depend on the location and the structure of the area where it intends to grow (e.g., superficial layer of dermis, hair, nails). Nevertheless, the exact mechanism that contributes to the incidence of a fungal disease of this kind in the host has not been elucidated in full yet. Furthermore, it is of great importance to broaden the knowledge regarding all the factors that are involved in pathogenesis (e.g., enzymes, adhesion mechanisms, ability to modulate defense mechanisms, etc.) (Baddley et al., 2008).

Dermatomycetes can cause irritation and inflammation of the underlying epithelial cells of the skin, leading that way in several cases to cells' death. Although dermatomycoses are not life threatening in general, under certain conditions can be chronic and even responsible for significant morbidity. Although there are a lot of antifungal drugs in the market, therapy can be limited in many cases due to the toxicity of the administered medicines (Morgan et al., 2005).

As a result of globalization, mass tourism, international sports activities, and migrations, less common or forgotten species can spread among humans (Lockhar et al., 2012). Classification of dermatomycetes depends on their habitat and presence in different ecological niches. Dermatomycetes can be classified into: zoophilic (they infect animals but can also be transmitted to people), geophilic (they occur naturally in soil, being also able to infect animals and people), and anthropophilic (they infect people and are not able be transferred to animals) (Lortholary et al., 2011; Shah et al., 2012).

Dermatomycetes are divided into three genera: *Epidermophyton*, *Microsporum*, and *Trichophyton* (Table 2.1). Several representatives of these three genera are reported in the following text, regarding their virulence and significance for human health.

Dermatomycoses are infections, which are caused by dermatomycetes such as ringworm. Ringworm may potentially affect the human skin in different areas (e.g., beard, body, feet, groin area, scalp), with the fungus causing it found in warm and moist places. Therefore, such infections might more often occur under wet conditions, like sweating, and as a result of injuries of the skin, scalp, or nails. To name that kind of infections is used the location of the lesion, following the Latin word *tinea* (Marković et al., 2016).

TABLE 2.1 Major dermatomycetes that infect people.

Geophilic	Zoophilic	Anthropophilic	
M. gypseum	*M. canis* (dogs, cats)	*M. audouinii*	*Microsporum* spp.
	M. equinum (horses)		
	M. nanum (pigs)		
	M. persicolor (rodents)	*M. ferrugineum*	
T. terrestre	*T. equinum* (horses)	*T. mentagrophytes* var. *interdigitale*	*Trichophyton* spp.
	T. mentagrophytes var. *mentagrophytes* (mice, rodents)		
	T. verrucosum (cattle)	*T. rubrum*	
		T. tonsurans	
		E. floccosum	*Epidermophyton* spp.

2.5 Community-acquired and nosocomial mycotic infections

Nosocomial fungal infections are invasive mycotic infections developed in health care and community-associated settings (Perlroth et al., 2007). The majority of immunocompromised patients can develop that kind of mycoses in the hospital (Benoit et al., 2000).

During the past decades, the incidence of fungal infections has dramatically increased. These infections have high morbidity and mortality range, probably as a consequence of advanced medical technologies, more intensive surgical therapies, immunotherapy, etc. These trends of modern society will continue to rise over the forthcoming decades and it is anticipated that the incidence of these infections will continue to increase (Perlroth et al., 2007).

Candida is until now the most common cause of nosocomial fungal infections (Wilson et al., 2002); it is also a potential cause of sepsis in the hospital community and with a high mortality rate (Pasqualotto et al., 2005).

Aspergillus species are the second most frequent factors that lead to nosocomial fungal infections (Perlroth et al., 2007). Unlike *Candida*, species of *Aspergillus* infect less often immunocompetent hosts with nosocomial risk (venous catheters, postsurgery, antibiotics therapy, etc.). *Aspergillus* species predominantly infect immunocompromised patients with serious health conditions (hematologic malignancies, transplant recipients, etc.) (Brakhage, 2005).

Infections caused by *Fusarium* are rarer then *Aspergillus* caused diseases. The most common fusariosis are caused by *F. moniliforme*, *F. solani*, and *F. oxysporum* (Morrison et al., 1994).

Micromycetes, such as *Acremonium*, *Mucorales*, *Paecilomyces*, *Cladophialophora*, *Cladosporium*, and many other hyalohyphomyces and dematiaceous fungi are not so common, but they can be noticed at some immunocompromised patients who were exposed and suppressed for a longer period (Perlroth et al., 2007).

Early diagnosis and antifungal treatment of invasive infections are not easy and they are not effective enough. A commonality shared by all nosocomial fungal infections is the difficulty in establishing the diagnosis. Despite many years of intensive studies, only a few diagnostic studies are available and reliable for these infections. Hence, emphasis has been placed on early empiric therapy for patients who are clinically suspected of having a fungal infection, or on prophylaxis for the patients with the highest risk.

2.6 Systemic mycoses

These are diseases that occur acutely but are, essentially, chronic diseases of the whole body. They are caused by exogenously or endogenously developed fungal mycelia. The exogenously live as saprobes in soil, or on leaves and enter the body by inhaling infected dust. Endogenous systemic mycoses develop from at any part of the body by changing the biological relationships in the infected organism and in that way these fungi become pathogenic.

The weakening of the immune system can lead to infections by fungi that are otherwise considered as nonpathogenic. Such infections can occur in people with impaired immune systems, such as HIV patients, diabetics, or patients with alerted immune function as a result of corticosteroid or antimetabolic therapy. Other factors that can affect the infection by fungi are long-term intravenous treatments, gastrointestinal surgeries, long antibiotic treatments, etc. (Shanahan, 1994).

Systemic mycoses are being caused by primary pathogens, which are inherently virulent, localized primarily in the lungs, and can infect other organs. Systemic mycoses caused by opportunistic pathogens are found in immunocompromised patients. Examples of this kind of infections are aspergillosis, candidiasis, and cryptococcosis.

There are several different fungal species that are capable to cause systemic mycoses. One of these species is *Histoplasma capsulatum*, which is endemic in southern America, causing many serious infections that often turn into diseases with a fatal outcome. The spores are developed first and cause lesions in the mouth, then spores spread, infecting the lungs, liver, kidneys, and spleen. Subsequently, they undergo division and the infected tissues become disorganized. Some infections can cause anemia, fever, and tissue ulcers with a fatal outcome. Tissue infection can occur in postoperative conditions, or after using broad-spectrum antibiotics, or steroid therapy. Fungal infections affecting the CNS are considered as the worst and can often lead to chronic meningitis. Infections of these species are in many cases associated with AIDS (Malcolm and Chin-Hong, 2013). In Brazil there is the highest AIDS prevalence in Latin America and also the highest death rates in patients infected with histoplasmosis, with mortality rates of 33%–42% (Damasceno et al., 2019).

Cryptococcus neoformans is a species that can cause meningitis. In Zaire, cryptococci were found in every third registered patient with AIDS and in India as many as in 50% of the cases. It should be mentioned at this point that cryptococcal meningitis is a major cause of death in patients with AIDS in Africa and Asia. Other types of this genus cause pneumonia, skin, and bone diseases (Bicanic and Harrison, 2014).

Species of the genus *Blastomyces* are also geographically restricted. Most infections have been reported in North America, slightly less in South America and Africa (Jiang et al., 2018). There is little information about direct research into this genus. It causes lesions on the skin, attacks the lungs and bones, and most commonly infects the urinary and genital tract, especially the prostate gland. For the most part, the course of infection by these fungi shares the following key features: inhalation of infectious propagules from the environment, pulmonary and extrapulmonary disease, a significant role of the acquired cellular immune system in the control of the infection and, in some cases, endogenous reactivation upon severe immune impairment of the host (De Hoog et al., 2016).

Coccidioidomycosis is a fungal disease caused by *Coccidioides immitis* and *C. posadasii*. Symptoms could vary, from mild to febrile diseases and even to life-threatening ones. The incidence of these mycoses increased in the past 20 years. The majority of the infections are noticed in California, Arizona, Mexico, and Central America. It is evident that elderly people, members of certain ethnic groups, immunodeficient and transplant patients, pregnant women, diabetics, and prisoners are groups that are considered to be vulnerable. These fungi infect the bones, lungs, skin, and can cause meningitis. Untreated infections lead to chronic conditions and may be fatal in the end (Brown et al., 2013).

Paracoccidioides brasiliensis is a species widespread in southern America and causes one of the most prevalent systemic fungal diseases in that area. It is male-specific and infects the lungs, the lymph, the reticuloendothelial organs, creating ulcers (Ryan, 1994). The susceptibility to paracoccidioidomycosis is genetically based (Borges-Walmsley et al., 2002).

Some of the *Aspergillus* species are capable to induce systemic mycoses. Aspergillosis is an illness resulting from infection caused by *Aspergillus* species, usually affects the respiratory system and secondly, other organs (Kousha et al., 2011). The incidence of systemic mycoses varies from 0.98% to 10.4%. The possibilities of therapy are still limited, so invasive aspergillosis is still associated with a high mortality rate, from 30% to 90% (Brakhage, 2005).

Fungal species from *Candida* genera are also causing agents of systemic mycoses. Invasive candidiasis is a serious illness due to infection of eyes, blood, heart, brain, bones, and other organs, and can also occur in hospitalized and immunosuppressed patients. One of the most serious types of this infection, caused by *Candida* species, is a bloodstream infection, named as candidemia and it usually it affects hospitalized patients. The epidemiology of this disease is rising and became a major cause of systemic infections, with increasing numbers of risk factors that are leading to candidiasis (Chakravarthi and Haleagrahara, 2011).

Systemic fungal infections are very serious, easily transmitted and difficult to treat. They are very virulent, the treatment is long-lasting, and they can even lead to death (Sokovic, 2001).

2.7 Specific fungal species

Aspergillus spp.

The genus *Aspergillus* contains 344 species, mostly saprophytic molds, but some of these can cause human diseases named aspergillosis (Samson et al., 2014), *A. fumigatus* represents the most important species, but several other in the section *fumigati* are also recognized as pathogenic for humans and animals with compromised immune system (Latgé, 1999).

Patients with compromised immune system are at high risk to be infected with these species. The appropriate knowledge of *Aspergillus* species is necessary and could be very valuable in the fight against the diseases they cause, while it is fundamental for further progress in the field (Sugui et al., 2015).

As early as in the end of 19th century were observed similarities in *Aspergillus* species that caused animal and human infections. The signs and symptoms of aspergillosis vary with the type of illness a patient develops. One of the weakest reactions is the allergic reaction. Patients with asthma or cystic fibrosis can develop allergic reaction to *Aspergillus* species. Symptoms could be allergic bronchopulmonary aspergillosis, fever, cough with blood or plugs of mucus, worsening asthma. More serious is considered aspergilloma, connected to chronic lung conditions (emphysema, tuberculosis, or advanced sarcoidosis) and forming of cavities in lungs. Fungus can infect these patients and fibers may enter cavities and form masses known as aspergillomas. Finally, the most serious symptoms are known as invasive aspergillosis. This form starts when the infection spreads fast from the lungs to the other organs (brain, heart, kidneys, or skin). Untreated, this form of aspergillosis may be fatal.

Invasive aspergillosis (IA) was described in a patient with aplastic anemia (Rankin, 1953). Very rapidly IA increased in the number of infected immunocompromised patients (cancer, AIDS, transplanted patients, and other patients receiving immunosuppressive treatments) (Kontoyiannis and Bodey, 2002; Marr et al., 2002). It became the most common fungal infection with approximately 5500 patients (Marr et al., 2002). Invasive aspergillosis (IA) is one of the most serious clinical invasive fungal infections, resulting in a high case-fatality rate among immunocompromised patients (Lin et al., 2001). In susceptible individuals, aspergillosis can cause arthritis (Peromingo et al., 2016), which can be manifested with fever, pain, swelling, tenderness, and restricted range of motion.

Aspergillus infections are the result of the inhalation of spores, but in healthy individuals, the immune system quickly deals with foreign bodies and infection is stopped at the very beginning. On the other hand, in immunocompromised patients this fungus easily can provoke and develop infection and induce further complications (Jariwal et al., 2018). These species are allergenic agents, causing nasal/sinus lesions. Compromised patients, like those that have asthma and tuberculosis, are very sensitive against the spores. In some cases, if immunocompromised persons are infected and if infection is not properly treated, spores could spread through the circulation in the lungs, brain, liver, and other organs, leading that way to systemic infection.

The results from studying the virulence factors in *Aspergillus* species isolated from human and from environment showed the capacity of invasiveness of the these species. It can be concluded that α-amylase, proteinase, hemolysin, phospholipase, and biofilm production are important virulence factors and the activity is in range from 42.85% to 100%. Isolates from environment showed lower presence of virulence factors than those in humans. It looks that species from human produce more virulence factors when they are at human tissues and when they causing tissue damage (Raksha et al., 2017).

Aspergillus fumigatus is the major causing agent of the fungal disease, responsible for approximately 90% of human infections, followed by *Aspergillus flavus*, *Aspergillus niger*, *Aspergillus terreus*, and *Aspergillus nidulans* (Latgé et al., 1997). It should be pointed out that approximately, 95% of the human diseases are linked to only these three species: *A. fumigatus*, *A. niger*, and *A. flavus* (Raksha et al., 2017).

Aspergillus fumigatus

A. fumigatus is an airborne saprophytic fungus with worldwide distribution. It could be found in the soil, compost, and plant surfaces. The spores of this fungus are present widely in the environment. Every conidial head produces thousands of conidia, which are released into the atmosphere. With a diameter of 2–3 μm, this fungus is small enough to reach the lung. It is assumed that several hundred conidia are inhaled during the day. In most cases, the immune system is able to fight off advances with innate immune mechanisms. Allergic bronchopulmonary aspergillosis and aspergilloma, uncommon clinical syndromes, are probably the only infections that could be observed in such hosts (Latgé, 1999).

A. fumigatus is capable to grow at temperatures up to 50°C, while its spores could survive at 70°C.

Among the 63 species in the section *Fumigati*, 17 species are described as opportunistic pathogens of animals. *A. fumigatus* is an opportunistic fungal pathogen, but as in the case with many other fungal pathogens, when the immune system is compromised, it becomes highly susceptible to *A. fumigatus* infection.

This species causes a lung disease (chronic pulmonary aspergillosis), infects patients with cell and organ transplantation, chronic neutropenia, patients with diabetes, and other immunocompromised patients, but also can cause systemic infection, which could be life threatening (invasive aspergillosis) (Frisvad and Larsen, 2016). In the last 20 years, *A. fumigatus* has become the most prevalent among fungal pathogens, infecting and causing usually fatal infections in immunocompromised hosts, especially in developed countries (Denning, 1998). More than 600,000 people die annually as a result of infections caused by *A. fumigatus*, with a mortality rate of 25%–90% (Dhingra and Cramer, 2017). Currently, treatments of this infection consist of antifungal medications but still remain ineffective (Nierman et al., 2005).

It is reported that *A. fumigatus* produces a variety of different extrolites, which may contribute to infection in humans and animals (Coleman et al., 2011): melanins from conidium (Langfelder et al., 2003), fusarinine C and triacetylfusarinine C (Petrik et al., 2014), proteins that protect conidia from being recognized by the immune system or give them the ability to break down barriers within the host to allow their entry (Aimanianda et al., 2009). It was found that proteinases secreted by the fungus are able to hydrolyze pulmonary basal lamina. Human serum in percentage up to 80% does not inhibit this fungus. Even more, this fungus probably uses the human serum as a source of nutrients. In contrast to other fungal species, which are using fatty acids and lipids as their primary carbon source during growth, *A. fumigatus* is using host's proteins as a major carbon source.

Aspergillus flavus

A. flavus is considered to be the second most common fungus that causes human infections, following *A. fumigatus*. The regular and optimal growth is achieved at 33°C, but it can stand in the range of 10–48°C, requires a moist environment, also has a wide tolerance for pH values, growing from 2 to 11 (Gamaletsou et al., 2017).

This species is growing on carpets, building materials, in different kinds of food (meat, milk and milk products, contaminated grain, dead plants and animals, and in soil), at moist environments, in general. Very often it infects corn, peanuts, cotton, and nut trees and also other important industrial and agricultural crops, leading to huge losses of millions of dollars (Rubens and Cardwell, 2005).

As other *Aspergillus* species, *A. flavus* is a producer of mycotoxins, aflatoxins B_1 and B_2. Cyclopiazonic acid is also produced by *A. flavus* and leads to degeneration and necrosis of organs (Amaike and Keller, 2011). Aflatoxin contamination is a major problem in the developing countries (Pitt and Hocking, 2009). These toxins induce cirrhosis, acute liver damage, and they are acting as immunosuppressive and carcinogenic agents. In some cases, mycotoxins are affecting people who have never been in contact with the corresponding fungi. Such is the case with *Aspergillus flavus*, perhaps the most significant fungal species from agricultural standpoint. Aflatoxin could be transferred through the food, eggs, meat, and milk products of animals to humans. Acute exposure to these mycotoxins can cause many problems in humans, starting from nausea and vomiting, abdominal pain, pulmonary and cerebral edemas, convulsions, to severe liver damage and death.

A. flavus spores can affect the eyes and ear canals, cause issues like asthma and tuberculosis, and can develop aspergillosis at alerted immune systems, while in healthy persons can only lead to a mild allergic reaction.

Aspergillus niger

Aspergillus niger is in general present in warmer climates, at field and stored foods. The black spores protect fungus from sun and UV irradiation (Hocking, 2006).

This fungus is responsible for postharvest infection of fresh fruits (apples, pears, peaches, citrus, figs, strawberries, mangoes, melons, and onions), but very often is isolated from nuts (peanuts, pistachios, hazelnuts, walnuts, coconut, and copra), from cereals and oilseeds and also from dried, smoked fish, and meat products. It is a common contaminant of food (Samson et al., 2001).

A. niger is spread via the air, soil, and water. It is capable to grow in a wide temperature range of 6–47°C and extremely wide pH range: 1.4–9.8. It can survive low water conditions, having an index of 0.88, which is relatively high compared to other *Aspergillus* species (Schuster et al., 2002).

Some strains of *A. niger* are producers of very potent mycotoxins called ochratoxins and fumonisin B_2 and aflatoxins (Samson et al., 2001; Schuster et al., 2002; Noonim et al., 2009; Al-Abdalall, 2009). Mycotoxins produced by *A. niger* are linked to quality deterioration, discoloration, reduction in commercial values of fruits and vegetables, but the most important fact is that they also cause damage in kidneys, liver, digestive tract, nervous system, muscles, skin, respiratory organs, genital organs, etc. This group of toxins is leading to nephrotoxicity and renal tumors in animals and they are hazardous to human health through the food chain (Muntanola, 1987; Truckesses and Scott, 2008).

Aspergillus niger is not so virulent as *A. fumigatus* and *A. flavus* in causing human diseases, but still cases of human illness caused by this species are being reported. In rare instances, due to a lung disease, or some other complication, it can provoke aspergillosis. This is frequent among farm workers who are very often exposed to inhalation of spores rich in *Aspergillus niger* spores. This fungus can also be found in the mummies of ancient Egyptian tombs and can be inhaled when they are disturbed (Handwerk, 2005).

A. niger is very often attributed with ear infections (otomycosis). It is consider that the most frequently isolated species responsible for otomycosis are *C. albicans* (30%), *C. parapsilosis* (20%), *A. niger* (20%), *A. flavus* (10%), *A. fumigatus* (5%), *C. tropicalis* (5%), *Trichosporon asahii* (5%), and *Scedosporium apiospermum* (5%) (Pontes et al., 2009).

Candida spp.

It is well known that several *Candida* species live on human body with some of them colonized in the first days after birth, mainly in the mucosa of the gastrointestinal tract (40%–50%), alongside with mouth, pharynx, and larynx. Moreover, they are found on the genitalia of up to 10% of the males, in the areas between the mucous membrane and the skin and in 5%–30% of nonpregnant females. Regarding pregnant women, *Candida* species colonization reaches 30%–40% as a result of alterations in the immune system, elevated glycogen levels, and modified pH (Achkar and Fries, 2010).

Infections can potentially be transmitted from other sources, such as contaminated catheters, syringes shared by drug users, or infected sexual partners. In infants, the infection may occur after inoculation in the birth canal, while in older children and adults might occur as a result of immunosuppression.

Some predisposing factors that could be involved in *Candida* infections are modified pH levels (e.g., of vagina and mouth), pregnancy, premature birth, diabetes, primary immune deficiency, acquired immune deficiency, use of oral contraceptives, surgeries, catheter placement, inflammatory skin diseases, nail trauma, persistent moisture on the hands, poor teeth, malnourishment, etc. (Jenkinson and Douglas, 2002).

Several strains of the genus *Candida* are well known among a series of commensal oral fungi, with *Candida albicans* being the most common among them. The non-*albicans* species, such as *C. glabrata*, *C. krusei*, *C. tropicalis*, and *C. dubliniensis*, are considered as less frequent, although there are facts revealing their presence in clinical cases of oral candidiasis (Samaranayake, 1991; Scully et al., 1994). When certain criteria are met, *Candida* spp. commensals turn into pathogens, causing different kinds of diseases ranging from mucous membrane infection (Gonsalves et al., 2007; Lam et al., 2012), to life-threatening systemic infections (Edmond et al., 1999; Samaranayake and MacFarlane, 1990; Seneviratne et al., 2008). Oral candidiasis is an opportunistic infection that can be asymptomatic, but it can also occur with a series of symptoms such as burning, painful sensation, change of taste, and difficulty in swallowing. It is usually treated by administering synthetic antifungal drugs that can be potentially harmful to the host tissues and also be responsible for emergence of resistance to existing agents.

Indications of oral candidiasis are the signs of inflammation of the mucous surfaces of the oral cavity and the presence of whitish deposits, the removal of which is followed by an unpleasant odor. There are several causes that are responsible for the development of oral candidiasis, such as irrational use of antibiotics, immunosuppressive drugs, poor oral hygiene, etc. (Scully et al., 1994). In invasive candidiasis many symptoms are witnessed, including fever, low blood pressure, rapid pulse, and even systemic rash or peeling of the skin in some of the cases. It should be pointed out that In the United States, candidemia is one of the most common infections of hospitalized patients, with approximately 10,000 deaths per year and annual treatment costs of over

USD $ 1 billion (Kuhn et al., 2004). More than 17 different *Candida* species are known to be responsible for invasive infections, including *C. albicans*, *C. glabrata*, *C. parapsilosis*, *C. tropicalis*, and *C. krusei* (Sardi et al., 2013). As it was previously stated, *C. albicans* is one of the most common pathogens at infection sites, urogenitary tract, oral cavity, etc., being able to cause several health issues, less or more severe (Mayer et al., 2012).

In recent international literature there can be found data connecting pathogenic fungi that are present in intestines, with the incidence of adenomas (Luan et al., 2015). *Candida* was one of the genera present in adenomas (7%) and adjacent tissues (1%). It should be noted at this point that more than 80% of sporadic colorectal cancer cases were induced by colorectal adenoma (Ullmann et al., 2012). Furthermore, it has been reported that more than 70% of all cancer deaths occurred in low- and middle-income countries with statistically poor oral hygiene (Cecchini et al., 2012).

Candida albicans

Since the 1980s, a growing number of hospital reports stated that fungal infections were increasing, with *Candida albicans* being more responsible compared to other strains for that phenomenon.

C. albicans can be found in yeast form (in the form of oval cells—blastospores, by budding), as elongated ellipsoidal cells (pseudohyphae) and in the form of true hyphae with parallel walls that together form a mycelial form (Jacobsen et al., 2012). The transition from yeast to mycelial forms (known as dimorphism) and vice versa is regulated by a complex network of pathways and signal transduction, in response to environmental stimuli. It is generally established that the mycelial form is invasive, as it is observed in most histological preparations, while yeast cell invasion reports are rare in general. Furthermore, dimorphism seems to be of high importance for the formation of biofilms with other resistant microorganisms (Molero et al., 1998).

C. albicans is a communal microorganism and it is a part of healthy human microbiota. It can also colonize some organs at human body, including gastrointestinal tract, female reproductive organs, oral cavity, etc. Any changes in balance, pH shifts or nutritional changes, use of antibiotics, immunodeficiency, etc. can lead *C. albicans* to proliferate in a high rate and cause infection (superficial and dermal diseases, vaginal infections, or even more serious infections with high mortality rates) (47% in some cases). This fungus causes hospital-acquired infections, 15% of all sepsis cases, 40% of bloodstream infections. These serious infections are more often in immunocompromised patients (HIV infected patients, cancer patients, transplanted people, etc.). *C. albicans* can be found and isolated from pacemakers, urinary and central venous catheters, heart valves, joint prostheses, contact lenses, and dentures. It is capable to form biofilm on all of the mentioned medical devices but also is very resistant to existing antifungal drugs and could lead to candidemia (bloodstream infections) (Gulati and Nobile, 2016).

Candida glabrata

This species lacks the ability to form dimorphic phases, so it can be found in the form of blastoconidia (1–4 μm) that are smaller compared to *C. albicans* (4–6 μm) (Kucharickova et al., 2011). As a part of local microbiota, it could be found in the oral cavity and gastrointestinal tract. For many years this species was considered as a nonpathogenic species. Recent research found that after *C. albicans*, this species is the most common in mucosal and systemic candidiasis (Brunke and Hube, 2012).

C. glabrata is responsible for up to 29% of total *Candida* bloodstream infections (Montagna et al., 2014). After *C. albicans* this species is the second most common species, which causes bloodstream *candidiasis* in Northern Europe and the United States (Guinea, 2014), but in the third place in Asia (Tan et al., 2015).

C. glabrata is more closely related to *Sacharomyces cerevisiae* by phylogeny than to *C. albicans*. Moreover, virulence factors detected in *C. albicans* are not conserved in *C. glabrata*. The similarity of these two species is in the easy adhesion to the host cells, where *C. glabrata* possesses a variety of subtelomeric genes, responsible for encoding cell surface proteins (Castaño et al., 2006).

New results of investigation of *Candida* epidemiology showed increased prevalence of non-*albicans Candida* species, with *C. glabrata* among them (Yapar, 2014). However, *C. glabrata* is significantly less pathogenic than *C. albicans*. Probably, the lack of hyphal forms responsible for invasion, non-utilization of hemoglobin, and activity of secreted proteases, contribute to the less observed pathogenicity of this species in comparison with *C. albicans*. *C. glabrata* shows resistance to fluconazole, which is implicated in infections with a high mortality rate. *C. glabrata* showed susceptibility to azoles and became resistant to echinocandins (Kumar et al., 2019).

Candida krusei

Morphologically, *C. krusei* is very different from *C. albicans* and *C. glabrata*, having elongated rice-shaped grains and white-yellow colonies. It is very tolerant to high temperatures, being able to stand even higher than 45°C, and is among the few microorganisms that grow at a medium without vitamins. It is isolated from fruits, soil, wine, and beer. Probably, due to overuse of antimycotics it becomes an opportunistic pathogen (Amaral-Lopes and Moura, 2012).

C. krusei was defined as a transient species that usually had no clinical significance in the past. In the beginning of 20th century was recognized that *C. krusei* may cause disease in human. Since then, this species was recognized as a commensal in animals with very low pathogenicity. However, the reports for *C. krusei* as a human pathogen have remarkably increased in last few decades. Later research described it as a pathogenic species that could induce a variety of clinical symptoms (arthritis, fungemia, endophthalmitis, and endocarditis), notably in immunocompromised patients in nosocomial occasions (Samaranayake and Samaranayake, 1994).

C. krusei could be recovered from different environmental sources. This yeast is a significant causing agent of vaginitis. It is the fifth most common cause of candidemia, but it is the most important species for its innate resistance to fluconazole and capability for reduced susceptibility to other drugs (Pelletier et al., 2005). Mostly, it is isolated from neutropenic patients receiving prophylactic fluconazole therapy. *C. krusei* is a fungus that has the lowest survival rates (53.6%) during 90 days of observing period, among all *Candida* spp. (Pfaller et al., 2012). The prevalence of this species is lower (2%) than that of the other yeast species (Pfaller et al., 2011). It is a fungal pathogen that usually infects patients with hematologic malignancies and transplant patients (Pfaller et al., 2008a,b).

C. krusei, like *C. glabrata*, has become important factor of resistance to azoles, in contrast to *C. albicans* isolates, which continue to demonstrate lower resistance levels (Cleveland et al., 2012). It has also shown resistance to fluconazole, flucytosine, voriconazole, and amphotericin B and according to these facts is considered to be a multiresistant species (Samaranayake, 1991). *C. krusei* is responsible for candidiasis with mortality rates of 30%–60% (Pfaller et al., 2008a,b) (Fig. 2.1).

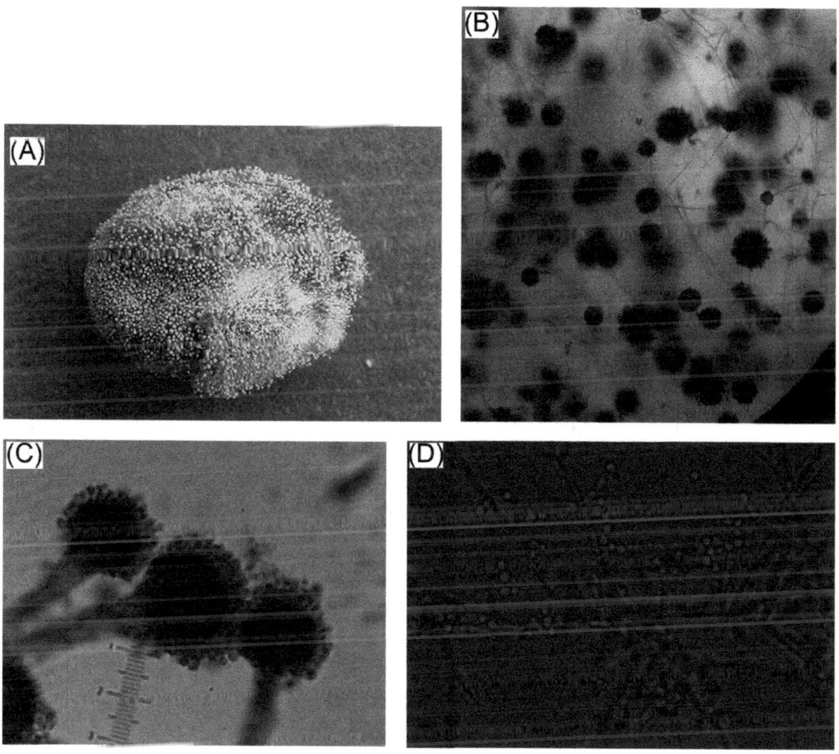

FIG. 2.1 Pictures of some fungi: (A) *Aspergillus flavus*, (B) *Aspergillus niger*, (C) *Aspergillus fumigatus*, (D) *Candida albicans*.

Fusarium spp.

Species of *Fusarium* genus are distributed in soil, plants, and many other organic substrates (Nelson et al., 1994), but also in water, where they can form biofilms (Elvers et al., 1998). Among them are plant and opportunistic human pathogens and also mycotoxin producers. Human diseases caused by *Fusarium* spp. are rising during the time. These infections are usually noticed in immunocompromised patients, but onychomycoses due to these species are not rare in healthy humans, since up to 10% of them are caused by *Fusarium* spp. (van Diepeningen et al., 2015). These fungi are causing agents of eye infections (van Diepeningen et al., 2014) and leukemic patients are especially at risk (Nucci and Anaissie, 2002). Supposedly, the places of entry of *Fusarium* species are different than those of other microfungi. They enter the body through wounds, via catheters and intravenously, through orthopedic traumas, or through eyes (Sutton and Brandt, 2011). *Fusarium* possesses high levels of resistance to most of the known antifungal drugs, which complicates the treatment of these infections. On that way, fusariosis became a life-threatening disease (Guarro, 2013).

F. solani and *F. oxysporum* are species that are in general implicated in human infections, since approximately 80% of all human infections are caused by members of the *F. solani* and *F. oxysporum* species (O'Donnell et al., 2009). The mortality rates with disseminated, skin, and pulmonary diseases were 50%, 40%, and 37.5%, respectively. *Fusarium solani* was the most frequent species, isolated from 49% of the literature cases. *F. solani* can damage the corneal membrane, while *F. dimerum*, is mostly connected with keratomycosis, especially in people with low hygiene.

Due to inhalation of conidia, the occurrence of sinusitis and pneumonia is evident. The role of skin as a portal of entry is supported by the development of infection following skin breakdowns due to trauma (automobile accidents, bamboo), burns, or onychomycosis in normal hosts (Nucci and Anaissie, 2002), and the development of cellulitis (typically at sites of tissue breakdown such as toes and fingers), which may remain localized or lead to disseminated infection in immunocompromised patients (Nucci et al., 2003).

Fusarium species can cause allergic diseases, sinusitis, or chronic noninvasive or invasive sinusitis in immunocompetent persons (Wickern, 1993) and mycotoxicosis in humans after ingestion of contaminated food by toxin produced by these fungi (Nelson et al., 1994).

Fusarium species are famous as mycotoxin producers. Among them, the most important are trichothecenes and zearalenone (deoxynivalenol, diacetoxyscirpenol, HT-2, T-2, and nivalenol). There are examples of human poisoning by these toxins: scabby grain intoxication in India and contaminated, by fumonisins, maize (Marasas, 2001; Voss et al., 2007).

Fusarium oxysporum

F. oxysporum represents soil-borne fungi, very common within the rhizosphere of plants, present all around the world. These fungi are saprophytic, with some of them being pathogenic (they are affecting approximately 120 different plants), being responsible for damage on many economically and industrial important plants.

F. oxysporum is a pathogen that can be dormant for 30 years before expressing its virulence and infecting a plant, animal, or human.

After *F. solani* it is the second most frequent species that can cause human infections (keratitis, endophthalmitis, onychomycosis, cutaneous and subcutaneous infections, sinusitis, arthritis and mycetoma, disseminated infections in immunosuppressed individuals with mortality close to 100%) (Tupaki-Sreepurna and Kindo, 2018). It is reported that *F. oxysporum* causes systemic fusariosis in immunocompromised patients, observed in clinical settings (Fisher et al., 2012).

This fungus is a highly resistant species. Mostly resistant to existing antifungal drugs, but in some cases, and against some strains, amphotericin B and voriconazole could be active. The rest of azoles and echinocandins are inactive against this species (Tupaki-Sreepurna and Kindo, 2018).

Fusarium solani

F. solani is a complex of different species, very similar and closely related, with at least 26 of them (Summerbell, 2003). This fungus is the second most frequent isolated fungi from soil and colonist of plants. Usually it is isolated from a number of industrially important plants (pea, cucurbits, and sweet potato). Nevertheless, it is connected with serious invasive mycoses in immunosuppressed patients. This species is also associated with opportunistic infections of humans and animals, causing systemic infections with high mortality. It can affect skin and other body parts, provoke infection of eyes cornea, osteomyelitis, disseminated disease, skin infection, fungemia, endophthalmitis, and keratitis (Kwon-Chung and Bennett, 1992). Keratitis is reported usually during harvest season in workers, due to dust or plant material, which influences corneal, where spores spread and grow (Pearlman et al., 2011). *F. solani* is the most frequent cause of fungal keratitis in AIDS-positive patients in Africa (Kwon-Chung and Bennett, 1992). This fungus is also found to form biofilm on the surface of contact lenses. The most serious infections caused by this fungus are reported in neutropenic patients, where superficial or subcutaneous initiation can develop to more serious infections that are, usually, fatal (Zhang et al., 2006).

Malassezia spp.

The genus *Malassezia* is represented by dimorphic fungi that are present on normal skin. In general, these fungi mostly live in coexistence with their hosts, but in some cases they are causing serious diseases in humans. This fungus needs fat for normal growth, so usually can be found on scalp, face, and upper part of the body (Gueho et al., 1996). Earlier investigation was focused on a debate about the ability of *Malassezia* to cause human disease. Recent research showed that there is no doubt about that and these species are accepted and proved to be the etiological agent of both cutaneous and systemic diseases.

Under specific conditions, in the presence of certain predisposing factors, these yeasts can cause superficial skin infection (Midgley et al., 1998). These species are able to cause cutaneous and systemic diseases in predisposed humans. The yeasts *Malassezia* spp. are connected with diseases affecting the human skin (atopic dermatitis, seborrheic dermatitis and dandruff, pityriasis versicolor, folliculitis, psoriasis, papillomatosis, onychomycosis, and transient acantholytic dermatosis) (DeAngelis et al., 2007; Gupta et al., 2004).

Immunocompromised patients that stay in intensive care units are at risk to be infected by *Malassezia* spp. Bloodstream infections are less common, but those individuals and premature infants are exposed to these infections.

As the yeasts metabolize the skin surface fat, they form break-down products that can potentially trigger harmful reactions on the skin. *Malassezia* could be involved in the development of skin cancer since it can activate known tumor-promoting pathways, reacting in ways similar to those triggered by sun light (Velegraki et al., 2015).

The occurrence of clinical disease by *Malassezia* depends on the factors permitting conversion of the saprophytic yeast phase of the organism to the mycelial phase (Hay, 2000). High sebum levels in puberty, excessive sweating, warmer season (May to September), application of oil, malnutrition, administration of systemic steroids or the presence of Cushing's disease, pregnant state or therapy with oral contraceptives, antibiotics, or immunosuppressives are some factors that facilitate massive growth of the fungus (Hay, 2000).

It is not clearly known whether the disease is initiated by alteration of the usual relationship between the human host and the resident yeast flora that facilitates mycelial conversion or by transmission from a source patient. Some species constituting normal flora may have a greater pathogenic potential and may be converted to the mycelial phase readily. The occurrence of diseases like pityriasis versicolor in several members of a family suggests a genetic susceptibility to the infection (Inamadar and Palit, 2003).

Seven species could be recognized within *Malassezia* genus: *M. furfur*, *M. sympodialis*, *M. obtusa*, *M. globosa*, *M. restricta*, *M. slooffiae*, and *M. pachydermatis* (Gueho et al., 1996). *M. pachydermatis* is not so common but it still causes canaliculitis, wound infection, and systemic disease in premature neonates (Gueho et al., 1996).

Malassezia globosa is a pathogen that causes tinea versicolor. Conversion of the yeast form into the invasive mycelial form makes it able to invade the stratum corneum. This leads to chronic condition, hyperpigmented lesions with minimal pruritus. *M. globosa* is a species that is isolated in patients with tinea versicolor. *Malassezia* species are responsible for other superficial diseases: acne vulgaris, dacryocystitis, seborrhoeic blepharitis, neonatal pustulosis, confluent and reticulated papillomatosis, onychomycosis, nodular hair infection, and psoriasis (Aractingi et al., 1991). Some more serious conditions caused by these species are also reported (mastitis, sinusitis, septic arthritis, and malignant otitis externa). Systemic disease, peritonitis, due to *M. furfur*, has been noticed in peritoneal dialysis patients (Johnson et al., 1996). *M. furfur* and *M. pachydermatis* have been also reported to cause systemic disease. *M. globosa* and *M. restricta* are, in general pathogenic, associated with tinea versicolor and tinea capitis or seborrheic dermatitis (Crespo et al., 2000). Hair follicles in ears could be also affected by these species (Ashbee et al., 2002).

M. sympodialis is a skin pathogen from tropical and subtropical climates. Individual species have specific host preferences (Kwon-Chung and Bennett, 1992). Usually it affects domestic animals, dogs, causing otitis externa, but some cases of human infections have been also reported (Crespo et al., 2000).

Penicillium spp.

The species of *Penicillium* genus are among the most common microfungi present in the nature. *Penicillium* contains 354 accepted species (Samson et al., 2014). They are present in soil, in decaying vegetation, and in the air. They can contaminate various substrates, food, and they are also mycotoxin producers.

These species are producers of a number of mycotoxins, including ochratoxin, penicillic acid, and others. The production of the toxin usually occurs in cereal grains in cold climates (Pitt, 2000).

In general, they are considered as opportunistic species, but they can be pathogenic in immunocompromised hosts and they can even cause death. They can lead to a variety of diseases known as penicilliosis (Lyratzopoulos et al., 2002), endophtalmitis, otomycosis, necrotizing esophagitis, pneumonia, endocarditis, peritonitis, and urinary tract infections (Deshpande and Koppikar, 1999). It has been noticed that corneal infections were posttraumatic (Deshpande and Koppikar, 1999).

Penicillium marneffei

This fungus was reported as a human pathogen in 1973. *P. marneffei* is a pathogenic fungus and particularly infects AIDS patients in endemic areas, Thailand, Taiwan, and India (Singh et al., 1999).

Infections can also be reported in patients with hematological malignancies and other immunosuppressive patients (Wong et al., 2001). Infection is able to spread by inhalation of spores and, as a result, patients can develop pulmonary infection, while some other routes are also reported, such as skin and digestive tract (Hilmarsdottir et al., 1993). This fungus can infect also liver, lymphatic system, spleen, kidneys, lungs, lymph nodes, spleen, skin, soft tissue, and bones. Unfortunately, *P. marneffei* infections are often fatal (Garbino et al., 2001), affect persons with some of the diseases that involve cell-mediated immunity (Hodgkin's disease, tuberculosis, corticosteroid therapy), but also immunocompetent individuals (Lobulago and Taylor, 1995). The most common symptoms are body weight loss, fever, and anemia.

P. marneffei species are very resistant and may stand high minimal inhibitory concentrations of antifungal drugs, such as flucytosine, fluconazole, and amphotericin B, but they are more sensitive to itraconazole, voriconazole, ketoconazole, and terbinafine (McGinnis et al., 2000).

Penicillium chrysogenum

P. chrysogenum, known as *P. notatum*, is the source of the first antibiotic, penicillin. It is a ubiquitous organism usually found in soil, decaying plants, and in construction sites. This species is isolated also from the human scalp. It was reported that it was also isolated from the Russian space station MIR and the International Space Station, which shows the amazing survival capabilities of this fungus (Makimura et al., 2001; Vesper et al., 2008). It is an inducer of cutaneous infections, pneumonia, pneumonitis, pulmonary fibrosis and pulmonary abscesses, endocarditis, brain abscesses, and endophthalmitis (Lyratzopoulos et al., 2002). *P. chrysogenum* has been also isolated from prosthetic aortic valve in women and also from patients with necrotizing esophagitis and necrotizing pneumonia (D'Antonio et al., 1997). MICs of drugs (amphotericin B, itraconazole, ketoconazole, and voriconazol) needed to inhibit *P. chrysogenum* are acceptably low (Wildfeuer et al., 1998).

Penicillium verrucosum

P. verrucosum is a widespread fungus, common in temperate and subtropical regions and could be found in food, grain, seeds, and decaying vegetation (Summerbell et al., 1994) and animal feed (Cabanes et al., 2010).

In general, this species is opportunistic, but it can cause some infections in humans, such as skin rashes and otomycosis, sinusitis, posttraumatic endophthalmitis, necrotizing esophagitis necrotizing pneumonia, intestinal invasion, and disseminated disease (Barcus et al., 2005; Howard, 2003).

It produces ochratoxin A (OTA), a very potent mycotoxin with immunosuppressive, teratogenic, genotoxic, and human carcinogenic effects (Pfohl-Leszkowicz and Manderville, 2007). It has been noticed that this species causes nephritis in northern and central Europe. Consumption of contaminated food has been notably reported in Bulgaria, minor Yugoslavia, and Romania in 1950s, with high rates of mortality (Pfohl-Leszkowicz and Manderville, 2007).

Inhalation of *P. verrucosum* spores could lead to allergy. This was case recorded at some workers in a cheese factory. Positive culture isolation from workers sputum and antibodies in the blood confirmed the infection and the allergic reaction. That way, it was found that intensive exposure to fungal spores could develop allergic response in humans (Guglielminetti et al., 2001).

2.8 Dermatomycetes

Epidermophyton spp.

This genus includes two species, of which the pathogenic one is *Epidermophyton floccosum*. *Epidermophyton* is distinct from *Microsporum* and *Trichophyton* due to the absence of microconidia (Dongyou, 2011).

E. floccosum is an anthropophilic species with a worldwide distribution. It often causes tinea cruris, tinea pedis (athlete's foot), tinea corporis, and onychomycosis. There are no data that it can infect hairs and only attacks the body, feet and nails. The infection is localized to the nonliving cornified layers of the epidermis, since it is not capable to penetrate the viable tissues of immunocompetent hosts (Aman et al., 2001). The fungus is communicable and usually transmitted by contact, particularly in public showers and gym facilities and can be distributed all around the earth (Christiansen and Svejgaard, 1976). It was the fourth most common cause of dermatomycoses in North America in 20th century (Julius and Belmont, 1997).

Novel results of clinical testing showed its capacity to infect eyes (Dongyou, 2011). This fungus infects humans and very rarely infects animals (Murray and Baron, 2003). *E. floccosum* is one of the most infectious dermatomycetes (Rippon, 1988). In rare cases causes chronic infections. It can cause invasive infections in immunocompromised patients, leading to skin lesions, onychomycosis, and subcutaneous nodules. Among the known drugs than can be used in clinical practice to treat infections caused by *E. floccosum* are itraconazole, terbinafine, and ketoconazole (Hay, 2000).

Microsporum spp.

There are 19 species encompassed in this genus, with only 9 of them being capable to infect humans (Deacon, 2005). *Microsporum* species degrade keratin and while being noninvasive, they can reside on skin and its appendages. They can produce enzymes such as keratinase, proteinases, and elastases that can promote virulence.

With the exception of *M. persicolor*, which does not infect hair, *Microsporum* spp. in most of the cases has the ability to infect the hair and skin. On the contrary, nail infections are considered rare. Special care should be provided to immunocompromised patients since they can be easily infected (Christiansen and Svejgaard, 1976) and should be also taken into account that asymptomatic carriage of the fungi is also possible. These species can be transmitted via contact with the soil (geophilic), via animals (zoophilic), or via humans (anthropophilic).

Taking into consideration their pathogenicity, the most clinically important *Microsporum* species are *M. canis* and *M. gypseum*.

M. canis usually infects cats and dogs. Symptoms that this species induce are tinea capitis and tinea corporis in humans, while mycetoma lesions very rare in immunocompromised hosts. *M. canis* is a widely distributed (Christiansen and Svejgaard, 1976) zoophilic species that is considered as the most common source of ringworm in humans. It can infect the body of adults and the skin of the scalp in children, while it rarely infects the nails. Animals, such as dogs and cats, are carriers of ringworm without necessarily demonstrating any signs of the disease. Asymptomatic animals and people can be carriers of this fungus and it is considered that spreaders of the disease in about 50% of the infected humans are animals. The infections are most usually transmitted via spores, but mycelial fragments in the skin and hair can potentially occur as well. It should be also mentioned that spores have the ability to remain viable for years in blankets, clothing, combs, etc. In older human, females are more susceptible than males. It should be mentioned that human-to-human infection has been frequently recorded (Mancianti et al., 2003).

Failure of treatment in infected patients is noticed in 25%–40% of treated patients, due to lack of compliance or drug penetration into tissue, medication bioavailability, and, in the end, resistance phenomena (Bueno et al., 2010). There is no standardized reference method for testing the antifungal susceptibility of this species, so there is no susceptibility profile for antifungal drugs to treat infections of this fungus.

M. gypseum is a geophilic species, frequently isolated from soil, that can also infect animals. It infects the skin and hair of humans (Christensen et al., 1977). *M. gypseum* can cause tinea capitis and tinea corporis, usually in immunocompetent hosts. Tinea corporis looks as erythematous patches with central clearing or large papulosquamous plaques. In immunocompromised patients, the lesions can form intensive hyperkeratosis. In some cases symptoms are atypical, intensive, and difficult to control (Pandey and Pandey, 2013). Dermatomycoses caused by this fungus are like an inflammatory mycosis that typically affects the glabrous skin and scalp, especially in children, sometimes even presented as onychomycosis (Romano et al., 2009). For a long period, griseofulvin was usually the preferable drug in clinical practice for treating infections caused by *Microsporum* spp. but nowadays, terbinafine and itraconazole are the most usually administered agents (Elewski, 2000).

Trichophyton spp.

This genus contains 22 species (Deacon, 2005), with *Trichophyton mentagrophytes*, *Trichophyton rubrum*, and *Trichophyton tonsurans* being the most pathogenic ones.

T. mentagrophytes is anthropophilic and zoophilic species. In ancient times, symptoms caused by this fungus were attributed to insect bites. It can be found in the soil and it infects the feet, body, nails, beard, scalp, arm, and groin, but the disease it causes more often is athlete's foot, the most common fungal skin infection. It is particularly prevalent among men, since nearly 70% of the population will meet the infection at some point. This infection is contagious and can spread by contact, from an infected person, or via contaminated surfaces (towels, floors, and shoes). An interesting fact is that the anthropophilic species cannot infect hair (Christiansen and Svejgaard, 1976). It should also be noted that *T. mentagrophytes* is widely distributed.

T. rubrum is the most usually isolated dermatomycete and the diseases caused by this fungus are among the most common human skin diseases. Approximately 80% of the treated patients respond well to topical treatment with antifungal drugs, but 20% develop a chronic form of dermatomycoses, becoming resistant to antifungal therapy (Blutfield et al., 2015). It is an anthropophilic species that can also be found in animals, but not in soil. It usually infects the feet, nails, body, groin, and rarely the scalp. This fungus is widely distributed (Gregory et al., 2009). This species cause an aggressive and invasive infection granuloma, invasive disease, limited to the extremities and subcutaneous infections. It affects immunocompetent and immunocompromised patients. It has been reported that, in general, *T. rubrum* isolates are found to be more susceptible to antifungals when compared to *T. mentagrophytes* (Salama et al., 2001).

T. tonsurans is an anthropophilic fungus responsible for infections of the scalp and glabrous skin or nails. The clinical images of infections caused by *T. tonsurans* are not very apparent initially (Gregory et al., 2009). *Trichophyton* is a keratinophilic filamentous fungus that can potentially attack keratinized tissues. It has several enzymes such as acid proteinases, elastase, and keratinases that are considered as its major virulence factors (Weitzman and Summerbell, 1995). *T. tonsurans* mostly infects children, but also can affect adults who are in contact with infected children (Hay et al., 1996). The clinical manifestations of tinea capitis caused by *T. tonsurans* vary. Infections induced by *T. tonsurans* are named "black dot" tinea capitis (Sombatmaithai et al., 2015). Tinea corporis due to *T. tonsurans* is demonstrated as small, erythematous, scaly plaques, 1–2 cm in diameter, with no central clearing, looking like eczema. Symptoms can be grouped into three types: the seborrheic (dandruff and crusts), the kerion celsi type, and the black dot type (Hiruma et al., 2015).

Commercial drugs ketoconazole, itraconazole, and clotrimazole exhibit antifungal potential against *Trichophyton* spp. under in vitro conditions.

2.9 Dematiaceous fungi

Dematiaceae family contains species that are found in the soil and in decomposing vegetables (Azulay and Azulay, 2004). Approximately 100 species and 60 genera are connected to a wide spectrum of human infections. They are divided into several groups depending their morphology: annellides (*Phaeoannellomyces*, *Exophiala*), phialides (*Phialophora*, *Wangiella*), adelophialides (*Phialemonium* without collarettes, *Lecythophora* with collarettes), differentiation of conidiophores (*Xylohypha* versus *Cladosporium*), and conidial hilum, septation and germination (*Bipolaris*, *Drechslera*, *Exserohilum*). These fungi are responsible for a variety of infections in humans. Many people are exposed to them, at least by inhalation or some kind of trauma and they are considered as the most common etiological agents. These fungi possess unique mechanisms of pathogenicity due to the presence of melanin in their cell walls, giving them dark color. It should be mentioned at this point that melanin is a virulence factor in fungi (Revankar, 2007).

Chromomycosis is a term for animal and human fungal infections caused by species of family Dematiaceae (known as dark brown or black fungi). These species can cause a spectrum of clinical diseases with four major categories: Superficial Chromomycosis-colored macule induced by fungus inhabiting a dead layer of the skin, Chromoblastomycosis-slow-growing nodule produced by fungi in the dermis, Chromohyphomycosis-A dermal nodule produced by fungi. Fungi can form infection in deeper tissues, muscles, bone, and brain or can infect eye (keratomycosis), Chromomycetoma-draining sinuses (Zaias, 1978). Fungi belonging to the *Dematiaceae* family are the main pathogens, after *Aspergillus* genus, in causing allergic fungal sinusitis. Pulmonary diseases caused by these species are allergic bronchopulmonary and nonallergic syndromes (Revankar, 2007).

In general, are affected populations in rural regions, often in tropical or subtropical regions of developing countries. Disease is progressing slowly, developing for over a mean of 10 years, and usually is connected with secondary bacterial infections (Sampaio and Rivitti, 2000).

Cladophialophora bantiana is one of the most often causing agents among other dematiaceae species associated with human disease. It was recorded that this fungus causes sinusitis and paranasal infections, but also brain abscess and fungal meningitis (Krishanko et al., 2019).

Exophiala is a genus of dermatiaceous hyphomycetes, which are widely distributed in nature. It is a rare fungal pathogen of humans, but in recent years the infections due to these species increased a lot, especially in immunocompromised patients. Diseases caused by these fungi include pheohyphomycosis, eumycotic mycetoma, chromoblastomycosis, and disseminated infection (García-Martos et al., 2002).

Bipolaris species are the most common dematiaceae fungi that are associated with sinusitis. *Ochroconis gallopava* is causing encephalitis in turkey but is also isolated from human brain. *Phialophora* species are connected with chromoblastomycosis and pheohyphomycosis. *Cladosporium bantianum* and *C. trichoides* are also fungi that often cause disease in humans. They are soil fungi, and the main route of infection is the respiratory tract (Dixon et al., 1987).

References

Achkar, J.M., Fries, B.C., 2010. *Candida* infections of the genitourinary tract. Clin. Microbiol. Rev. 23 (2), 253–273.

Aimanianda, V., Bayry, J., Bozza, S., Kniemeyer, O., Perrucio, K., Elluru, S.R., Clavaud, C., Paris, S., Brakhage, A.A., Kaveri, S.V., Romani, L., Latgé, J.P., 2009. Surface hydrophobin prevents immune recognition of airborne fungal spores. Nature 460, 1117–1121.

Al-Abdalall, A.H.A., 2009. Production of aflatoxins by *Aspergillus flavus* and *Aspergillus niger* strains isolated from seeds of pulses. J. Food Agric. Environ. 7, 33–39.

Alexander, B.D., Johnson, M.D., Pfeiffer, C.D., Jiménez-Ortigosa, C., Catania, J., Booker, R., Castanheira, M., Messer, S.A., Perlin, D.S., Pfaller, M.A., 2013. Increasing echinocandin resistance in *Candida glabrata*: clinical failure correlates with presence of FKS mutations and elevated minimum inhibitory concentrations. Clin. Infect. Dis. 56, 1724–1732.

Amaike, S., Keller, N.P., 2011. Aspergillus flavus. Annu. Rev. Phytopathol. 49, 107–133.

Aman, S., Haroon, T.S., Hussain, I., Bokhari, M.A., Khurshid, K., 2001. Tinea unguium in Lahore, Pakistan. Med. Mycol. 39, 177–180.

Amaral-Lopes, S., Moura, A., 2012. Neonatal fungal sepsis by *Candida krusei*: a report of three cases and a literature review. Med. Mycol. Case Rep. 1 (1), 24–26.

Aractingi, S., Cadrand, S., Reygagne, P., Wallach, D., 1991. Non follicular pustules induced by *Malassezia furfur* in a newborn. Ann. Dermatol. Venereol. 118, 856–858.

Ashbee, H.R., Leck, A.K., Puntis, J.W., Parsons, W.J., Evans, E.G., 2002. Skin colonization by *Malassezia* in neonates and infants. Infect. Control Hosp. Epidemiol. 23, 212–216.

Azulay, R.D., Azulay, D.R., 2004. Dermatologia. Guanabara Koogan, Rio de Janeiro, pp. 287–288. Capitulo 23, Cromomicose.

Baddley, J.W., Patel, M., Bhavnani, S.M., Moser, S.A., Andes, D.R., 2008. Association of fluconazole pharmacodynamics with mortality in patients with candidemia. Antimicrob. Agents Chemother. 52, 3022–3028.

Barcus, A.L., Burdette, S.D., Herchline, T.E., 2005. Intestinal invasion and disseminated disease associated with *Penicillium Chrysogenum*. Ann. Clin. Microbiol. Antimicrob. 4, 21.

Benoit, D., Peleman, R., Claeys, G., Hertveldt, K., Vandewoude, K., 2000. Mixed community-acquired fungal infection in an apparently healthy patient. Eur. J. Clin. Microbiol. Infect. Dis. 19 (8), 642–643.

Bicanic, T.A., Harrison, T.S., 2014. Systemic fungal infections. Fung. Infect. 42, 26–30.

Blutfield, M.S., Lohre, J.M., Pawich, D.A., Vlahovic, T.C., 2015. The immunologic response to *Trichophyton Rubrum* in lower extremity fungal infections. J. Fungi (Basel) 1 (2), 130–137.

Borges-Walmsley, M.I., Chen, D., Shu, X., Walmsley, A.R., 2002. The pathobiology of *Paracoccidioides brasiliensis*. Trends Microbiol. 10 (2), 80–87.

Brakhage, A.A., 2005. Systemic fungal infections caused by *Aspergillus* species: epidemiology, infection process and virulence determinants. Curr. Drug Targets 6 (8), 875–886.

Brown, J., Benedict, K., Park, B.J., Thompson, G.R., 2013. Coccidioidomycosis: epidemiology. Clin. Epidemiol. 5, 185–197.

Brunke, S., Hube, B., 2012. Two unlike cousins: *Candida albicans* and *C. glabrata* infection strategies. Cell. Microbiol. 15 (5), 701–708.

Bueno, J.G., Martinez, C., Zapata, B., Sanclemente, G., Gallego, M., Mesa, A.C., 2010. *In vitro* activity of fluconazole, itraconazole, voriconazole and terbinafine against fungi causing onychomycosis. Clin. Exp. Dermatol. 35 (6), 658–663.

Cabanes, F.J., Bragulat, M.R., Castella, G., 2010. Ochratoxin a producing species in the genus *Penicillium*. Toxins 2 (5), 1111–1120.

Castaño, I., Cormack, B., De Las Peñas, A., 2006. Virulence of the opportunistic pathogen mushroom *Candida glabrata*. Rev. Latinoam. Microbiol. 48 (2), 66–69.

Cecchini, C., Silvi, S., Cresci, A., Piciotti, A., Caprioli, G., Papa, F., Sagratini, G., Vittori, S., Maggi, F., 2012. Antimicrobial efficacy of *Achillea ligustica* All. (Asteraceae) essential oils against reference and isolated oral microorganisms. Chem. Biodivers. 9, 12–24.

Chakravarthi, S., Haleagrahara, N., 2011. A comprehensive review of the occurrence and management of systemic candidiasis as an opportunistic infection. Microbiol. J. 1, 1–7.

Christensen, M.S., Hargens, C.W., Nacht, S., Gans, E.H., 1977. Viscoelastic properties of intact human skin: instrumentation, hydration effects, and the contribution of the stratum corneum. J. Invest. Dermatol. 69, 282–286.

Christiansen, A.H., Svejgaard, E., 1976. Studies of the antigenic structure of *Trichophyton rubrum, Trichophyton mentagrophytes, Microsporon canis* and *Epidermophyton floccosum* by crossed immuno-electrophoresis. Acta. Pathol. Microbiol. Scand. 84, 337–341.

Cleveland, A.A., Farley, M.M., Harrison, L.H., Stein, B., Hollick, R., Lockhart, S.R., Magill, S.S., Derado, G., Park, B.J., Chiller, T.M., 2012. Changes in incidence and antifungal drug resistance in candidemia: results from population-based laboratory surveillance in Atlanta and Baltimore, 2008–2011. Clin. Infect. Dis. 55, 1352–1361.

Coleman, J.J., Ghosh, S., Okoli, I., Mylonakis, E., 2011. Antifungal activity of microbial secondary metabolites. PLoS ONE 6, e25321.

Crespo, E.V., Ojeda, M.A., Vera, C.A., Crespo, E.A., Sanchez, F.E., 2000. *Malassezia globosa* as the causative agent of pityriasis versicolor. Br. J. Dermatol. 143, 799–803.

D'Antonio, D., Violate, B., Farina, C., Sacco, R., Angelucci, D., Masciulli, M., 1997. Necrotizing pneumonia caused by *Penicillium chrysogenum*. J. Clin. Microbiol. 35 (12), 3335–3337.

Damasceno, L.S., de Teixeira, M., Barker, B.M., Almeida, M.A., de Muniz, M., Pizzini, C.V., Mesquita, J.R.L., Rodrigez-Arellanes, G., Ramírez, J.A., Vite-Garín, T., Leitão, M.J.S., Taylor, M.L., Almeida-Paes, R., Zancopé-Oliveira, R.M., 2019. Novel clinical and dual infection by *Histoplasma capsulatum* genotypes in HIV patients from Northeastern, Brazil. Sci. Rep. 9, 11789.

De Hoog, G.S., Guarro, J., Gené, J., Figueras, M.J., 2016. Atlas of Clinical Fungi, E-version 4.1.4. CBS-KNAW Fungal Biodiversity Centre, Utrecht.

Deacon, J.W., 2005. Fungal Biology, fourth ed. Oxford, Wiley-Blackwell.

DeAngelis, Y.M., Saunders, C.W., Johnstone, K.R., Reeder, N.L., Coleman, C.G., Kaczvinsky Jr., J.R., Gale, C., Walter, R., Mekel, M., Lacey, M.P., Keough, T.W., Fieno, A., Grant, R.A., Begley, B., Sun, Y., Fuentes, G., Youngquist, R.S., Xu, J., Dawson Jr., T.L., 2007. Isolation and expression of a *Malassezia globosa* lipase gene, LIP1. J. Invest. Dermatol. 127 (9), 2138–2146.

Denning, D.W., 1998. Invasive aspergillosis. Clin. Infect. Dis. 26, 781–805.

Deshpande, S.D., Koppikar, G.V., 1999. A study of mycotic keratitis in Mumbai. Indian J. Pathol. Microbiol. 42, 81–87.

Dhingra, S., Cramer, R.A., 2017. Regulation of sterol biosynthesis in the human fungal pathogen *Aspergillus fumigatus*: opportunities for therapeutic development. Front. Microbiol. 8, 92.

Dixon, D.M., Merz, W.G., Elliott, H.L., Macleay, S., 1987. Experimental central nervous system phaeohyphomycosis following intranasal inoculation of *Xylohypha bantiana* in cortisone-treated mice. Mycopathologia 100, 145–153.

Dongyou, L., 2011. Molecular Detection of Human Fungal Pathogens. CRC Press, Boca Raton, FL, pp. 421–422.

Edmond, M.B., Wallace, S.E., McClish, D.K., Pfaller, M.A., Jones, R.N., Wenzel, R.P., 1999. Nosocomial bloodstream infections in United States hospitals: a three-year analysis. Clin. Infect. Dis. 29, 239–244.

Elewski, B.E., 2000. Tinea capitis: a current perspective. J. Am. Acad. Dermatol. 42, 1–24.

Elvers, K.T., Leeming, K., Moore, C.P., Lappin-Scott, H.M., 1998. Bacterial-fungal biofilms in flowing water photo-processing tanks. J. Appl. Microbiol. 84, 607–618.

Farrer, R.A., Weinert, L.A., Bielby, J., Garner, T.W., Balloux, F., Clare, F., Bosch, J., Cunningham, A.A., Weldon, C., du Preez, L.H., Anderson, L., Pond, S.L., Shahar-Golan, R., Henk, D.A., Fisher, M.C., 2011. Multiple emergences of genetically diverse amphibian-infecting chytrids include a globalized hypervirulent recombinant lineage. Proc. Natl. Acad. Sci. U. S. A. 108, 18732–18736.

Fisher, M.C., Henk, D.A., Briggs, C.J., Brownstein, J.S., Madoff, L.C., McCraw, S.L., Gurr, S.J., 2012. Emerging fungal threats to animal, plant and ecosystem health. Nature 484, 186–194.

Frisvad, J.C., Larsen, T.O., 2016. Extrolites of *Aspergillus fumigatus* and other pathogenic species in *Aspergillus* section *Fumigati*. Front. Microbiol. 6, 1485.

Gamaletsou, M.N., Rammaert, B., Bueno, M.A., Sipsas, N.V., Moriyama, B., Kontoyiannis, D.P., Roilides, E., Zeller, V., Taj-Aldeen, S.J., Henry, M., Petraitis, V., Denning, D.W., Lortholary, O., Walsh, T.J., 2017. International osteoarticular mycoses consortium. *Aspergillus* arthritis: analysis of clinical manifestations, diagnosis, and treatment of 31 reported cases. Med. Mycol. 55 (3), 246–254.

Garbino, J., Kolarova, L., Lew, D., Hirschel, B., Rohner, P., 2001. Fungemia in HIV-infected patients: a 12-year study in a tertiary care hospital. AIDS Patient Care STDs 15, 407–410.

García-Martos, P., Márquez, A., Gene, J., 2002. Infecciones humanas por levaduras negras del género Exophiala. Rev. Iberoam. Micol. 19, 72–79.

Gonsalves, W.C., Chi, A.C., Neville, B.W., 2007. Common oral lesions. Part II: masses and neoplasia. Am. Fam. Physician 75, 509–512.

Gregory, P.J., Johnson, S.N., Newton, A.C., Ingram, J.S.I., 2009. Integrating pests and pathogens into the climate change/food security debate. J. Exp. Bot. 60, 2827–2838.

Guarro, J., 2013. Fusariosis, a complex infection caused by a high diversity of fungal species refractory to treatment. Eur. J. Clin. Microbiol. Infect. Dis. 32, 1491–1500.

Gueho, E., Midgley, G., Guillot, J., 1996. The genus *Malassezia* with description of four new species. Antonie Van Leeuwenhoek 69 (4), 337–355.

Guglielminetti, M., Valoti, E., Cassini, P., Taino, G., Caretta, G., 2001. Respiratory syndrome very similar to extrinsic allergic alveolitis due to *Penicillium verrucosum* in workers in a cheese factory. Mycopathologia 149, 123–129.

Guinea, J., 2014. Global trends in the distribution of *Candida* species causing candidemia. Clin. Microbiol. Infect. 20, 5–10.

Gulati, M., Nobile, C.J., 2016. *Candida albicans* biofilms: development, regulation, and molecular mechanisms. Microbes Infect. 18 (5), 310–321.

Gupta, A.K., Batra, R., Bluhm, R., Boekhout, T., Dawson Jr., T.L., 2004. Skin diseases associated with *Malassezia* species. J. Am. Acad. Dermatol. 51 (5), 785–798.

Hajjeh, R.A., Sofair, A.N., Harrison, L.H., Lyon, G.M., Arthington-Skaggs, B.A., Mirza, S.A., Phelan, M., Morgan, J., Lee Yang, W., Ciblak, M.A., Benjamin, L.E., Sanza, L.T., Huie, S., Yeo, S.F., Brandt, M.E., Warnock, D.W., 2004. Incidence of bloodstream infections due to *Candida* species and *in vitro* susceptibilities of isolates collected from 1998 to 2000 in a population-based active surveillance program. J. Clin. Microbiol. 42, 1519–1527.

Handwerk, B., 2005. Egypt's "King Tut Curse" caused by tomb toxins? Natl. Geogr. https://www.nationalgeographic.com/science/2005/05/news-tutankhamun-carnarvon-mold-bacteria-toxins/ (Accessed 20 February 2020).

Hay, R.J., 2000. Dermatophytosis and other superficial mycoses. In: Mandell, G.L., Bennett, J.E., Dolin, R. (Eds.), Principles and Practice of Infectious Diseases, fifth ed. Mandell, Douglas, and Bennett's, London.

Hay, R.J., Clayton, Y.M., De Silva, N., Midgley, G., Rossor, E., 1996. Tinea capitis in south-East London-a new pattern of infection with public health implications. Br. J. Dermatol. 135 (6), 955–958.

Hilmarsdottir, I., Meynard, J.L., Rogeaux, O., Guermonprez, G., Datry, A., Katlama, C., Brucker, G., Coutellier, A., Danis, M., Gentilini, M., 1993. Disseminated *Penicillium marneffei* infection associated with human immunodeficiency virus: a report of two cases and a review of 35 published cases. J. Acquir. Immune Defic. Syndr. 6, 466–471.

Hiruma, J., Ogawa, Y., Hiruma, M., 2015. *Trichophyton tonsurans* infection in Japan: epidemiology, clinical features, diagnosis and infection control. J. Dermatol. 42 (3), 245–249.

Hocking, A.D., 2006. Aspergillus and related teleomorphs. In: Food Spoilage Microorganisms, pp. 451–487.

Howard, D.H., 2003. Pathogenic Fungi in Humans and Animals. Marcel Dekker, New York, pp. 329–333.

Hyde, K.D., Al-Hatmi, A.M.S., Andersen, B., Boekhout, T., Buzina, W., Dawson Jr., T.L., Eastwood, D.C., Jones, E.B.G., de Hoog, S., Kang, Y., Longcore, J.E., McKenzie, E.H.C., Meis, J.F., Pinson-Gadais, L., Rathnayaka, A.R., Richard-Forget, F., Stadler, M., Theelen, B., Thongbai, B., Tsui, C.K.M., 2018. The world's ten most feared fungi. Fungal Divers. 93, 161–194.

Inamadar, A.C., Palit, A., 2003. The genus *Malassezia* and human disease. Indian J. Dermatol. Venereol. Leprol. 69, 265–270.

Summerbell, R.C., Staib, F., Ahearn, D.G., Ando, M., Ajello, L., Crow, S.A., Fung, D., Gregor, T., Noble, J., Price, D.L., Simmons, R.B., Tarlo, S.M., Woychuk, W., 1994. Household hyphomycetes and other indoor fungi. J. Med. Vet. Mycol. 32 (Suppl. 1), 277–286.

Jacobsen, D., Wilson, D., Wächtler, B., Brunke, S., Naglik, J.R., Hube, B., 2012. *Candida albicans* dimorphism as a therapeutic target. Expert Rev. Anti Infect. Ther. 10 (1), 85–93.

Jariwal, R., Heidari, A., Sandhu, A., Patel, J., Shoaepour, K., Natarajan, P., Cobos, E., 2018. Granulomatous invasive *Aspergillus flavus* infection involving the nasal sinuses and brain. J. Investig. Med. High Impact. Case Rep. 6, 2324709618770473.

Jenkinson, H.F., Douglas, L.J., 2002. Interactions between *Candida* species and bacteria in mixed infections. In: Brogden, K.A., Guthmiller, J.M. (Eds.), Polymicrobial Diseases. ASM Press, Washington, DC (Chapter 18).

Jiang, Y., Dukik, K., Muñoz, J.F., Sigler, L., Schwartz, I.S., Govender, N.P., Kenyon, C., Feng, P., van den Ende, B.G., Stielow, J.B., Stchigel, A.M., Lu, H., de Hoog, S., 2018. Phylogeny, ecology and taxonomy of systemic pathogens and their relatives in *Ajellomycetaceae* (Onygenales): *Blastomyces*, *Emergomyces*, *Emmonsia*, *Emmonsiellopsis*. Fungal Divers. 90, 245–291.

Johnson, A.S., Bailey, E., Wright, P.A., Solomon, L., 1996. *Malassezia furfur*: a possible cause of culture negative CAPD-peritonitis. Perit. Dial. Int. 16, 187–188.

Julius, K., Belmont, C.A., 1997. Laboratory Handbook of Dermatophytes: A Clinical Guide and Laboratory Handbook of Dermatophytes and Other Filamentous Fungi From Skin, Hair, and Nails. Star Pub, ISBN: 978-0898631579.

Kao, A.S., Brandt, M.E., Pruitt, W.R., Conn, L.A., Perkins, B.A., Stephens, D.S., Baughman, W.S., Reingold, A.L., Rothrock, G.A., Pfaller, M.A., Pinner, R.W., Hajjeh, R.A., 1999. The epidemiology of candidemia in two United States cities: results of a population-based active surveillance. Clin. Infect. Dis. 29, 1164–7110.

Kogoj, F., Čajkovac, S., 1959. Medicinskaenciklopedija. vol. 3 JLZ, Zagreb, pp. 75–91.

Kontoyiannis, D.P., Bodey, G.P., 2002. Invasive aspergillosis in 2002: an update. Eur. J. Clin. Microbiol. Infect. Dis. 21, 161–172.

Kousha, M., Tadi, R., Soubani, A.O., 2011. Specific fungal species, *Pulmonary aspergillosis*: a clinical review. Eur. Respir. Rev. 20, 156–174.

Krishanko, D., Nandini, C., Shantasil, P., Sourin, M., 2019. Phaeohyphomycosis meningitis in an immunocompetent person: a case report and review of literature. Int. J. Curr. Res. 12 (1), 9870.

Kucharickova, S., Tournu, H., Lagrou, K., Van Dijck, P., Bujdakova, H., 2011. Detailed comparison of *Candida albicans* and *Candida glabrata* biofilms under different conditions and their susceptibility to caspofungin and anidulafungin. J. Med. Microbiol. 60, 1261–1269.

Kuhn, D.M., Mukherjee, P.K., Clark, T.A., Pujol, C., Chandra, J., Hajjeh, R.A., Warnock, D.W., Soll, D.R., Ghannoum, M.A., 2004. *Candida parapsilosis* characterization in an outbreak setting. Emerg. Infect. Dis. 10 (6), 1074–1081.

Kumar, K., Askari, F., Sagar Sahu, M., Kaur, R., 2019. *Candida glabrata*: a lot more than meets the eye. Microorganisms 7 (2), 39. 1–22.

Kwon-Chung, K.J., Bennett, J.E., 1992. Infections caused by *Malassezia* species. In: Medical Mycology. Lea and Febiger, Philadelphia, pp. 70–182.

Lam, O.L.T., Bandara, H.M.H.N., Samaranayake, L.P., Mcgrath, C., Li, L.S.W., 2012. Oral health promotion interventions on oral yeast in hospitalised and medically compromised patients: a systematic review. Mycoses 55 (2), 123–142.

Langfelder, K., Streibel, M., Jahn, B., Haase, G., Brakhaage, A.A., 2003. Biosynthesis of fungal melanins and their importance for human pathogenic fungi. Fungal Genet. Biol. 38, 143–158.

Latgé, J.P., 1999. *Aspergillus fumigatus* and aspergillosis. Clin. Microbiol. Rev. 12, 310–350.

Latgé, J.P., Paris, S., Sarfati, J., Debeaupuis, J.P., Beauvais, A., Jaton-Ogay, K., Monod, M., 1997. Infectivity of *Aspergillus fumigatus*. In: Vanden Bossche, H., Stevens, D.A., Odds, F. (Eds.), Host-Fungus Interplay. National Foundation for Infectious Diseases, Bethesda.

Lin, S.J., Schranz, J., Teutsch, S.M., 2001. Aspergillosis case-fatality rate: systematic review of the literature. Clin. Infect. Dis. 32, 358–366.

Lobulago, K.F., Taylor, J.W., 1995. Phylogeny and PCR identification of the human pathogenic fungus *Penicillium marneffei*. J. Clin. Microbiol. 33, 85–89.

Lockhar, S.R., Iqbal, N., Cleveland, A.A., Farley, M.M., Harrison, L.H., Bolden, C.B., Baughman, W., Stein, B., Hollick, R., Park, B.J., Chiller, T., 2012. Species identification and antifungal susceptibility testing of *Candida* bloodstream isolates from population-based surveillance studies in two U.S. cities from 2008 to 2011. J. Clin. Microbiol. 50 (11), 3435–3442.

Lortholary, O., Desnos-Ollivier, M., Sitbon, K., Fontanet, A., Bretagne, S., Dromer, F., French Mycosis Study Group, 2011. Recent exposure to caspofungin or fluconazole influences the epidemiology of candidemia: a prospective multicenter study involving 2,441 patients. Antimicrob. Agents Chemother. 55, 532–538.

Luan, C., Xie, L., Yang, X., Huifang, M., Na, L., Ruifen, Z., Xue, X., Yongfei, H., Yulan, L., Na, W., Yuanmin, Z., Baoli, Z., 2015. Dysbiosis of fungal microbiota in the intestinal mucosa of patients with colorectal adenomas. Sci. Rep. 5, 7980.

Lyratzopoulos, G., Ellis, M., Nerringer, R., Denning, D.W., 2002. Invasive infection due to *Penicillium* species other than *P. marneffei*. J. Infect. 45 (3), 184–195.

Madhani, H.D., Fink, G.R., 1998. The control of filamentous differentiation and virulence in fungi. Trends Cell Biol. 8 (9), 348–353.

Magill, S.S., Edwards, J.R., Bamberg, W., Beldavs, Z.G., Dumyati, G., Kainer, M.A., Lynfield, R., Maloney, M., McAllister-Hollod, L., Nadle, J., Ray, S.M., Thompson, D.L., Wilson, L.E., Fridkin, S.K., 2014. Emerging infections program healthcare-associated infections and antimicrobial use prevalence survey team. Multistate point-prevalence survey of health care-associated infections. New Eng. J. Med. 370, 1198–1208.

Makimura, K., Hanazawa, R., Takatori, K., Tamura, Y., Fujisaki, R., Nishiyama, Y., Abe, S., Uchida, K., Kawamura, Y., Ezaki, T., Yamaguchi, H., 2001. Fungal flora on board the Mir-Space Station, identification by morphological features and ribosomal DNA sequences. Microbiol. Immunol. 45 (5), 357–363.

Malcolm, T.R., Chin-Hong, P.V., 2013. Endemic mycoses in immunocompromised hosts. Curr. Infect. Dis. Rep. 15 (6), 536–543.

Mancianti, F., Nardoni, S., Corazza, M., D'achille, P., Ponticelli, C., 2003. Environmental detection of *Microsporum canis* arthrospores in the households of infected cats and dogs. J. Feline Med. Surg. 5, 323–328.

Marasas, W.F., 2001. Discovery and occurrence of the fumonisins: a historical perspective. Environ. Health Perspect. 109, 239–243.

Marković, T., Nikolić, M., Glamočlija, J., Ćirić, A., Emerald, M., Radanović, D., Zheljazkov, V., Soković, M., 2016. Essential oils for the prevention and treatment of human opportunistic fungal diseases. In: Medicinal and Aromatic Crops: Production, Phytochemistry, and Utilization, pp. 247–277.

Marr, K.A., Carter, R.A., Crippa, F., Wald, A., Corey, L., 2002. Epidemiology and outcome of mould infections in hematopoietic stem cell transplant recipients. Clin. Infect. Dis. 34, 909–917.

Mayer, F.L., Wilson, D., Hube, B., 2012. *Candida albicans* pathogenicity mechanisms. Virulence 4 (2), 119–128.

McGinnis, M.R., Nordoff, N.G., Ryder, N.S., Nunn, G.B., 2000. *In vitro* comparison of terbinafine and itraconazole against *Penicillium marneffei*. Antimicrob. Agents Chemother. 44 (5), 1407–1408.

Midgley, G., Gueho, E., Guillo, J., 1998. Diseases caused by Malassezia species. In: Ajello, L., Hay, R.J. (Eds.), Topley and Wilson's Microbiology and Microbial Infections (Medical Mycology), ninth ed. Arnold, London, pp. 201–214.

Molero, G., Díez-Orejas, R., Navarro-García, F., Monteoliva, L., Pla, J., Gil, C., Sánchez-Pérez, M., Nombela, C., 1998. *Candida albicans*: genetics, dimorphism and pathogenicity. Int. Microbiol. 1 (1), 95–106.

Montagna, M.T., Lovero, G., Borghi, E., Amato, G., Andreoni, S., Campion, L., Lo Cascio, G., Lombardi, G., Luzzaro, F., Manso, E., et al., 2014. Candidemia in intensive care unit: a nationwide prospective observational survey (GISIA-3 study) and review of the European literature from 2000 through 2013. Eur. Rev. Med. Pharmacol. Sci. 18, 661–674.

Morgan, J., Meltzer, M.I., Plikaytis, B.D., et al., 2005. Excess mortality, hospital stay, and cost due to candidemia: a case-control study using data from population-based candidemia surveillance. Infect. Control Hosp. Epidemiol. 26, 540–700.

Morrison, V.A., Haake, R.J., Weisdorf, D.J., 1994. Non-*Candida* fungal infections after bone marrow transplantation: risk factors and outcome. Am. J. Med. 96, 497–503.

Muntanola, M., 1987. General Mycology. NIRO. Knjizevne Novine, Beograd, pp. 257–269.

Murray, P.R., Baron, E.J., 2003. Manual of clinical microbiology. In: American Society for Microbiology, eighth ed. ASM Press, Washington, DC.

Nelson, P.E., Dignani, M.C., Anaissie, E.J., 1994. Taxonomy, biology, and clinical aspects of *Fusarium* species. Clin. Microbiol. Rev. 7, 479–504.

Nierman, W.C., et al., 2005. Genomic sequence of the pathogenic and allergenic filamentous fungus *Aspergillus fumigatus*. Nature 438 (22), 1152–1156.

Noonim, P., Mahakarnchanakul, W., Nielsen, K.F., Frisvad, J.C., Samson, R.A., 2009. Fumonisin B_2 production by *Aspergillus niger* in Thai coffee beans. Food Addit. Contam. Part A 26, 94–100.

Nucci, M., Anaissie, E., 2002. Cutaneous infection by *Fusarium* species in healthy and immunocompromised hosts: implications for diagnosis and management. Clin. Infect. Dis. 35, 909–920.

Nucci, M., Anaissie, E.J., Queiroz-Telles, F., Martins, C.A., Trabasso, P., Solza, C., Mangini, C., Simoes, B.P., Colombo, A.L., Vaz, J., Levy, C.E., Costa, S., Moreira, V.A., Oliveira, J.S., Paraguay, N., Duboc, G., Voltarelli, J.C., Maiolino, A., Pasquini, R., Souza, C.A., 2003. Outcome predictors of 84 patients with hematologic malignancies and Fusarium infection. Cancer 98, 315–319.

O'Donnell, K., Gueidan, C., Sink, S., Johnston, P.R., Crous, P.W., Glenn, A., Riley, R., Zitomer, N.C., Colyer, P., Waalwijk, C., Lee, T., Moretti, A., Kang, S., Kim, H.S., Geiser, D.M., Juba, J.H., Baayen, R.P., Cromey, M.G., Bithell, S., Sutton, D.A., Skovgaard, K., Ploetz, R., Corby Kistler, H., Elliott, M., Davis, M., Sarver, B.A., 2009. A two-locus DNA sequence database for typing plant and human pathogens within the *Fusarium oxysporum* species complex. Fungal Genet. Biol. 46, 936–948.

Pandey, A., Pandey, M., 2013. Isolation and characterization of dermatophytes with tinca infections at Gwalior (MP), India. Int. J. Pharm. Sci. Invent. 2, 5–8.

Pasqualotto, A.C., Nedel, W.L., Machado, T.S., Severo, L.C., 2005. A comparative study of risk factors and outcome among outpatient-acquired and nosocomial candidaemia. J. Hosp. Infect. 60, 129–134.

Pearlman, E., Leal, S., Tarabishy, A., Sun, Y., Szczotka-Flynn, L., Imamura, Y., Mukherjee, P., Chandra, J., Momany, M., Hastings-Cowden, S., Ghannoum, M., 2011. Pathogenesis of fungal keratitis. In: Dartt, D.A., D'Amore, P., Dana, R., Niederkorn, J.Y. (Eds.), Immunology, Inflammation and Disease of the Eye. Academic Press, San Diego, p. 110.

Pelletier, R., Alarie, I., Lagacé, R., Walsh, T.J., 2005. Emergence of disseminated candidiasis caused by *Candida krusei* during treatment with caspofungin: case report and review of literature. Med. Mycol. 43 (6), 559–564.

Perlroth, J., Choi, B., Spellberg, B., 2007. Nosocomial fungal infections: epidemiology, diagnosis, and treatment. Med. Mycol. 45 (4), 321–346.

Peromingo, B., Rodríguez, A., Bernáldez, V., Delgado, J., Rodríguez, M., 2016. Effect of temperature and water activity on growth and aflatoxin production by *Aspergillus flavus* and *Aspergillus parasiticus* on cured meat model systems. Meat Sci. 122, 76–83.

Petrik, M., Haas, J.H., Laverman, P., Schrettl, M., Franssen, G.M., Blatzer, M., et al., 2014. ^{68}Ga-triacetylfusarinine C and ^{68}Ga-ferrioxamine E for *Aspergillus* infection imaging: uptake specificity in various microorganisms. Mol. Imaging Biol. 16, 102–108.

Pfaller, M.A., Diekema, D.J., Ostrosky-Zeichner, L.J., Rex, H.B., Alexander, D., Andes, D., Brown, S.D., Chaturvedi, V., Ghannoum, M.A., Knapp, C.C., Sheehan, D.J., Walsh, T.J., 2008a. Correlation of MIC with outcome for *Candida* species tested against Caspofungin, Anidulafungin, and Micafungin: analysis and proposal for interpretive MIC breakpoints. J. Microbiol. 40 (8), 2620–2629.

Pfaller, M.A., Diekema, D.J., Gibbs, D.L., Newell, V.A., Nagy, E., Dobiasova, S., Rinaldi, M., Barton, R., Veselov, A., 2008b. *Candida krusei*, a multidrug-resistant opportunistic fungal pathogen: geographic and temporal trends from the ARTEMIS DISK antifungal surveillance program, 2001 to 2005. J. Clin. Microbiol. 46 (2), 515–521.

Pfaller, M.A., Messer, S.A., Moet, G.J., Jones, R.N., Castanheira, M., 2011. *Candida* bloodstream infections: comparison of species distribution and resistance to echinocandin and azole antifungal agents in intensive care unit (ICU) and non-ICU settings in the SENTRY antimicrobial surveillance program (2008–2009). Int. J. Antimicrob. Agents 38, 65–69.

Pfaller, M., Neofytos, D., Diekema, D., Azie, N., Meier-Kriesche, H.U., Quan, S.P., Horn, D., 2012. Epidemiology and outcomes of candidemia in 3648 patients: data from the Prospective Antifungal Therapy (PATH Alliance®) registry, 2004–2008. Diagn. Microbiol. Infect. Dis. 74, 323–331.

Pfohl-Leszkowicz, A., Manderville, R., 2007. Ochratoxin A: an overview on toxicity and carcinogenicity in animals and humans. Mol. Nutr. Food Res. 51 (1), 61–99.

Pitt, J.I., 2000. Toxigenic fungi: which are important? Med. Mycol. 38, 17–22.

Pitt, J.I., Hocking, A.D., 2009. *Aspergillus* and related teleomorphs. In: Fungi and Food Spoilage. Springer, Boston, MA, pp. 467–468.

Pontes, Z.B.V.S., Silva, A.D.F., Lima, E.O., Guerra, M.H., Oliveira, N.M.C., Carvalho, M.F.F.P., Guerra, F.S.Q., 2009. Otomycosis: a retrospective study. Braz. J. Otorhinolaryngol. 75 (3), 367–370.

Raksha, Singh, G., Urhekar, A.D., 2017. Virulence Factors detection in *Aspergillus* isolates from clinical and environmental samples. J. Clin. Diagn. Res. 11 (7), 13DC–18DC.

Rankin, N.E., 1953. Disseminated aspergillosis and moniliasis associated with agranulocytosis and antibiotic therapy. Br. Med. J. 183, 918–919.

Revankar, S.G., 2007. Dematiaceous fungi. Mycoses 50 (2), 91–101.

Rippon, J.W., 1988. Medical Mycology: The Pathogenic Fungi and the Pathogenic Actinomycetes, third ed. W.B. Saunders Co., Philadelphia.

Romano, C., Massai, L., Gallo, A., Fimiani, M., 2009. *Microsporum gypseum* infection in the Siena area in 2005–2006. Mycoses 52 (1), 67–71.

Rubens, J., Cardwell, K.F., 2005. The cost of mycotoxin management in the United States. Aflatoxin Food Saf. 1, 1–13.

Ryan, J.K., 1994. Medical Microbiology. An Introduction to Infectious Diseases. III. Appleton and Lange, Norwaik, CT, pp. 571–615.

Salama, S.M., Atwal, H., Gandhi, A., Simon, J., Poglod, M., Montaseri, H., Khan, J.K., Furukawa, T., Saito, H., Nishida, K., Higashitani, F., Uji, T., Unemi, N., Daneshtalab, M., Micetich, R.G., 2001. *In vitro* and *in vivo* activities of Syn2836, Syn2869, Syn2903, and Syn2921: new series of triazole antifungal agents. Antimicrob. Agents Chemother. 45, 2420–2426.

Samaranayake, L.P., 1991. Superficial oral fungal infections. Curr. Opin. Dent. 1, 415–422.

Samaranayake, L.P., MacFarlane, T.W. (Eds.), 1990. Oral Candidosis. Wright Butterworth, London, Great Britan, pp. 124–132.

Samaranayake, Y.H., Samaranayake, L.P., 1994. *Candida krusei*: biology, epidemiology, pathogenicity and clinical manifestations of an emerging pathogen. J. Med. Microbiol. 41, 295–310. The pathological society of Great Britain and Ireland Rev. Article: Clinical Mycology.

Sampaio, S.A.P., Rivitti, E.A., 2000. Dermatologia, second ed. Artes Médicas, São Paulo, pp. 545–547.

Samson, R.A., Houbraken, J., Summerbell, R.C., Flannigan, B., Miller, J.D., 2001. Common and important species of fungi and actinomycetes in indoor environments. In: Microorganisms in Home and Indoor Work Environments. CRC, pp. 287–292.

Samson, R.A., Visagie, C.M., Houbraken, J., Hong, S.-B., Hubka, V., Klaasen, C.H.W., et al., 2014. Phylogeny, identification and nomenclature of the genus *Aspergillus*. Stud. Mycol. 78, 141–173.

Sardi, J.C.O., Gullo, F.P., Pitanguo, N.S., Fusco-Almeida, A.M., Mendes-Giannini, M.J.S., 2013. *In vitro* antifungal susceptibility of *Candida albicans* isolates from patients with chronic periodontitis and diabetes. Clin. Microbiol. 2, 103.

Schmit, J.P., Mueller, G.M., 2007. An estimate of the lower limit of global fungal diversity. Biodivers. Conserv. 16, 99–111.

Schuster, E., Dunn-Coleman, N., Frisvad, J.C., van Dijck, P.W.M., 2002. On the safety of *Aspergillus niger*—a review. Appl. Microbiol. Biotechnol. 59, 426–435.

Scully, C., el-Kabir, M., Samaranayake, L.P., 1994. Candida and oral candidosis: a review. Crit. Rev. Oral Biol. Med. 5, 125–157.

Seneviratne, C.J., Jin, L., Samaranayake, L.P., 2008. Biofilm lifestyle of Candida: a mini review. Oral Dis. 14 (7), 582–590.

Shah, D.N., Yau, R., Lasco, T.M., et al., 2012. Impact of prior inappropriate fluconazole dosing on isolation of fluconazole-nonsusceptible *Candida* species in hospitalized patients with candidemia. Antimicrob. Agents Chemother. 56, 3239–3243.

Shanahan, P., 1994. Microbiological Techniques, IX. Mosby, pp. 734–786.

Singh, P.N., Ranjana, K., Singh, Y.I., Singh, K.P., Sharma, S.S., Kulachandra, M., Nabakumar, Y., Chakrabarti, A., Padhye, A.A., Kaufman, L., Ajello, L., 1999. Indigenous disseminated *Penicillium marneffei* infection in the state of Manipur, India: report of four autochthonous cases. J. Clin. Microbiol. 37, 2699–2702.

Sokovic, D.M., 2001. Antifungal Activities of Essential Oils of Selected Aromatic and Medical Plants *In Vitro* and *In Vivo* (Ph.D. thesis). Faculty of Biology, University of Belgrade, Yugoslavia.

Sombatmaithai, A., Pattanaprichakul, P., Tuchinda, P., Surawan, T., Muanprasart, C., Matthapan, L., Bunyaratave, S., 2015. Tinea capitis caused by *Trichophyton tonsurans* presenting as an obscure patchy hair loss due to daily antifungal shampoo use. Dermatol. Pract. Concept 5 (2), 133–135.

Sugui, J.A., Kwon-Chung, K.J., Juvvadi, P.R., Latgé, J.-P., Steinbach, W.J., 2015. *Aspergillus fumigatus* and related species. Cold Spring Harb. Perspect. Med. 5, 1–17.

Summerbell, R., 2003. Ascomycetes: *Aspergillus*, *Fusarium*, *Sporothrix*, *Piedraia*, and their relatives. In: Howard, D.H. (Ed.), Pathogenic Fungi in Humans and Animals, second ed. Marcel Dekker, New York, pp. 400–425.

Sutton, D.A., Brandt, M.E., 2011. *Fusarium* and other opportunistic hyaline fungi. In: Versalovic, J., Carroll, K., Funke, G., Jorgensen, J.H., Landry, M.L. (Eds.), Manual of Clinical Microbiology, 10th ed. ASM Press, Washington, DC, pp. 1853–1879.

Tan, B.H., Chakrabarti, A., Li, R.Y., Patel, A.K., Watcharananan, S.P., Liu, Z., Chindamporn, A., Tan, A.L., Sun, P.L., Wu, U.I., et al., 2015. Incidence and species distribution of candidaemia in Asia: a laboratory-based surveillance study. Clin. Microbiol. Infect. 21, 946–953.

Truckesses, M.W., Scott, P.M., 2008. Mycotoxins in botanicals and dried fruits: a review. Food Addit. Contam. 25, 181–192.

Tupaki-Sreepurna, A., Kindo, A.J., 2018. *Fusarium*: the versatile pathogen. Indian J. Med. Microbiol. 36 (1), 8–17.

Ullmann, A.J., Akova, M., Herbrecht, R., Viscoli, C., Arendrup, M.C., Arikan-Akdagli, S., Bassetti, M., Bille, J., Calandra, T., Castagnola, E., Cornely, O.A., Donnelly, J.P., Garbino, J., Groll, A.H., Hope, W.W., Jensen, H.E., Kullberg, B.J., Lass-Flörl, C., Lortholary, O., Meersseman, W., Petrikkos, G., Richardson, M.D., Roilides, E., Verweij, P.E., Cuenca-Estrella, M., 2012. ESCMID* guideline for the diagnosis and management of *Candida* diseases adults with haematological malignancies and after haematopoietic stem cell trans plantation (HCT) for the ESCMID Fungal Infection Study Group (EFISG). Clin. Microbiol. Infect. 18 (s7), 53–67.

Vallabhaneni, S., Cleveland, A., Farley, M., et al., 2015. Epidemiology and risk factors for echinocandin nonsusceptible *Candida glabrata* bloodstream infections: data from a large multisite population-based candidemia surveillance program, 2008–2014. Open Forum Infect. Dis. 2 (4), ofv163.

van Diepeningen, A.D., Al-Hatmi, A.M.S., Brankovics, B., de Hoog, G.S., 2014. Taxonomy and clinical spectra of *Fusarium* species: where do we stand in 2014? Curr. Clin. Microbiol. Rep. 1, 10–18.

van Diepeningen, A.D., Feng, P., Ahmed, S., Sudhadham, M., Bunyaratavej, S., de Hoog, G.S., 2015. Spectrum of *Fusarium* infections in tropical dermatology evidenced by multilocus sequencing typing diagnostics. Mycoses 58, 48–57.

Velegraki, A., Cafarchia, C., Gaitanis, G., Iatta, R., Boekhout, T., 2015. *Malassezia* infections in humans and animals: pathophysiology, detection, and treatment. PLoS Pathog. 11 (1), e1004523.

Vesper, S.J., Wong, W., Kuo, C.M., Pierson, D.L., 2008. Mold species in dust from the International Space Station identified and quantified by mold-specific quantitative PCR. Res. Microbiol. 159, 432.

Voss, K.A., Smith, G.W., Haschek, W.M., 2007. Fumonisins: toxicokinetics, mechanism of action and toxicity. Anim. Feed Sci. Technol. 137, 299–325.

Weitzman, I., Summerbell, R.C., 1995. The dermatophytes. Clin. Microbiol. Rev. 8 (2), 240–259.

Wickern, G.M., 1993. *Fusarium* allergic fungal sinusitis. J. Allergy Clin. Immunol. 92, 624–625.

Wildfeuer, A., Seidl, H.P., Paule, I., Haberreiter, A., 1998. *In vitro* evaluation of voriconazole against clinical isolates of yeasts, moulds and dermatophytes in comparison with itraconazole, ketoconazole, amphotericin B and griseofulvin. Mycoses 41, 309–319.

Wilson, L.S., Reyes, C.M., Stolpman, M., et al., 2002. The direct cost and incidence of systemic fungal infections. Value Health 5, 26–34.

Wong, S.S.Y., Wong, K.H., Hui, W.T., Lee, S.S., Lo, J.Y.C., Cao, L., Yuen, K.Y., 2001. Differences in clinical and laboratory diagnostic characteristics of penicilliosis marneffei in human immunodeficiency virus (HIV)- and non-HIV-infected patients. J. Clin. Microbiol. 39, 4535–4540.

Wu, B., Hussain, M., Zhang, W., Stadler, M., Liu, X., Xiang, M., 2019. Current insights into fungal species diversity and perspective on naming the environmental DNA sequences of fungi. Mycology 10 (3), 123–140.

Yapar, N., 2014. Epidemiology and risk factors for invasive candidiasis. Ther. Clin. Risk Manag. 10, 95–105.

Zaias, N., 1978. Chromomycosis. J. Cutan. Pathol. 5 (4), 155–164.

Zhang, N., O'Donnell, K., Sutton, D.A., Nalim, F.A., Summerbell, R.C., Padhye, A.A., Geiser, D.M., 2006. Members of the *Fusarium solani* species complex that cause infections in both humans and plants are common in the environment. J. Clin. Microbiol. 44 (6), 2186–2190.

Chapter 3

Antifungal therapy

Marina Soković[a] and Konstantinos Liaras[b]
[a]*Mycological Laboratory, Institute for Biological Research "Siniša Stanković", University of Belgrade, National Institute of Republic of Serbia, Belgrade, Serbia,* [b]*Department of Pharmaceutical Chemistry, School of Pharmacy, Aristotle University of Thessaloniki, Thessaloniki, Greece*

Chapter outline

3.1 Antifungal therapy in the past

In the far past and particularly in the Middle Ages, the cure of diseases depended mainly on the evaluation of the effects of medicines on patients' survival.

Before the beginning of 20th century, treatments for infections were based primarily on folklore by employing in most of the cases plant or fungal materials. From the scientific legacy of Pasteur, Koch, and Fleming, it was established that compounds can be screened for their ability to inhibit the growth of bacterial cultures. Since the introduction of penicillin in the market (1941), natural products kept a crucial role in antibiotics discovery, due to the fact that most of the antibacterial agents were derivatives of natural products and their analogues. Fungi as a cause of infectious diseases were considered to be more neglected than bacteria, as was also antifungal therapy (Fig. 3.1).

The beginning of medical mycology is placed in the early 19th century, when tinea favosa is discovered (Espinel-Ingroff, 1996) and later on, when a human case of blastomycosis (1894) is reported in the US. In that time, antifungal therapy was limited in using phenol, potassium iodide, potassium permanganate, methyl violet, bromine, and turpentine oil. To be exact, antifungal chemotherapy began in 1903, when potassium iodide (KI) was successfully used for the treatment of sporotrichosis (Stevens, 1995). With the number of infections increasing, the needs for using orally or intravenously antifungal drugs also emerged.

Antifungal Compounds Discovery. https://doi.org/10.1016/B978-0-12-815824-1.00003-2

FIG. 3.1 Antifungal therapy in the past.

In the next 50–60 years new dimorphic fungi are being recognized as human pathogens that can cause systemic fungal diseases.

Giseofulvin was one of the first developed safe antifungal drugs at Oxford, in 1939, followed by benzimidazole as the first azole that was found in 1944. Nystatin, which was discovered in 1950, enabled the effective control of fungal diseases for the first time in history (Ostrosky-Zeichner et al., 2010) and that could also be considered as the beginning of a new era for antifungal therapy. Later on, amphotericin B (AmB) was discovered in 1955 and a few years later griseofulvin that could be administered orally. Later, in 1964 the orally administered drug, 5-fluorocytosine, was developed and was used for the treatment of *Candida albicans* and *Cryptococcus neoformans* infections, followed by the production of miconazole, in 1969, that could be used in formulations for topical application. The new class of antifungal drugs, imidazoles, exhibited broad-spectrum activity against dermatomycetes but also toward other fungi. Ketoconazole was developed in 1981 and still is being used for the treatment of systemic fungal infections. In the years between 1983 and 1987, only 16 systemic antimicrobial agents had been approved. In line with that, it has to be admitted that the threat, which antimicrobial resistance brings, has been underestimated and strict actions should be taken to avoid repetition of mistakes of the past. It would also be interesting to point out that Food and Drug Administration (FDA) approved only two new antimicrobials for antibacterial infections from 2009 till 2014 (Spellman, 2014) (Fig. 3.2).

With new laboratory techniques that are essential for the identification of fungi, fungal diseases attract the scientific interest and are being observed

FIG. 3.2 Industrial production of drugs in the 1950s and 1960s.

with more attention, establishing their prominent place in medicine by the formation of a separate branch known as medical mycology (Gubbins and Anaissie, 2009).

Discoveries of fluconazole and the echinocandins were a great success for antifungal therapy in the last decades (Gubbins and Anaissie, 2009). In this period great attention was given to the activity of antifungal drugs, but also to their toxicity. In 1992, itraconazole was introduced for the treatment of *Candida* and a variety of filamentous fungi infections. Moreover, new and safer formulations of AmB and nystatin were developed. Novel classes of antifungal drugs were represented by echinocandins, caspofungin in 2001, and voriconazole, which was introduced as drug with increased potential against fluconazole-resistant *Candida* species and molds (Gubbins and Anaissie, 2009).

3.2 Antifungal therapy nowadays

Nowadays, research and development in this field is conducted by small and mid-sized companies, with relatively limited resources, making it more difficult to discover and produce new agents. It becomes a trend among pharmaceutical companies to avoid antibiotics and antimycotics development, probably due to the rise in antimicrobial resistance and lower perceived returns on investments, turning that way their interest to more commercially attractive, therapeutics (Spellman, 2014).

At the moment, natural products are still remaining a pivotal source for potential drug candidates. Development of compound and extracts libraries

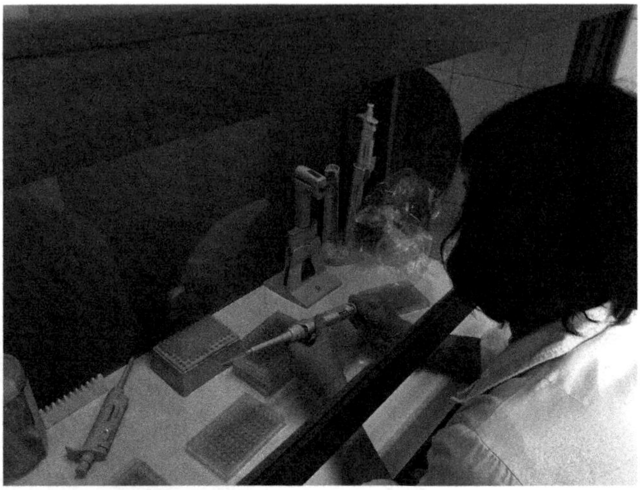

FIG. 3.3 Testing of antifungal activity by microdilution method in Mycological Laboratory, IBISS, University of Belgrade.

containing all available chemical structures and properties from various natural sources, could turn to be of great significance in the discovery of new antimicrobials (Fig. 3.3).

The advances in medicine during the last decades led to introduction of modern technologies and innovative therapies, especially in transplantations, chemotherapy, and care in specialized units, parenteral nutrition and assisted ventilation, among others, having as a result the increase of the populations of hospitalized and immunocompromised patients. In the growing numbers of immunosuppressed individuals, AIDS patients are considered to be in danger of acquiring nosocomial infections of high severity, from fungi that were not considered pathogenic in the past. These kinds of infections might progress rapidly and in many cases it is difficult for physicians to diagnose. For the treatment of such infections, a deep knowledge of the nature of fungi is in need, because as eukaryotic microorganisms they are more complex than bacteria and thus, proper patient care should always be a priority (Fig. 3.4).

Nowadays, infectious diseases still possess the second place among the conditions that can cause death and this issue is enhanced by the emerging multidrug resistance of common pathogens that leads to development of new kinds of infections. As a result of this problem, for infections caused by certain pathogens, only a few or no treatment options exist.

Invasive fungal infections represent a serious global health problem, with very high incidence, killing approximately 1.4 million people per year. This number is comparable to the number of deaths caused by tuberculosis. The increasing numbers of immunocompromised patients who are extremely susceptible to fungal pathogens, but also insufficient diagnostics and lack of appropriate treatment are reasons for such a high incidence (Brown et al., 2012).

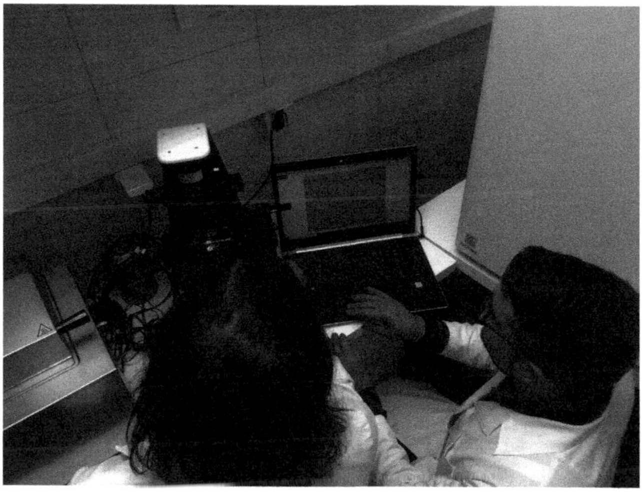

FIG. 3.4 Computer analysis of images captured by microscope.

Approximately more than 300 million people are suffering from some kind of fungal infection. Fungal species from genera *Cryptococcus*, *Candida*, and *Aspergillus* are responsible for more than 90% of the reported fungal-associated deaths (Cowen et al., 2018).

Recognition of the importance of fungal infections has led to a dramatic rise in the application of antifungal agents for their treatment and prevention. Unfortunately, the treatment options are highly limited, as there are only a few chemical classes that are represented by the existing antifungal drugs.

There are more fundamental reasons why only a few new classes of antifungal compounds have been produced and found their place in market during the past 50 years. At first, it is very reasonable that it is more difficult to produce new broad-spectrum compounds nowadays than it was in the golden age of antibiotic/antifungal discovery. The case of antifungals discovery is a little different compared to that of antibiotics. It has been suggested that the main antibacterial classes of compounds that act on specific metabolic pathways have been already developed and placed in the market and that there are no more of them available (Becker et al., 2006). However, it is possible that the potential classes of broad-spectrum antifungals have not been discovered, so far, in the same extent with those of the antibacterial compounds. This could provide a chance to not only discover, but also to place in the market new classes of antifungal agents.

Another factor that could obstruct the antifungal discovery progress is the large investments needed of either science or pharmaceutical industry in the genomics approach that could deliver new classes of compounds into the market. Using DNA sequences data to predict enzymatic pathways, which can be inhibited by novel classes of compounds, could be very useful in the fight against fungi, but comes at a high cost (Brotz-Oesterhelt and Sass, 2010) (Fig. 3.5).

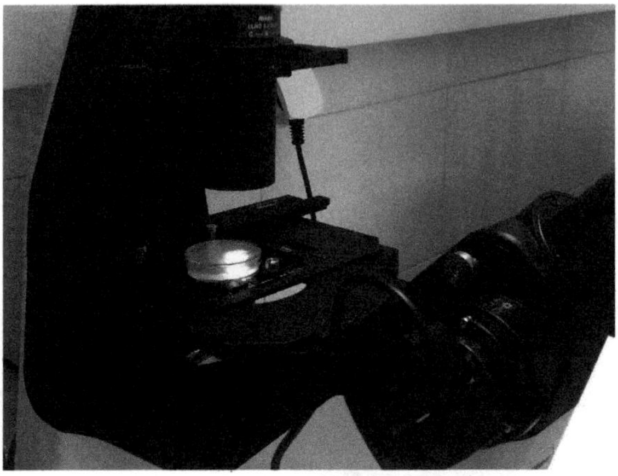

FIG. 3.5 Fungal culture under the microscope.

It should also be taken into consideration that scientists, small and large pharmaceutical companies come across many difficulties in the process of registration of antibiotics/antifungals and their enthusiasm can be suppressed by long and expensive procedures to enter the stage of producing novel drugs that will be ultimately be used in clinical practice for the treatment of microbial infections (Coates et al., 2011).

Current strategies for developing antifungal agents based on biotechnology and related novel approaches are very promising, but we have to be aware that these techniques can be time consuming and expensive.

Finally, all of the cited studies reveal a great potential for development of new drugs, but most of them have been performed by testing molecules, extracts, and oils. Further research regarding the toxicity of these compounds must be conducted, especially concerning their pharmacodynamic characteristics and the potential implications of pharmacogenomics in predicting responses to these treatments. In other words, the development of new options for antifungal drugs is mandatory in view of the increase in fungal infections and their mortality rates. This process is a challenge to researchers and the pharmaceutical industry, but the recent application of biotechnological approaches has facilitated new discoveries.

Fungal infections in combination with resistance to antimycotics are relatively poorly investigated and this issue could continue to evolve. The spread of these infections and resistance has to be investigated in different aspects and new approaches are necessary to be introduced. In particular, the high mortality of fungal infections and the limited efficacy of antifungal agents turned the interest to the development of new drugs and new formulations of existing antimycotics for infections, which are difficult to treat. The comprehensive data presented here should support future investigations concerning severe fungal infections. The use of the available in vitro and in vivo data and clinical studies in patients should move toward more useful results, with less potential risks and more benefits for the patients. Therefore, future focus in this important field seems to be mandatory.

3.3 Antifungal resistance

The evolution of resistant strains is a natural phenomenon that occurs when microorganisms replicate themselves erroneously or when resistance traits are exchanged between them. The use and misuse of antimicrobial drugs accelerates the emergence of drug-resistant strains. Poor infection control practices, inadequate sanitary conditions, and inappropriate food handling encourage the further spread of antimicrobial resistance (AMR) (WHO, 2014).

The term "resistance" is used to describe a relative insensitivity of a microbe to an antimicrobial drug as tested in vitro and compared with other isolates of the same species. In contrast, clinical failure describes failure of an appropriate therapy for a certain indication to result in a clinical response. The reason for clinical failure may be, for example, antifungal resistance, but other causes, such as an impaired immune function, poor bioavailability of the drug given, or an accelerated metabolism of the drug, are possible causes of treatment failure.

Resistant microorganisms (including bacteria, fungi, viruses, and parasites) are able to withstand attack by antimicrobial drugs, such as antibiotics, antifungals, antivirals, and antimalarials, so that standard treatments become ineffective and infections persist, increasing the risk of spreading to others.

The inadequate and excessive administration and application of antimicrobial drugs leads to microbe resistance to medicines. Antimicrobial resistance develops when microorganisms adapt to the applied antimicrobial drugs and continue to grow in the presence of drugs. It is well established that the resistance of microorganisms depends on how often antimicrobial drugs are being used. Usually, many antimicrobials belong to the same class of drugs and resistance to one antimicrobial agent leads to resistance against all related classes of drugs. One of the important facts is that developed resistance in one organism can also spread to others very fast through the exchange of genetic material between different microorganisms. In that case, possibilities and options for treating microbes and curing the diseases that they could cause are extensively reduced and limited. Resistant microorganisms could circulate among humans and animals, through the environment (water, food, etc.). Infections and transmission of microorganisms could spread by both human and animal migration.

The direct consequences of antimicrobial resistance could be witnessed in aspects of human health and can influence the whole society. Resistant microorganisms can affect the protection of patients undergoing surgeries and all the other medical procedures in hospitals, which influence longer illnesses, prolonged stays in hospital, increased mortality, and higher costs in all aspects of medicine (WHO, 2015, 2017).

The United Nations General Assembly in New York in September 2016 confirmed with an official declaration the international agreement to begin a coordinated effort to confront the basis of resistance in human health, animal health, agriculture, and other sectors. The main role of WHO in this effort is the support toward Member States in their projects regarding the development of national action plans to address the antimicrobial resistance issue, taking into account the parameters and standards of the global action plan (WHO, 2018).

In recent years, the role of environment as a promoter of resistance has become prominent. Fungicidal use in agriculture and industry differs vastly between different regions, with the United States using about a tenth of the amounts used in Europe. Furthermore, azole-based fungicides are used in grape and cereal production in European countries and it should be noted that a strong link was found between countries where azole-based fungicides are being used and the incidence of antifungal resistance.

Several factors have contributed to the increase of AMR: medical progress that led to an expanding population of susceptible hosts, with impaired immunological defenses against infections in the community and hospitals; the expansion of severely ill and/or immunocompromised patient population, including HIV-infected patients, patients with cancer who have chemotherapy-induced neutropenia and transplant recipients who are receiving immunosuppressive therapy; other factors that have also contributed, include the frequent use of more invasive medical procedures, treatment with broad-spectrum antifungals and glucocorticoids, receipt of parenteral nutrition, and receipt of peritoneal dialysis or hemodialysis (Perea and Patterson, 2002). The majority of nosocomial fungal infections are caused by *Candida* species, with *Candida albicans* being the most common etiological agent of fungal bloodstream infections. Together with *Candida* infections, *Aspergillus* infections, which mostly affect granulocytopenic and other immunocompromised patients, account for 90% of all nosocomial fungal infections. *Cryptococcus neoformans* has become also a major opportunistic fungal pathogen in immunocompromised individuals, mainly HIV-infected patients and those receiving immunosuppressive treatment for cancer, organ transplantations, and other serious medical conditions. Several other less common fungi, many of which are intrinsically resistant to the available antifungal drugs, are becoming increasingly recognized as sources of deep fungal infections, such as Zygomycetes (*Rhizopus arrhizus*, *Absidia corymbifera* and *Rhizomucor pusillus*), *Fusarium* species, *Trichosporon beigelii*, *Blastoschizomyces capitatus*, *Scedosporium* species, *Acremonium* species, and dematiaceous fungi (Perea and Patterson, 2002).

Antifungal drugs and fungicides are being used nowadays to confront fungal infections in humans and plants. Nevertheless, this effort is obstructed by the rapid development of resistance of fungi to marketed compounds due to their highly plastic genomes and their ability to reproduce rapidly. Especially for humans, chronic prophylaxis for patients who are in critical condition can cause the emergence of resistance to antifungals. Resistance of clinical pathogens to all marketed antifungal agents has been reported, but with various rates of emergence among the known classes of drugs. This issue is intensified by the additional threat of withdrawal of some compounds or classes of compounds due to regulatory changes in jurisdictions (e.g., in European Union) (Fisher et al., 2018).

The control of pathogenic fungi by using chemical agents has progressed from the use of inorganic compounds to the application of organic surface chemicals and subsequently to the administration of systemic antifungals. The amount of antifungals that are used for confronting crop infections is more or less nine-fold higher than that of the agents used for the treatment of such diseases in animals. Marketed compounds for humans are limited to only a few classes of drugs, with azoles being the most widely used in clinical practice. Most of the chemical fungicides for crops have as a target the mitochondrial function, the cytoskeleton, or ergosterol biosynthesis, although some other agents exhibit their activity on other targets. Nevertheless, the azole derivatives are still the main class of compounds that are used for the treatment of fungal infections in crops, animals, and humans.

The factors responsible for sensitivity of host to fungal infections can be immunological or nonimmunological. Fungal resistance could be microbiological or clinical. Different fungal characteristics (gene amplification, gene transfer, gene deletion, point mutations, loss of cis- and trans- acting regulatory elements, and transcriptional activation) are connected with pathogenesis, virulence, and resistance to drugs and all together influence microbiological resistance. On the other hand, clinical resistance depends on host-fungal interaction, drug-fungal interaction, and drug-related factors. Knowing all these facts and considering the very limited number of available antifungals, one of the ways to reduce and avoid resistance is application of combinational antifungal treatment. This could be one of the strategies for improving therapy in mycoses in order to overcome antifungal resistance. One main advantage of combination therapy is the increased activity due to synergistic action of single drugs.

As it was stated before, microbiological resistance means that a fungus is not susceptible to an antifungal agent. The minimum inhibitory concentration (MIC) of the drug exceeds the susceptibility concentration for that fungus. When the fungi are resistant to some drug before exposure we can discuss about primary (intrinsic) resistance, while secondary (acquired) resistance is when fungi develop resistance when are exposed to antimicrobials as a response to that treatment and that usually is connected to altered gene expression (Wiederhold, 2017).

Clinical resistance is when therapy with certain drugs fails to stop a fungal infection. This highlights the importance of a new approach in antifungal treatment based on combination of all necessary factors (fungi, host characteristics, and drug pharmacology) connected with fungal resistance in association with each other. It depends on multiple host- and drug-related factors (host immunosuppression, delay in initiation of therapy, treatment with low antifungal dosage, long-term therapy) (Wiederhold, 2017).

AMR has been increased in the last years and still continues to develop in a great number of pathogens, leading in many cases to higher morbidity and mortality rates. Coordination among different countries and all sectors in human and animal health and environment is necessary. Global Action Plan (GAP) adopted by the World Health Organization (WHO, 2015) and political declaration made at the General Assembly of the UN (UN, 2016) recognized AMR as important biological, but also social problem for human and animal health. It was also pointed out the lack of development of new drugs as the emergent problem. Poor public health conditions in many developing countries and only limited access to high-quality medicine, make this problem even worse (Laxminarayan and Chaudhury, 2016). To support implementation of the GAP made by WHO and UN (WHO, 2017) we need coordination of governments, new national plans, and to expand monitoring of responsible commitments through uncommon and integrated approaches to measure and fight AMR (Wernli et al., 2017).

Fungi and resistance

Invasive fungal infections cause substantial morbidity and mortality and are costly, common problems in healthcare settings that in combination with antifungal resistance to existing antifungals have as a result even bigger economic burden.

Antifungal resistance represents a major clinical challenge to clinicians responsible for treating invasive fungal infections due to the limited arsenal of systemically available antifungal agents. In addition, current drugs may be limited by drug-drug interactions and serious adverse effects/toxicity that prevent their prolonged use or dosage escalation.

Fungi are extremely adaptable microorganisms and capable to overcome the effects of existing drugs. They have been present for around 1500 million years with more than 1.5 million species, out of which only about 300 species are known to cause human infections. Nowadays, at least 700,000 patients die per year of infections by resistant strains. If there will not be an appropriate response to this threat (warns the UN Ad hoc Interagency Coordinating Group on Antimicrobial Resistance), this number will increase to 10 million deaths per year by 2050. It should be taken also into account that the corresponding damage to the economy will be enormous and comparable to that of the 2008–2009 global financial crisis. By the year 2030, it is estimated that the antimicrobial resistance could lead up to 24 million people to extreme poverty (WHO, 2019).

Anthropophilic fungi are causing 70% of dermatomycoses at humans that could lead to a chronic infection, with slow progression. This means that these pathogenic fungi could easily adapt to the human host (White et al., 2008).

Traditionally, *Candida* and *Aspergillus* species accounted for the majority of the infections. Candidemia is the fourth leading cause of blood-stream infections and carries 35%–55% mortality. The incidence of mold infections has also increased in the recent past and especially infections caused by *Aspergillus* spp. where the mortality rate crosses 50% in such patients (Groth and Dodds-Ashley, 2006).

Concerning clinical resistance, there are a few challenges. One of them is the increasing resistance of non-*C. albicans* species to azole and echinocandins and azole resistance in *A. fumigatus* when it is exposed to these agents. Some pathogenic fungi are capable to develop resistance against many available antifungal agents. There are new antifungal agents that have advantages compared to current drugs, with less adverse effects and possibility to develop antifungal resistance. However, more extensive clinical and preclinical investigation is needed.

As it was previously stated, the achievements of modern medicine are put at risk by AMR. Without effective antimicrobials for prevention and treatment of infections, the success of organ transplantation, cancer chemotherapy, and major surgery would be compromised.

Resistance of Candida spp

Testing susceptibilities of *Candida* spp. aids the selection of the most promising treatments, since a number of the species have already developed resistance against widely used antifungal drugs. The increased resistance of these species to antimycotics in combination with the fact that many of them can be toxic and cause side effects turns the development of new antifungal agents into a subject of great interest (Table 3.1).

Recently though, antifungal resistance to antimycotics was determined to be acquired by 0.4%–40% of the isolates, including also patients who haven't used the drug before (Arendrup 2013). This could be an indication that resistance of fungi to commonly used agents is an emerging and underdiagnosed condition.

Before the year of 2000, resistance of *Aspergillus* and *Candida* species to the most commonly used antifungals was practically unknown. These species are prevailing among pathogenic fungi that can infect patients with suppressed immune system, increasing in that way mortality, suffering, and cost of treatment (WHO, 2014, 2017).

Bloodstream infection caused by *Candida* spp. (candidemia) usually brings 3 to 13 days of hospitalization for patients, which means around $6000 to $29,000 in healthcare costs (Morgan, 2005).

TABLE 3.1 Pan-European data about antimicrobial resistance.

Infection type	Estimated resistance levels[a]
Candida albicans	15%–20%
Candida auris	5%–40%
Candida glabrata	0.4%–1%
Aspergillus fumigatus	4%–6%

[a] *Resistance level: percentage of fungi that display resistance to common antifungals (WHO, 2014).*

Lockhart et al. (2012) showed that resistance incidence for fluconazole is 7%, for echinocandin resistance is 1%, and 0.4% for both echinocandin and fluconazole (8 of 9 isolates were *Candida glabrata*). Some types of *Candida* species are becoming extremely resistant to first- and second-line antifungals (fluconazole and echinocandins, respectively) (Vallabhaneni et al., 2015, 2017).

Candida auris infections can also be hospital-acquired infections, like all the other *Candida* caused infections. The mortality rate of these infections was 60%, similar to other reported cases: 44% in India, 46% in South Africa, 52% in Pakistan, and 72% in South America (Chowdhary et al., 2013a,b). Infections caused by *C. auris* were detected in Japan, South Korea, Kuwait, and Venezuela, from 26 patients with other invasive diseases and 45 patients with candidemia. *C. auris* is a multidrug-resistant pathogen, difficult to identify with traditional biochemical methods and capable to cause invasive fungal infections, especially among hospitalized patients. These reports proved salient characteristics of *C. auris* invasive infections, which occur as part of nosocomial outbreaks. The majority of isolates were fluconazole-resistant, while resistance rates for amphotericin B and echinocandins were approximately 30%–40% and 5%–10%, respectively. Almost 50% of the isolates showed multidrug resistance and just a few were pandrug-resistant strains. Mortality rates were found to be very high, 70% for candidemia (Clancy and Nguyen, 2018).

Resistance of *Aspergillus* spp

Alongside *Candida* species, antifungal resistance to antimycotics has been noticed for other fungal species, such as *Aspergillus*. It is still difficult to present the complete extent of the problem, but it is known that the global prevalence of azoles-resistance rate in *Aspergillus* is approximately 3%–6% (Arendrup, 2013).

Aspergillus infections are also connected to a high level of mortality, the same as *Candida* species. It should be mentioned again that infections caused by resistant species could be developed in patients who were already exposed to antifungals (Howard et al., 2009).

Furthermore, some reports showed that resistance in *Aspergillus* spp. could be also caused, partially, by the azoles used in agriculture for crops protection from fungi (Mortensen et al., 2010; Verweij et al., 2009). The antifungal drug voriconazole, used for treatment for invasive aspergillosis, has been marked in 2002 as a primary cause of antifungal resistance. Since then, it is necessary to highlight the emergence of azoles resistance against *Aspergillus fumigatus*. In 10 years, already, some registries of European countries are making entries concerning the increasing amount of azoles-resistant *A. fumigatus* strains. Evermore, in some hospitals in Europe, prevalence of *Aspergillus*-resistance incidence turns to be as high as 30%. It was stated that patients infected with these resistant isolates showed high mortality, almost 90% rate, while for patients suffered from aspergillosis caused by susceptible strains, mortality reaches approximately 40% rate. Resistant strains were also isolated from countries in the Middle East, Asia, Africa, South America, and more recently in the United States (Mortensen et al., 2010; Verweij et al., 2009).

One of the eminent fungal pathogens is *Cryptococcus neoformans* and this species has become also a major opportunistic fungal pathogen in immunosuppressed patients (Mitchell and Perfect, 1995). Among others, some of the Zygomycetes (*Rhizopus arrhizus*, *Absidia corymbifera*, and *Rhizomucor pusillus*), *Fusarium* species, *Trichosporon beigelii*, *Blastoschizomyces capitatus*, *Scedosporium* species, *Acremonium* species, and dematiaceous fungi have become intrinsically resistant to the available antifungal drugs and they are recognized as causes of deep fungal infections (Fridkin and Jarvis, 1996). Species from *Scedosporium* and *Fusarium* genera show also reduced susceptibility or pan-resistance to the available antifungal drugs. Infections provoked by *Penicillium* spp. as well as histoplasmosis and coccidioidomycosis are considered important in some geographical regions for both immunocompetent and immunodeficient patients (Loeffler and Stevens, 2003).

Although there was an increase in the spectrum of the already-available medicines, there are some limitations mostly because of the emergence of more resistant fungal strains, slow mycological diagnosis, diverse drug bioavailabilities in immunocompromised patients, toxicity of drugs, lack of either oral or intravenous formulations, drug interactions, and the development of breakthrough infections.

The growth of human population, urbanization, and the generally improved economic conditions brought in the surface the need for larger amounts and different kinds of food. Advances in agriculture have also contributed to meet the previously mentioned need by trying to reach maximum productivity in crops with the help of broad-scale pesticide use, compromising that way the efficacy of plants' own defenses. Moreover, the human populations, which are considered at risk for developing fungal infections are still rising, due to a number of factors such as aging, medical interventions, or HIV infections. Furthermore, it should be stated that advances in medicine, which have as a result greater survival rates for patients with severe illnesses or had organ transplantations, could potentially leave them susceptible to infections by opportunistic fungi, leading that way to an increased administration of antifungal agents.

Mechanisms of resistance to synthetic antifungals

A good way to understand the mechanisms of resistance is by comparing how it is developed, the similarities and the differences between fungi and bacteria. For that purpose, the differences in the structures and functions between these two kinds of microorganisms should be taken into account. For example, several antibacterial compounds target peptidoglycan of the cell wall, while the majority of antifungal agents act on ergosterol, an essential component of the cell membrane of fungi.

The main difference in resistance evolution between fungi and bacteria is that regarding fungal strains, in most of the cases occurs independently by the use of drugs, while in bacteria more often takes place via plasmids and phages transfer. Nevertheless, there are also data indicating horizontal gene transfer between fungi, which takes longer time to occur compared to bacteria and the dynamics of resistance occurring by this route is considered insignificant.

There are several types of mechanisms that promote the emergence of resistance to antifungals. The basic resistance mechanisms to antifungals include three types: (a) decrease of concentration of active drug, e.g., increased drug efflux, or number of targets, drug sequestration, poor conversion of drug; (b) alterations of drug target; (c) bypasses in metabolisms (Ghannoum and Rice, 1999; Sanglard, 2016) (Fig. 3.6).

The presence of orthologs in target proteins of human cells seems to be a common source of inspiration in many modern efforts to develop effective antifungal agents. The targets in the majority of conventional antifungals are unique in fungi and this brings a new concept by trying to investigate the sometimes small structural differences between host and fungal proteins and find compounds, which would be selective for the pathogen's protein and, thus, with low toxicity for the patient. It should be added that these innovative approaches could be used in combination with the conventional ones to address the issue of resistance and successful antifungal treatment in general (Butts and Krysan, 2012).

Four main resistance factors are connected and interacting between each other: fungal factors (reduction of accumulation of drug within the cell, decrease of affinity of drug for the appropriate target, modification of metabolism, cellular stress, biofilm formation); host factors (site of infection, immunosuppression degree, kind of infection, and provided treatment); drug factors (dosage, drug interaction, pharmacokinetics); and environmental factors (O'Neill, 2014) (Fig. 3.7).

Nevertheless, these facts could be of great help for researchers to elucidate the fungal strains that are candidates for developing resistance to the drugs that are used in clinical practice.

In general, development of resistance in fungi takes time, while it can occur faster in bacteria. Taking into consideration that there was no known mechanism of horizontal resistance gene transfer in fungi, it was thought that when a great number of viable fungi are exposed to considerably high levels of antifungal drugs in the oropharyngeal candidiasis for example, might lead to resistance in these drugs.

Fungi, like bacteria, can also produce a biofilm in vitro and there have been conducted studies for its effect in infections. For instance, changes in extracellular matrix and especially boosted beta glucan synthesis was observed, aiming to obstruct penetration of antifungals (e.g., azoles and polyenes). There have been also conducted clinical trials showing the importance of the biofilm for several pathological conditions, e.g., in candidemia (Pappas et al., 2009). However, it

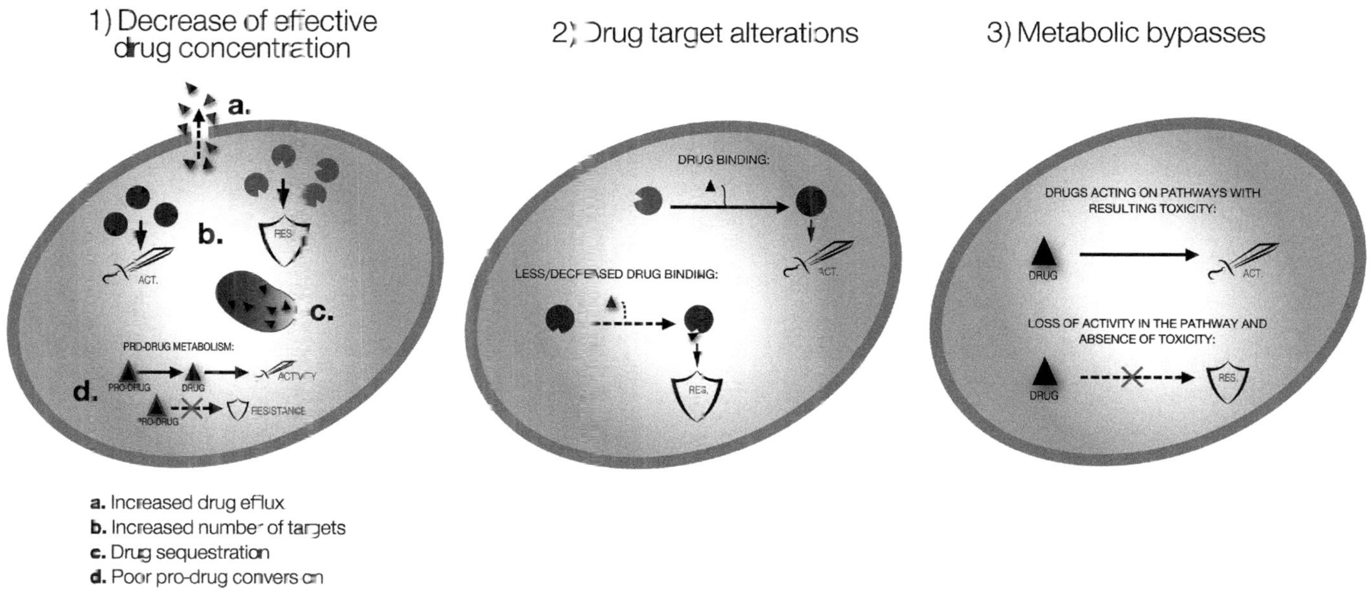

FIG. 3.6 Basic mechanisms of fungal resistance to drugs.

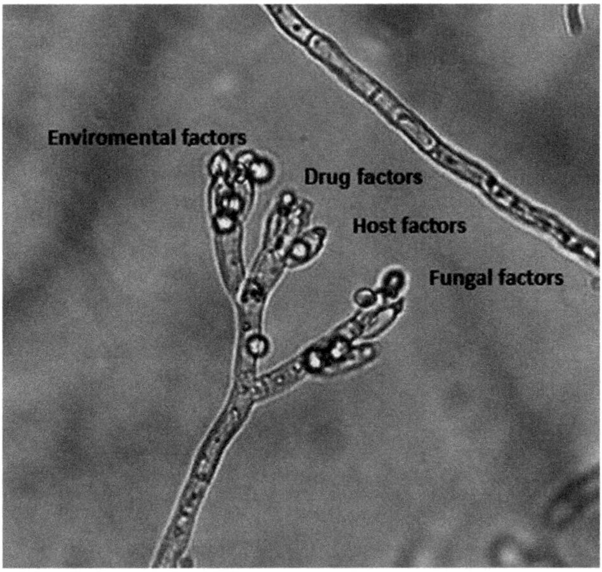

FIG. 3.7 The main resistance factors.

should be noted that there no well-standardized systems to measure and evaluate resistance parameters in fungal strains, making its study and evaluation more difficult for scientists.

New resistance mechanisms emerge and spread globally, threatening our ability to treat common infectious diseases, resulting in death and disability of individuals who until recently could continue a normal course of life. Infections caused by resistant microorganisms often fail to respond to the standard treatment, resulting in prolonged illness, higher health care expenditures, and a greater risk of death.

References

Arendrup, M.C., 2013. Update on antifungal resistance in *Aspergillus* and *Candida*. In: ESCMID Conference Invasive Fungal Infections, Italy.

Becker, D., Selbach, M., Rollenhagen, C., Ballmaier, M., Meyer, T.F., Mann, M., Bumann, D., 2006. Robust salmonella metabolism limits possibilities for new antimicrobials. Nature 16, 303–307.

Brotz-Oesterhelt, H., Sass, P., 2010. Postgenomic strategies in antibacterial drug discovery. Future Microbiol. 5, 1553–1579.

Brown, G.D., Denning, D.W., Gow, N.A., Levitz, N.M., Netea, M.G., White, T.C., 2012. Hidden killers: human fungal infections. Sci.Transl. Med. 4 (165). 165rv13.

Butts, A., Krysan, D.J., 2012. Antifungal drug discovery: something old and something new. PLoS Pathog. 8 (9), e1002870.

Chowdhary, A., Sharma, C., Duggal, S., Agarwal, K., Prakash, A., Singh, P.K., Jain, S., Kathuria, S., Randhawa, H.S., Hagen, F., Meis, J.F., 2013a. New clonal strain of *Candida auris*, Delhi, India. Emerg. Infect. Dis. 19, 1670–1673.

Chowdhary, A., Kathuria, S., Xu, J., Meis, J.F., 2013b. Correction: emergence of azole-resistant *Aspergillus fumigatus* strains due to agricultural azole use creates an increasing threat to human health. PLoS Pathog. 9 (10), e1003633.

Clancy, C.J., Nguyen, M.H., 2018. Diagnosing invasive candidiasis. J. Clin. Microbiol. 56 (5), 1–9.

Coates, A.R.M., Halls, G., Hu, Y., 2011. Novel classes of antibiotics or more of the same? Brit. J. Pharmacol. 163, 184–194.

Cowen, L.E., Sanglard, D., Howard, S.J., Rogers, P.D., Perlin, D.S., 2018. Mechanisms of antifungal drug resistance. Cold Spring Harb. Perspect. Med. 5 (7), 1–22.

Espinel-Ingroff, A., 1996. History of medical mycology in the United States. Clin. Microbiol. Rev. 9 (2), 235–272.

Fisher, M.C., Hawkins, N.J., Sanglard, D., Gurr, S.J., 2018. Worldwide emergence of resistance to antifungal drugs challenges human health and food security. Science 360 (6390), 739–742.

Fridkin, S.K., Jarvis, W.R., 1996. Epidemiology of nosocomial fungal infections. Clin. Microbiol. Rev. 9 (4), 499–511.

Ghannoum, A.M., Rice, B.R., 1999. Antifungal agents: mode of action, mechanisms of resistance, and correlation of these mechanisms with bacterial resistance. Clin. Microbiol. Rev. 12 (4), 501–517.

Groth, C.M., Dodds-Ashley, E.S., 2006. Fungal infections in the ICU. In: Infection Critical Care, pp. 1–23.

Gubbins, O.P., Anaissie, J.E., 2009. Antifungal therapy. In: Clinical Mycology, Second Ed. Elsevier Inc.

Howard, S.J., Cerar, D., Anderson, M.J., Albarrag, A., Fisher, M.C., Pasqualotto, A.C., Laverdiere, M., Arendrup, M.C., Perlin, D.S., Denning, D.W., 2009. Frequency and evolution of azole resistance in *Aspergillus fumigatus* associated with treatment failure. Emerg. Infect. Dis. 15, 1068–1076.

Laxminarayan, R., Chaudhury, R.R., 2016. Antibiotic resistance in India: drivers and opportunities for action. PLoS Med. 13 (3), e1001974.

Lockhart, S.R., Iqbal, N., Cleveland, A.A., Farley, M.M., Harrison, L.H., Bolden, C.B., Baughman, W., Stein, B., Hollick, R., Park, B.J., Chiller, T., 2012. Species identification and antifungal susceptibility testing of *Candida* bloodstream isolates from population-based surveillance studies in two U.S. cities from 2008 to 2011. J. Clin. Microbiol. 50, 3435–3442.

Loeffler, J., Stevens, D.A., 2003. Antifungal drug resistance. Clin. Infect. Dis. 15 (3), S31–S41.

Mitchell, T.G., Perfect, J.R., 1995. Cryptococcosis in the era of AIDS-100 years after the discovery of *Cryptococcus neoformans*. Clin. Microbiol. Rev. 8, 515–548.

Morgan, J., 2005. Global trends in Candidemia: review of reports from 1995-2005. Curr. Infect. Dis. Rep. 7 (6), 429–439.

Mortensen, K.L., Mellado, E., Lass-Flörl, C., Rodriguez-Tudela, J.L., Johansen, H.K., Arendrup, M.K., 2010. Environmental study of azole-resistant *Aspergillus fumigatus* and other Aspergilli in Austria, Denmark and Spain. Antimicrob. Agen. Chemother. 54 (11), 4545–4549.

O'Neill, J., 2014. Antimicrobial resistance: tackling a crisis for the health and wealth of nations. The review on antimicrobial resistance. HM Government, Welcome Trust.

Ostrosky-Zeichner, L., Casadevall, A., Galgiani, J.N., Odds, F.C., Rex, J.H., 2010. An insight into the antifungal pipeline: selected new molecules and beyond. Nat. Rev. Drug Discov. 9, 719–727.

Pappas, G.P., Kauffman, C.A., Andes, D., Benjamin, K.D., Calandra, T.F., Edwards, J.E., Filler, S.G., Fisher, J.F., Kullberg, B.-J., Ostrosky-Zeichner, L., Reboli, A.R., Rex, J.H., Walsh, T.J., Sober, J.D., 2009. Clinical practice guidelines for the management of candidiasis: 2009 update by the Infectious Diseases Society of America. Clin. Infect. Dis. 48 (5), 503–535.

Perea, S., Patterson, S.T., 2002. Antifungal resistance in pathogenic fungi. Clin. Infect. Dis. 35, 1073–1080.

Sanglard, D., 2016. Emerging threats in antifungal-resistant fungal pathogens. Front. Med. 3 (11).

Spellman, B., 2014. Alliance for the prudent use of antibiotics, new antibiotic development: barriers and opportunities in 2012. APUA Clin. Newsl. 30 (1).

Stevens, D.A., 1995. Therapy for opportunistic fungal infections: past, present and future. Indian J. Cancer 32, 1–9.

UN, 2016. Political declaration of the high-level meeting of the general assembly on antimicrobial resistance: draft resolution. In: President; UN. General Assembly (71st sess.: 2016-2017). High-Level Plenary Meeting on Antimicrobial Resistance, New York.

Vallabhaneni, S., Cleveland, A., Farley, M., Harrison, L.H., Schaffner, W., Beldavs, Z.G., Derado, G., Pham, C.D., Lockhart, S.R., Smith, R.M., 2015. Epidemiology and risk factors for echinocandin non-susceptible Candida glabrata bloodstream infections: data from a large multisite population-based candidemia surveillance program, 2008–2014. Open Forum Infect. Dis. 2 (4), 163.

Vallabhaneni, S., Kallen, A., Tsay, S., Chow, N., Welsh, R., Kerins, J., Kemble, S.K., Pacilli, M., Black, S.R., Landon, E., Ridgway, J., Palmore, T.N., Zelzany, A., Adams, E.H., Quinn, M., Chaturvedi, S., Greenko, J., Fernandez, R., Southwick, K., Furuya, E.Y., Calfee, D.P., Hamula, C., Patel, G., Barrett, P., Lafaro, P., Berkow, E.L., Moulton-Meissner, H., Noble-Wang, J., Fagan, R.P., Jackson, B.R., Lockhart, S.R., Litvintseva, A.P., Chiller, T.M., 2017. Investigation of the first seven reported cases of *Candida auris*, a globally emerging invasive, multidrug-resistant fungus-United States, May 2013–August 2016. Am. J. Transplant. 17 (1), 296–299.

Verweij, P.E., Snelders, E., Kema, G.H., Mellado, E., Melchers, W.J., 2009. Azole resistance in *Aspergillus fumigatus*: a side-effect of environmental fungicide use? Lancet Infect Dis. 9, 789–795.

Wernli, D., Jorgensen, P.S., Harbarth, S., Carroll, S.P., Laxminarayan, R., Levrat, N., Røttingen, J.A., Pittet, D., 2017. Antimicrobial resistance: The complex challenge of measurement to inform policy and the public. PLoS Med. 14 (8), e1002378.

White, T.C., Oliver, B.G., Graser, Y., Henn, M.R., 2008. Generating and testing molecular hypotheses in the dermatophytes. Eukaryot. Cell. 7, 1238–1245.

WHO, 2014. Antimicrobial resistance. Fact sheet N°194. World Health Organization.

WHO, 2015. Global action plan on antimicrobial resistance. World Health Organization, Geneva.

WHO, 2017. Library of national action plans. World Health Organization.

WHO, 2018. Antimicrobial resistance. World Health Organization.

WHO, 2019. New report calls for urgent action to avert antimicrobial resistance crisis.

Wiederhold, N.P., 2017. Antifungal resistance: current trends and future strategies to combat. Infect. Drug Resist. 10, 249–259.

Chapter 4

Natural products as antifungals

Marina Soković[a] and Konstantinos Liaras[b]

[a]*University of Belgrade, Institute for Biological Research "Siniša Stanković", National Institute of Republic of Serbia, Mycological Laboratory, Belgrade, Serbia,* [b]*Aristotle University of Thessaloniki, School of Pharmacy, Department of Pharmaceutical Chemistry, Thessaloniki, Greece*

Chapter outline

During the last decades, the issue of increase in the incidence of fungal infections was magnified by the simultaneous rise of the resistance emergence of some species of fungi against various compounds regularly administered for medical use. Fungi are considered to belong to neglected pathogens in comparison with bacteria.

The dramatic rise in the incidence of life threatening systemic fungal infections brought up the need for developing innovative strategies in order to successfully treat fungal infections and especially opportunistic ones in HIV-infected patients and other immunocompromised individuals who are under cancer chemotherapy and/or overuse of antibiotics. Most of the antifungals that find application in clinical practice are far from being characterized as ideal drugs in terms of toxicity, efficacy, and cost, not to mention that their frequent use has

Antifungal Compounds Discovery. https://doi.org/10.1016/B978-0-12-815824-1.00004-4

been one of the main causes of the emergence of resistant strains. Moreover, in order to minimize environmental and public health risks concerning use of synthetic fungicides in agriculture, public pressure was exerted toward alternative methods of dealing with fungal strains.

Although the synthetic chemistry, as a method to discover and produce drugs, was in dominant position in recent years, scientists realized the potential of bioactive compounds from natural sources and/or their extracts, and that nature could provide new products for prevention and disease treatment (Raskin et al., 2002). The discovery of antibiotics/antimycotics undoubtedly decreased the spread of a wide variety of diseases, but as a result of their uncontrolled use, the activity of many commercial antimicrobial agents was threatened by the emergence of microbial resistance to existing drugs (Cowan, 1999).

Therefore, there is still a great need for developing novel and structurally diverse antifungals, that would be selective, more potent, and with fewer or no side effects. A possible approach toward novel drugs discovery, could include testing of plants, mushrooms, and other natural sources that have been traditionally used for their antifungal applications. Therapeutic use of natural products is important only to the patients that traditional medicine is the sole choice for their treatments, or to the individuals who have chosen use of natural products as everyday lifestyle. Moreover, these medicinal plants, mushrooms, and other natural products can also serve as a source of new pharmaceuticals. Natural products, either in the form of pure compounds, or as standardized plant extracts, can provide us with a plethora of opportunities concerning novel drug leads discovery, mainly due to the availability of vast chemical diversity.

Organic chemists have been investigating different natural compounds since 1850s. All these intensive studies of natural obtained compounds initiated the development of new methodologies (separation techniques, spectroscopic approaches to structure elucidation, and synthetic methodologies). These new methodologies brought many possibilities for detection of new biological activities of numerous natural products, and marked this field as one of the most important ones in the continuous search for new antibiotics, drugs, insecticides, herbicides, etc.

While the total number of known natural compounds (plant, fungal, and microbial metabolites) is around one million, the need of mankind for innovative drugs is always increasing, making the discovery of natural compounds more than welcome (Berdy, 2005). The ongoing investigation of natural products in the hunt for discovering novel biologically active compounds to address problems in veterinary sciences, agriculture, and medicine, has also led to significant technological progress in the field. The biggest challenge is the optimization and successful application of novel discoveries in order to solve the targeted issues (Berdy, 2005). Only a few of the discovered natural compounds that possess innovative modes of action manage to reach clinical practice stage. This is one of the factors of the recently renewed scientific interest in pharmaceutical natural compounds application. In particular, the use of such natural products by ethnopharmacology practitioners is another eminent reason for encouraging that kind of research.

During the last years has been observed a raise in the use of natural products for therapeutic reasons. The main causes of that result are their sufficient availability around the globe, their fewer or none side effects, and their general decreased toxicity and higher biodegradability compared to conventional synthetic drugs. Natural therapeutic agents attract the scientific interest because of a variety of advantages that they possess such as lower price, safety, and being ecologically friendly.

Nature has been a source of medicinal products for millennia. Drug discovery from microbial sources opened this period and develop different techniques in the 1970s. It is universally known that nature has been a main source of therapeutic compounds for thousands of years. Of the 250,000 species of higher plants, approximately 5%–15% are chemically and pharmacologically investigated (Lahlou, 2013). Of the 252 drugs considered as basic and essential by the World Health Organization (WHO), 11% are exclusively of plant origin, and a significant number are synthetic drugs originating from natural precursors (Rates, 2001).

Natural substances derived from different origin (animals, plants, and microbes) were the lead compounds for the pharmaceutical industry between 1981 and 2002, and contributed greatly to the history and landscape of new molecular entities (NMEs). Approximately 877 small molecules were introduced, among them 49% pure natural compounds, analogues of semisynthetic products, or newly synthetic compounds based on natural-product components (Koehn and Carter, 2005). Nearly one-half of these are derived from mammals, one-quarter from microbes, and one quarter from plants (Patridge et al., 2016).

Combinatorial chemistry was developed at late 1980s and moved the focus from Nature to the laboratory bench. A new approach has risen: screening of natural compounds and their structures with the aim to develop novel drugs. Among 175 molecules, 131 (75%) were nonsynthetic, and 85 molecules (49%) were natural-based products (Newman et al., 2003). The establishing of the new approach, to use natural compounds in different areas, especially in the treatment of infectious diseases, showed that many natural obtained drugs used as antibacterial and antifungal agents have proved to be helpful in the treatment of infected patients (Singh and Barrett, 2006).

Presently, natural compounds have drawn the interest of pharmaceutical companies concerning the development of novel and efficient antimicrobials (Butler, 2005; Lam, 2007). Active metabolites deriving from natural sources are a great group of structurally diverse compounds and, thus, they can be responsible for different kinds of activities. For instance, secondary plant metabolites are important for their defensive mechanisms against predators and pathogens. These metabolites can act on sites similar to where the activity of pathogens is observed, providing in that way beneficial effects for protection of humans from them as well (Briskin, 2000; Treutter, 2005).

Natural products have complex molecular structures, cyclic semirigid scaffolds, several chiral centers, more than ten H-bond acceptors and five H-bond donors, five rotatable C–C bonds, a molecular weight above 500, and a large polar surface area. With these characteristics natural products may lead to good levels of bioavailability and could achieve the required efficacy (Jacob, 2009). Great structural diversity makes them as an incredible source of novel compounds against newly discovered therapeutic targets.

Natural products complement synthetic molecules, but they also could exhibit drug relevant characteristics unsurpassable by any synthetic compound. About 40% of the chemical scaffolds found in natural products could not be recognized in medicinal chemistry, and therefore complementary to synthetically produced molecules. Biosynthesis of natural products involves repeated interaction with modulating enzymes and the actual biological function of many natural products comprises binding to other proteins, which is an important advantage of these products. Many natural products possess binding characteristics more advanced compared to synthetics (Lutz, 2003).

In respect of all mentioned facts, traditional medicine is holding an eminent position among treatment strategies in both developing as well as developed countries, where natural products are becoming more popular as they present a safe and efficient therapeutic profile for the public. Ethnopharmacology is currently recognized by scientists as an efficient tool of great importance in treating a variety of severe diseases.

This book aims to examine the recent efforts toward discovering novel antifungal drugs of natural origin, from different sources, plant, fungi, insects, invertebrates, and new synthetic or semisynthetic antifungals. The information has been organized into easily accessible and comparable sections, with reference to the crude extracts of plants, fungi (micro and macro), insects, and other natural sources, their fractions and compounds, as well as new synthetic antifungal agents. Furthermore, bioactive natural products may have different mechanisms from existing commercial drugs and could be of significant clinical importance and promising in terms of health care improvement (Raskin et al., 2002).

4.1 Fungal metabolites

Fungi are organisms, which can be found practically everywhere and they have always been considered as extremely important for life on earth, but also for human society and everyday life. The vast majority of known fungal species are able to produce various primary and secondary metabolites that exhibit a wide spectrum of biological activities.

The medical use of antimycotics of microfungal origin has been the subject of extensive studies as far as the international literature is concerned (Berdy, 2005). However, their use in clinical practice remained limited because of concerns for resistance emergence among a number of pathogens that can infect humans (Soković et al., 2013).

The idea that substances derived from one organism might affect another is very old with many applications in traditional and conventional medicine. A number of the fungal-derived secondary metabolites can be found only in specific species, while others are more widespread. Therefore, it is observed that screening more fungal species can lead to discovering novel secondary metabolites that can potentially be biologically active.

Fungi are well known for their diversity, they are easy to obtain, making them an important source of natural products that can be further screened for their biological activity. Eminent natural products producers among fungal species are various microscopic (filamentous) fungi (fungi imperfecti) producing more or less 6400 compounds. Other 950, 900, and 300 compounds were isolated from well-known ascomycetes from *Aspergillus*, *Penicillium*, and *Fusarium* species, respectively. Moreover, many other filamentous and endophytic *Trichoderma*, *Phoma*, *Alternaria*, *Acremonium*, and *Stachybotrys* species are also considered to be good producers being in a position to produce several hundreds of biologically active compounds each. As far as higher fungal species are concerned (basidiomycetes or mushrooms), from species of *Ganoderma*, *Lactarius*, or *Agaricus*, around 2000 bioactive compounds have been isolated, while from yeasts only 140 and from Myxomycetes (slime molds) species about 60 bioactive metabolites have been characterized (Soković et al., 2013). Although compounds deriving from fungi are considered to be relatively simple in terms of chemistry, they frequently are found to exhibit diverse biological effects. Notable compounds of scientific and therapeutic interest in human history that are produced by fungi are beta lactams (penicillins, cephalosporins), cyclosporine, and some statins (compactin, lovastatin).

Recently, a growing interest for all types of fungal species has been observed and mainly for endophytic and the so-called marine fungi as promising sources for new pharmacologically active molecules. The development of newer and quicker screening processes led to increasing numbers of "unidentified" fungi that are considered as producers of bioactive metabolites (over 250 new metabolites in the last 2 years). This background proves the high speed of isolation, and identification processes as far as new fungal products are concerned, alongside with the long time needed for taxonomical identification of most new fungal species (Berdy, 2005). There is special interest in the discovery of innovative antimicrobials that represent new chemical classes and exhibit their activity by unique modes of action compared to existing agents and, consequently, lack cross-resistance to conventional drugs.

Natural compounds from microfungi

Microfungi are recognized nowadays as producers of compounds with diverse biological activities that could be currently used as pharmaceuticals. The golden period of antibiotics isolation from microfungi started with discovering bioactive molecules from *Penicillium* spp. Since then, many types of *Penicillium* have probably been scanned and various species are still being tested nowadays, indicating their importance as a source of new drugs to be used by the pharmaceutical industry. Since most fungi produce metabolites, molds are one of the most promising sources of new promising compounds with potential application in biotechnology and pharmaceutical industry.

Penicillium species produce many different known and unknown compounds that might be tested for their biological activities in many fields, including antifungal potential. *Penicillium ochrochloron* is one of the intensively tested species and known as a source of useful compounds (**penitrem A**, **B**, **C**, **D**, **E**, and **F**, **bromopenitrem A** and **F**, and **dehydro-penitrem D**) (Nielsen and Smedsgaard, 2003) (Figs. 4.1 and 4.2).

Kurobane et al. (1981) reported that *Penicillium brefeldianum* produces **fulvic acid** (Fig. 4.3). This compound is found to exhibit antiviral, antifungal, antioxidant, and antibiotic activities.

Moreover, Maskey et al. (2003) isolated two active substances, **8-*O*-methylaverufin** and **1,8-*O*-dimethylaverantin** (Fig. 4.4) from *Penicillium chrysogenum* as novel agents with antifungal properties.

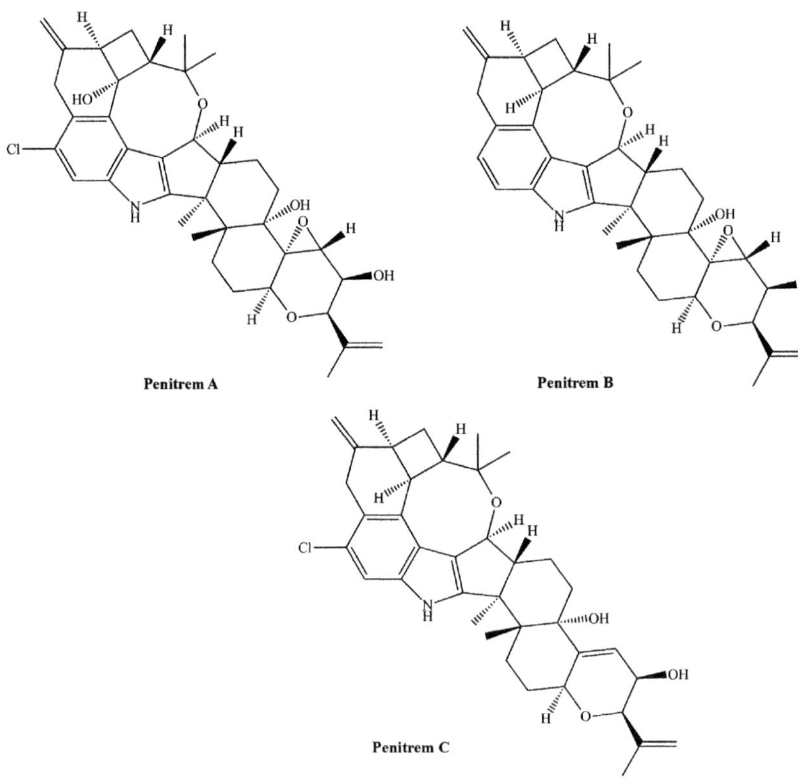

FIG. 4.1 Chemical structures of **penitrems A**, **B**, and **C**.

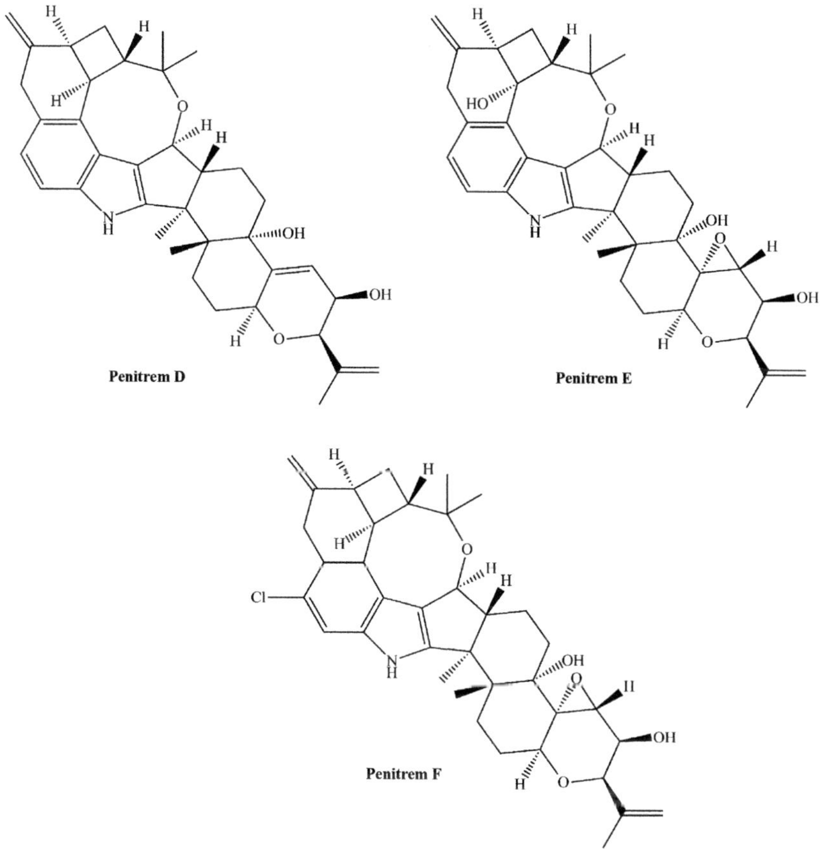

FIG. 4.2 Chemical structures of **penitrems D**, **E**, and **F**.

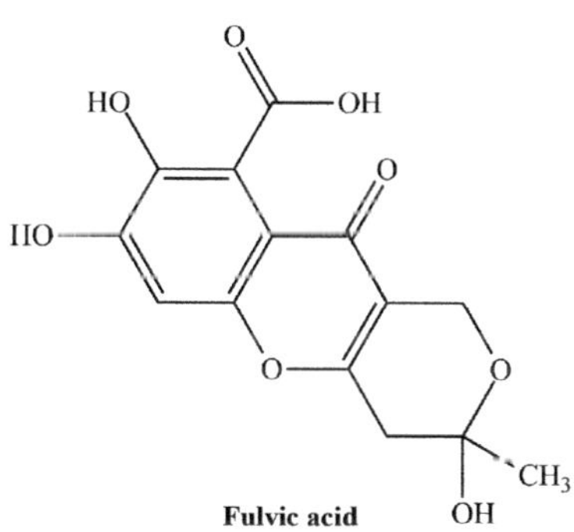

FIG. 4.3 Chemical structure of **fulvic acid**.

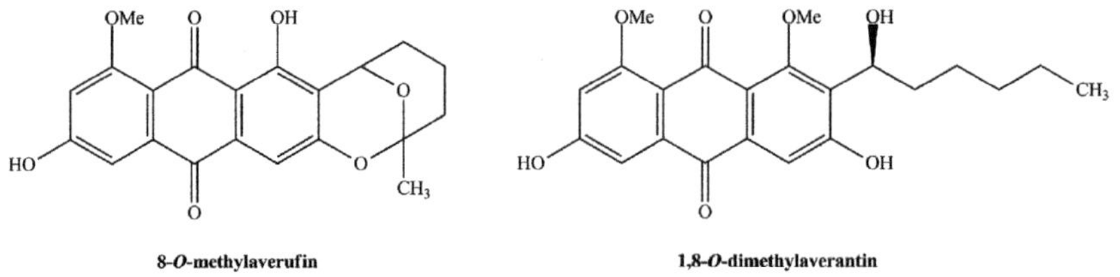

FIG. 4.4 Chemical structures of **8-*O*-methylaverufin** and **1,8-*O*-dimethylaverantin**.

Nam et al. (2000) stated that 8-*O*-methylsclererotiorinamine isolated from *Penicillium multicolor* possesses antimicrobial activity, while **funiculosin** (Fig. 4.5), found in *Penicillium funiculosum* exhibits antibiotic activity (Ando et al., 1969).

Extracts of seventeen microfungi were investigated for their effects against several fungi (*Alternaria alternata, Cladosporium cladosporioides, C. fulvum, Fusarium sporotrichioides, F. trincintum, Paecilomyces variotii, Penicillium ochrochloron, P. funiculosum, Phoma magdonaldii, Phomopsis helianthi, Stachybotrys chartarum, Trichoderma viride*) and five dermatomycetes, *Epidermophyton floccosum, Microsporum canis, Trichophyton mentagrophytes, T. rubrum*, and *T. tonsurans* were tested against *Candida albicans* with the use of bioautographic assay test on TLC plates (Fig. 4.6) (Rancic et al., 2006). The analysis of P. ochrochloron through the isolation of two secondary metabolites (**2,3,4-trihydroxybutanamide** and **erythritol**, Fig. 4.7) also showed antifungal potential of the extract and particularly these compounds against *C. albicans* (Rancic et al., 2006).

The potential demonstrated by *Penicillium* species as a source of interesting metabolites (tocopherols or phenolic compounds) was the reason to continue the search for their antimicrobial activities concerning the prospect of

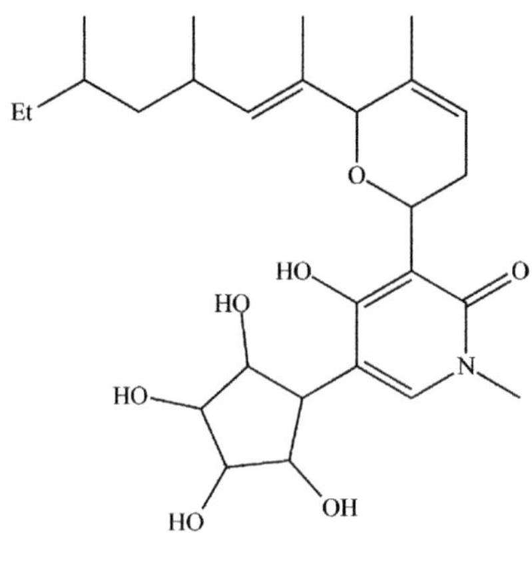

FIG. 4.5 Chemical structure of **funiculosin**.

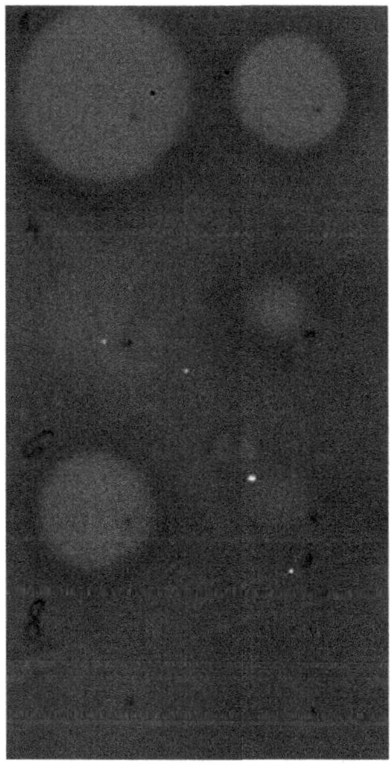

FIG. 4.6 TLC bioautography of microfungi extracts, inhibition zones.

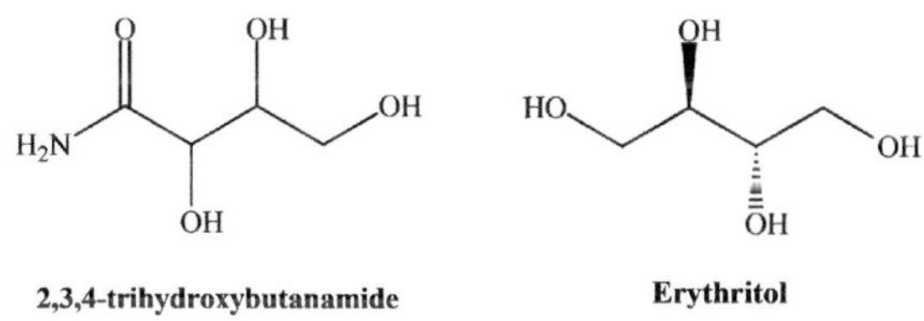

FIG. 4.7 Chemical structures of **2,3,4-trihydroxybutanamide** and **erythritol**.

further applications. The studied species (*P. ochrochloron*, *P. funiculosum*, and *P. verrucosum* var. *cyclopium*) were grown in culture media—Czapek and Malt media—in order to compare the produced compounds and their activity against *Candida albicans*.

Mycelia and culture media were tested for production of different compounds by all three fungi and the release of metabolites to the culture medium was monitored. Antimicrobial activity of phenolic acids, organic acids, and tocopherols extracts was analyzed. From sugars content, rhamnose, xylose, fructose, and trehalose we recorded in all three mycelia and culture media. Oxalic and fumaric acids were prevailing organic acids, and protocatechuic and p-hydroxybenzoic acids were the most abundant phenolic acids. Vitamin E isoform was represented by γ-tocopherol (Reis et al., 2013). Oxalic and fumaric acids exhibited antifungal activity against *C. albicans*. Mycelia of *P. ochrochloron* extract, grown on Cz medium and *P. verrucosum* var. *cyclopium* grown on

MA medium, possessed anticandidal activity. p-coumaric and cinnamic acids were also effective against *C. albicans* at all tested concentrations. Mycelial extracts of all three tested species, grown on Cz medium, were active toward *C. albicans*, while culture media extracts did not reveal anticandidal activity, except one of *P. funiculosum*. All tocopherol isoforms (α, ß, γ, and δ) showed effect on *C. albicans*. It is interesting to notice that the main detected compounds were oxalic and fumaric acid. Oxalic acid is usually associated with pathogenicity of species (Reis et al., 2013). Namely, in a process of pathogenesis in fungi, this acid is usually produced in order to make the host more susceptible (Nagarajkumar et al., 2005). It could be assumed that oxalic acid is produced as a response to oxidative stress, since the fungi were under in vitro conditions and in the presence of *C. albicans* it may be influenced and produced as a defense compound to inhibit the growth of this strain. The important fact is also that the organic acids are lipid-soluble compounds (when they are in the undissociated form) and then they could enter into the microbial cell. After entrance in cells, the acid releases the proton, creating a more acidic environment, which could result in decrease of intracellular pH. Fungal cells are forced in that condition to use energy and release protons, leading to an intracellular accumulation of acid anions. The production of organic acids by fungi and release of them in the culture media may be a consequence and answer to the stress conditions. When molds were not in their usual habitat, or this is modified due to some reasons or presence of a foreign agent, the possible way of defense could be inhibition of pathogens growth through the production of these molecules (Reis et al., 2013).

The interest in microfungi as promising and new sources of bioactive compounds increased in past 10 years. Even the opportunistic or pathogenic fungi, *Penicillium, Aspergillus*, and some other could be used as a source of bioactive molecules as they may possess beneficial activities. Research has to be focused not only on mycelia of microfungi, but also on production of bioactive compounds and release of valuable compounds in the culture media by them. Even further modifications and combinations in extraction methodology could be considered as approaches of great importance and could give better results in searching and detecting certain molecules.

Natural compounds from macrofungi

Macrofungi represent a large group of fungi of the phylum Basidiomycota (basidomycetes) and Ascomycota, which are a major source of pharmacologically active substances. Among this group 75% of tested species have shown high antimicrobial activities (Daniel et al., 2012).

Alongside with the common use of medicinal herbs in the Western Hemisphere, introduction of mushrooms in clinical practice saw a slight increase also in Europe during the last decades, following the long tradition of Asian countries in the field. Despite the extensive research concerning the medicinal potential of mushrooms, their true prospect is far from being fully revealed.

Mushrooms and the compounds they contain have been thoroughly studied, especially in traditional Chinese Medicine, for their immunomodulatory effects and generally their activity as adjuvants for autoimmune diseases, infectious diseases, cancer, allergies, etc. In the recent years, more and more in western allopathic medicine, food supplements containing mushrooms, and their ingredients are being used by patients or people who are following particular nutrition protocols and habits. *Ganoderma lucidum*, *Trametes versicolor*, *Lentinula edodes*, and *Agaricus brasiliensis* (*Agaricus blazei*) are some of the most well-known medicinal mushrooms that are being used and studied (Soković et al., 2006).

Nevertheless, it is observed a restraint in accepting mushrooms and their components, mainly due to the limited availability of double-blind clinical assays data and also because of the lack of pure active ingredients originating from mushrooms.

The majority of the studies concerning the pharmacological activity of mushrooms have taken place using beta glucans that take place in the formation of fungal cell wall. Beta glucans have the ability to activate leukocytes, among other cells of immune system, and thus to stimulate their cytotoxic, phagocytic, and generally their antimicrobial activity (Brown, 2006). They also take part in the stimulation and production of pro- and anti-inflammatory cytokines and chemokines, acting in that way as modulators of innate and acquired immunity and also cell growth (Smiderle et al., 2011).

The research studies on the immunomodulatory effects of nonpolysaccharide compounds that originate from mushrooms have been limited to date. Furthermore, another interesting group of bioactive mushroom components is that of the antioxidants. Despite the great amount of studies on the activity of this category of compounds, there is not much evidence of their preventive and/or therapeutic activity (Kozarski et al., 2015). Flavonoids and triterpenoids, among other polyphenols, were thoroughly tested and found to take part in the redox state of different cells and in that way to prevent oxidative damage. Prevention of neuropathological effects and problems in circulation were also suggested as possible fields of influence for the mentioned categories of mushroom compounds (Suabjakyong et al., 2015), while their antimicrobial activity has also been revealed from corresponding extensive studies.

It has been stated in literature that generation of reactive oxygen species in mushrooms can cause proinflammatory activity and the reason for this effect was linked to the presence of polyphenols rather than to β-glucans alone (Wei et al., 2008). There are also compounds that bind as ligands to cell membrane receptors, affecting in this way diverse cell functions and consequently health. In mammals, nitric oxide (NO) is such a signaling substance that possesses different kinds of roles in many processes that are connected to human health (Toledo Jr. and Augusto, 2012). Production of NO can induce damage of tissues than can be connected to acute and chronic inflammation (Taira et al., 2009). It is well known that inflammation takes part in the progress of many diseases, such as asthma, rheumatoid arthritis, atherosclerosis, inflammatory bowel diseases, various human cancers, (Saiki et al., 2017) as well as in emergence of infections, including the ones that are caused by fungi.

There are several medicinal and edible mushrooms such as *Agaricus brasiliensis*, *A. bochusii*, *A. campestris*, *Coprinus comatus*, *Coriolus versicolor*, *Ganoderma lucidum*, *Laetiporus sulphureus*, *Lentinula edodes*, *Phellinus linteus*, and *Armillaria mellea* among others, that have been used in traditional therapeutics as supplements for the prevention and cure of different kinds of diseases, such as atherosclerosis, cancer, chronic hepatitis, autoimmune diseases, infection diseases, and diabetes. These preventive and therapeutic effects of mentioned mushrooms and the compounds that derive from them have been extensively studied using in vitro and in vivo models as well as cancer cell lines (Heleno et al., 2013; Kostić et al., 2017; Petrović et al., 2014; Reis et al., 2012; Siljegovic et al., 2011a; Stojkovic et al., 2013b, 2014). The results of these studies led to a plethora of scientific facts about the biological activity of mushroom extracts and their corresponding modes of action. By now, it is widely recognized that mushroom extracts contain a variety of compounds, such as polysaccharides (i.e., glucans), small proteins, lectins, and polyphenols, each of which can possess its own biological or pharmacological activity.

Some of the most studied immunomodulatory effects of mushrooms can be attributed to β-(1 → 3)-(1 → 6)-glucans, which have been tested in depth (Smiderle et al., 2011). *Trametes versicolor*, *Laetiporus sulphureus*, and *Ganoderma lucidum* are mushrooms that are known to possess this kind of potential, turning them into ideal sources for desperately needed novel antimycotics (Zjawiony, 2004). *Laetiporus sulphureus* is a wood-rotting basidiomycete that contains various active compounds and that is the reason of its use as by humans. More particularly, articles concerning the antimicrobial activity of *L. sulphureus* extracts are scarce (Turkoglu et al., 2007; Zjawiony, 2004). The potential of using medicinal mushrooms on daily basis as therapy depends on the way the mushroom is consumed. It should also be mentioned that research on fungi as potential antimicrobial agents mostly concerns ethanolic and methanolic extracts of the fungal fruit body (Barros et al., 2007; Turkoglu et al., 2007), explaining why consumption of products under these conditions has no practical use. On the other hand, an aqueous extract obtained from *L. sulphureus* was tested for its antimicrobial properties using a microdilution assay in vitro against seven fungi (four *Aspergillus*, two *Penicillium* species and *Trichoderma viride*). It should be noted that this extract exhibited strong activity against the tested fungi in a dose-dependent manner (Siljegovic et al., 2011a,b). Antifungal activity of *L. sulphureus* n-hexane and chloroform extracts showed MFC in range of 3.0–5.0 mg/mL and 6.25–17.0 mg/mL, respectively. The best antifungal potential was obtained toward *T. viride* (n-hexane extract) and against *P. ochrochloron* (chloroform extract), while the lowest fungicidal effect of hexane extract was expressed on *P. verrucosum* and chloroform extract possessed the worst activity on *A. fumigatus* (Sinanoglou et al., 2015).

All tested extracts of *L. sulphureus* exhibited good antimicrobial effect, with small differences in their activity. Research showed that extraction methodology is one of the limitation factors for antimicrobial potential. Namely, extracts obtained by the ultrasound method showed higher antifungal potential. Ultrasound extraction is faster than the classical one, with minimal risk of light exposure, heating of extracts, denaturation of some valuable compounds, altering/affecting the composition of the mushroom, etc. Thongson et al. (2004) stated that a key step in the purification of bioactive compounds is solvent extraction, meaning that in conventional extraction of bioactive materials, a suitable solvent is required and appropriate heat has to be followed (due to degradation of active compounds is possible), or agitation to increase the solubility of materials (Petrović et al., 2014).

Aqueous and ethanolic extracts of *Inonotus obliquus* (Chaga) collected from different localities (Finland, Russia, and Thailand) showed considerable antifungal effects on all tested fungi. Inhibitory effect of ethanolic extracts was 0.40–1.50 mg/mL and fungicidal effect was achieved at 0.75–3.00 mg/mL. Aqueous extracts exhibited inhibition at 0.20–1.50 mg/mL, and fungicidal activity of these extracts was observed at 0.40–3.00 mg/mL. The most sensitive fungus was *Trichoderma viride* and *Aspergillus niger*, *A. fumigatus* were the most resistant species. The antifungal activity of all tested extracts against *Penicillium ochrochloron* and *Trichoderma viride* was higher than the potential of ketoconazole (Glamočlija et al., 2015a) (Fig. 4.8).

Ethanolic and methanolic extracts of *Agaricus bisporus*, *A. bitorquis*, *A. campestris*, and *A. macrosporus* showed antifungal activity against eight fungal

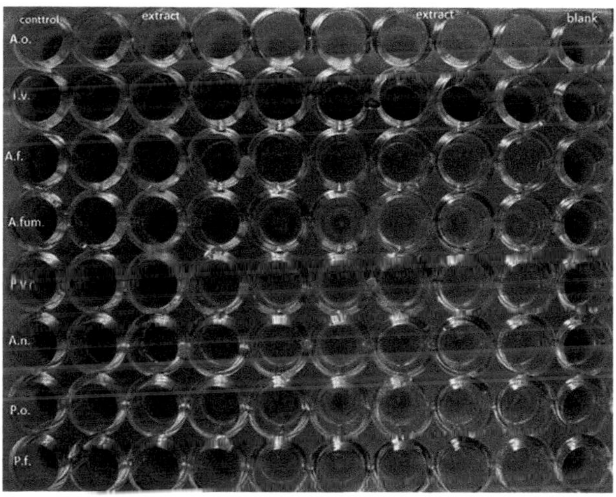

FIG. 4.8 Antifungal activity of *Inonotus obliquus* against *Aseprgillus ochraceus—A.o.*, *Trichoderma viride—T.v.*, *A. flavus—A.f.*, *A. fumigatus—A.fum.*, *Penicllium verucosum—P.v.c.*, *A. niger—A.n.*, *Penicillium ochrochloron—P.o.*, *P. Funiculosum—P.f.*

species. *A. bisporus* ethanolic extract exhibited lower effect than methanolic on *A. ochraceus* and *T. viride*, while *A. macrosporus* ethanolic extract showed lower activity than methanolic on *P. funiculosum* and *P. ochrochloron*. *A. macrosporus* extracts exhibited the highest antifungal potential against all microfungi (Glamočlija et al., 2015b).

Chemical and bioactive properties of wild and cultivated *Ganoderma lucidum* from Serbia (GS) and China (GCN) were tested. Both samples possessed strong antifungal activity. The results showed that antimycotic standards, ketoconazole and bifonazole, possess lower activity than extracts. Bioactive properties of *G. lucidum* are highly dependent on the origin of the samples (Stojkovic et al., 2013b).

Chemical analysis and biological potential of *Armillaria mellea* collected from nature was tested by Kostić et al. (2017). Methanolic extract of this mushroom showed good antifungal activity against all tested fungi. This was also the first record regarding antiquorum-sensing regulated potential (biofilm formation, pigment production) of the methanolic extract of *A. mellea*.

These facts might be of great importance in the near future of the fight against AMR. Mushroom-derived compounds could also be of particular importance, as they possess good antifungal potential and due to their natural origin, they could be a source of new antifungals with lesser side effects than the synthetic ones.

4.2 Plant extracts and metabolites

Nature has always been the source for biologically active molecules and the case of antifungal compounds could be no exception. Isolation and study of numerous natural compounds was and still is in the center of scientific attention for both prevention and treatment of fungal diseases.

Plants have been used for medicinal purpose for thousands of years. They have adapted to different conditions over millions of years, resisted pathogens (bacteria, insects, fungi) and produced a variety of structurally diverse and unique secondary metabolites. They have been used as a primary source of medicines for early drug discovery due to their ethnopharmacological properties (Daniel et al., 2012).

With the acceptance of medicinal plants as an alternative way for health care increasing year by year, the continuous search for active components in plants turns out to be of great importance. Such plants have been employed through the centuries of human history and are extensively mentioned in ancient folklore. Following the emergence of antibiotics in the 1950s, the therapeutic use of plants and their derivatives as antimicrobials became almost nonexistent, only to be rediscovered alongside with other alternative medical practices in the late 1990s (Cowan, 1999). Several different approaches for choosing sources of natural products in order to discover potential antifungal compounds have been followed by the scientists. One of these approaches consists of investigating extracts of whole plants to obtain biological material, which has not previously been studied and could be fractionated. Another strategy could be to use ethnopharmacology and/or chemical ecology information for the selection of plants that could serve as sources for samples (Duke et al., 2000).

There can been found numerous reports in literature concerning the antifungal activity of plant extracts, but only a few of these will be mentioned herein. Taking into account all the stated information, it turns out that fungal infections can be an excellent opportunity in order to gain greater expertise and improve understanding on the methods of using herbal medications in clinical practice.

Applying phytochemical screening on numerous plants has shown that secondary metabolites could exhibit a wide spectrum of biological activities, including antibacterial, antifungal, antioxidant, antidiabetic, anticancer, anti-inflammatory, etc. Consequently, plant extracts can be characterized as ideal material for research in the hunt for development of innovative formulations to treat infections caused by fungi.

An ethanolic extract of *Phlomis fruticosa* (Jerusalem sage) (Labiatae) has been tested against microfungi, inhibiting *Aspergillus niger*, *Penicillium ochrochloron*, *Trichoderma viride*, *Fusarium tricinctum*, and *Phomopsis helianthi*. In addition, this extract was found to possess fungicidal activity against *Cladosporium cladosporioides* and *Aspergillus ochraceus* at very low concentrations (10–20 μg/mL) (Ristić et al., 2000). A dealcoholized extract of leaves of *Cassia tora* exhibited strong antifungal activity against *Aspergillus niger* (Mukherjee et al., 1996), while Sato et al. (2000) tested 29 plants extracts against *Arthrinium sacchari* and *Chaetonium funicola*. Studying ethanol extracts of 15 different plants revealed antifungal activity, with *Acer nikoense* (Nikko maple), *Glycyrrhiza glabra* and *Thea sinensis* (Tsa) being the most effective among them and in very low concentrations. Whole, fresh involucral bracts of cardoon, *Cynara cardunculus* L. (Compositae), were extracted with ethanol (EtOH) and an aqueous suspension of this extract was partitioned successively with $CHCl_3$, EtOAc, and n-BuOH, leaving residual water extract. Each extract was studied concerning its antifungal activity, using a microdilution technique toward *Aspergillus niger*, *A. ochraceus*, *A. flavus*, *Penicillium ochrochloron*, *P. funiculosum*, *Trichoderma viride*, *Fusarium tricinctum*, and *Alternaria alternata*. All cardoon extracts revealed antifungal activity comparable to that of standard antimycotics (Kukić et al., 2008).

Branched centaury *Centaurium pulchellum* (Gentianaceae) extracts and secoiridoid glycosides deriving from them were evaluated for their antifungal potential against five fungal species. Methanolic extracts from both aerial parts and roots were found to possess excellent antifungal (0.1–2 mg/mL) potency, while pure secoiridoid glycosides isolated from these extracts revealed also very strong (0.001–0.1 mg/mL) antifungal activity (Šiler et al., 2010). Methanolic extracts of three different Labiatae species (Catmint), *Nepeta rtanjensis*, *N. sibirica*, and *N. nervosa* (grown in vitro) were tested for their activity against eight fungal species showing significant potency, with that from, *N. rtanjensis* being the strongest among them (Nestorović et al., 2010).

Fungal species from *Aspergillus*, *Penicillium*, *Cladosporium*, *Trichoderma*, *Alternaria* genera were found to be sensitive to hydromethanolic extract of *Sapindus saponaria*. The extract inhibited all microfungi at 0.075–0.6 mg/mL (MIC) and completely stopped the growth of fungi (MFC) at 0.3–1.25 mg/mL. The most sensitive were *Penicillium* species, while *Aspergillus fumigatus* was the most resistant species (Rashed et al., 2013). Extracts of dried fruits pericarp of *S. saponaria* possessed antifungal activity against clinical isolates of yeasts *Candida albicans* and *C.* non-*albicans*. The butanol (n-BuOH) and one of its fractions showed strong activity against all *Candida* isolates tested.

The antifungal potential of the *Achilea* species extracts could be presented as follows: *A. teretifolia* (MICs of 0.006–0.4 mg/mL, MFCs of 0.0125–0.8 mg/mL) > *A. biebersteinii* (MICs of 0.015–0.4 mg/mL, MFCs of 0.025–0.8 mg/mL) > *A. millefolium* (MICs of 0.0125–0.8 mg/mL, MFCs of 0.025–0.8 mg/mL). The water extracts showed the lowest antifungal effect and inhibited *A. versicolor*, *P. ochrochloron*, *P. verrucosum* and also showed fungicidal activity on *T. viride*. The ethyl acetate extracts exhibited a stronger effect than previous ones, while methanolic extracts were the most potent toward all fungi (Zengin et al., 2017).

Silene species (*S. alba*, *S. conoidea*, *S. dichotoma*, *S. italica*, *S. supina*, and *S. vulgaris*) were tested for their antifungal effect (8 strains). All six extracts *Silene* exhibited strong antifungal activity (0.003–0.4 mg/mL), even higher then bifonazole and ketoconazole (Zengin et al., 2018).

Humulus lupulus L. is a perennial plant cultivated mainly in brewing industry. The phenolic composition of seeds was tested for cytotoxic, antioxidant, and antimicrobial effects. Compounds identified in seeds of this plant, (+)-Catechin and (−)-epicatechin, represented 17.6 mg/g of total extract. The extract showed great antimicrobial properties, against all tested fungi of the genus *Penicillium* (Esteban et al., 2019).

Galium aparine L. (Rubiaceae) known as clivers or bedstraw is traditionally used as a medicinal plant. Infusions and hydromethanolic extracts of air- and freeze-dried samples are tested for antimicrobial, antioxidant, and cytotoxic properties. The freeze-dried extract possessed the highest content of phenolic compounds, followed by the infusion and the air-dried hydromethanolic extract. **5-O-Caffeoylquinic acid** (Fig. 4.9) was the major compound identified in this extracts. Antimicrobial properties were in dependence with phenolic content. The results suggested that extracts of this plant are a good source of phenolic compounds with antimicrobial properties (Senio et al., 2018).

Centauries erythema Raven extract was tested to notice the differences in biological activities in metabolomics profiles between nonhydrolyzed (ME) and hydrolyzed methanol extract (HME). ME and HME showed very strong antimicrobial activities against a variety of pathogenic fungi from *Aspergillus*, *Penicillium*, *Trichoderma*, and *Candida* species, with stronger effects than those of the reference antimycotics. UHPLC/DAD/+HESI−qqqMS analysis of secoiridoid glucosides (SGs) and their aglycones showed that gentiopicral and

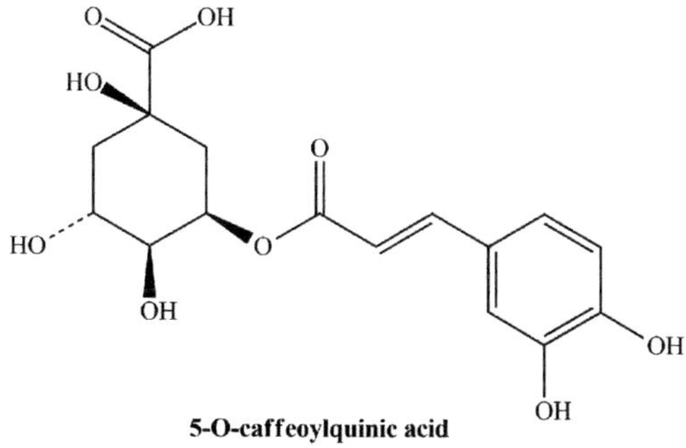

FIG. 4.9 Chemical structure of **5-O-caffeoylquinic acid**.

erythrocentaurin were major aglycones. Both extracts and secoiridoids exhibited the highest effect against all tested *Penicillium* species. It is noticed that *P. funiculosum* is capable to eliminate sublethal concentrations of secoiridoid glucosides by biotransformation and digestion of these compounds. Obtained results could be very helpful in clarification of the role of the secoiridoids from *C. erythraea* after ingestion of plant and of high significance for human health (Božunović et al., 2018).

Natural compounds from Bryophytes

Genuine mosses form a large group of nonvascular higher plants, containing about 14,000 species. Bryophytes, in general, cannot get damaged by microorganisms, insects, snails, slugs, and small mammals. Although they consist the second biggest group of land plants (app. 25,000 species), the reported information concerning the biological activity of extracts and compounds deriving from bryophytes are relatively scarce, labeling this group as neglected (Sabovljevic and Sabovljevic, 2008). Nevertheless, new compounds discovered in bryophytes led to their classification as a new source of biologically active molecules, with some of them being very potent (Pejin et al., 2013).

More than several hundreds of new compounds have been isolated and structurally determined so far from bryophytes (Veljic et al., 2009). Although a number of secondary metabolites of various mosses have been sufficiently determined in chemical terms, most of the species are insufficiently studied or even unknown in some cases. Some of the secondary metabolites that derive from mosses and have been determined to date include terpenoids, flavonoids, and bibenzyls and also derivatives of fatty acids. Mosses rich in flavonoids have been found to reveal strong antimicrobial potential (Veljic et al., 2010). An ethanolic extract of *Bryum argenteum* (silver moss), exhibited antifungal activity against two fungi (*A. niger* and *P. ochrochloron*) (Sabovljevic et al., 2006). Moreover, there are published results also demonstrating that methanolic extracts of particular genuine mosses (*Pleurozium schreberi*, *Palustriella commutata, Homalothecium philippeanum, Anomodon attenuatus, Rhytidium rugosum, Hylocomium splendens, Dicranum scoparium*, and *Leucobryum glaucum*) exhibit antimicrobial activity when tested by using the microdilution method (Veljic et al., 2008). By analyzing the antifungal potential of methanol extracts of the mosses *Fontinalis antipyretica* var. *antipyretica, Hypnum cupressiforme*, and *Ctenidium molluscum* was found that the first species possess strongest activity against the micromycetes *Trichoderma viride, Penicillium funiculosum, P. ochrochloron, Aspergillus fumigatus, A. flavus*, and *A. niger* (Veljic et al., 2009). Extracts of three bryophyte species, two mosses (*Atrichum undulatum, Physcomitrella patens*) and a liverwort (*Marchantia polymorpha* ssp. *ruderalis*), grown under natural conditions and in axenic culture, were tested using the microdilution method against five fungi, in order to determine their antifungal potential. All tested extracts were found to be active against all tested species. It was observed that extracts made from material grown in laboratory (in vitro) conditions reveal stronger antifungal potential than the ones made from material of natural origin. It should be mentioned that some of the tested fungal species had similar response against both mentioned types of extracts (Sabovljevic et al., 2011). Determination of antimicrobial activity against *Aspergillus versicolor, A. fumigatus, Penicillium funiculosum, P. ochrochloron*, and *Trichoderma viride* of a dimethyl sulfoxide extract of the moss *Rhodobryum ontariense* took place by using microdilution method. The mentioned extract was found to be potent against all tested microorganisms but at diverse rates. This finding implies that *R. ontariense* could be considered as a promising material for natural antifungal products (Pejin et al., 2012). Further studies by the same authors showed that this extract was more active against the tested fungi (MICs of 0.20–1.00 mg/mL and MFCs of 1.0–5.0 mg/mL). The best antifungal effect was achieved on *C. albicans* (MIC 0.20 mg/mL and MFC 1.00 mg/mL). Antifungal activity of dimethyl sulfoxide extract of this moss showed higher potential than the water extract. The tested extract was characterized as highly rich in saccharides. The results showed novelty in comparison with previous chemical profile of same species, when the presence of short-chain fructooligosaccharides (FOS; 1-kestose, nystose, and 1F-ß-fructofuranosylnystose) was noticed. These compounds are well known as prebiotics (Pejin et al., 2013).

In vitro grown bryophytes were found to possess antifungal potential. The only problem is how to obtain the necessary amount of material of bioactive species. One of the possible solutions is to produce biomass trough biotechnological processes, which could give enough material of these plants and also help to prevent extinction of some species that could be used in ex situ conservation, too (Rowntree et al., 2010). Bryophyte axenic culture give us a possibility to obtain enough material for some of the compounds that showed good activity and can be used for the treatment of fungal infections in environment-friendly manner and without toxic effects (Sabovljevic et al., 2011).

All the reports concerning testing of antifungal activity of mosses revealed that mosses could be sources of new antibacterial and, especially, antifungal agents. The isolation of biologically active substances from moss extracts could be useful in further investigations of agents providing protection against pathogenic microorganisms.

Plant secondary metabolites

Due to an increase in demand for developing new formulations and use of natural products in antimicrobial therapy, World Health Organization (WHO) has taken a broader step of including phytotherapy (Willis, 2017) in the health care programs and published new basic guidelines and procedures for validation of drugs of plant origin.

Plants are capable to produce a wide spectrum of secondary metabolites, many of which possess antifungal activity. A number of these molecules can be found in healthy plants in their biologically active forms while others, for instance cyanogenic glycosides and glucosinolates, can be usually found in the form of their inactive precursors and can be activated under special conditions, such as in response to tissue damage or pathogen attack. Activation usually involves specific plant enzymes that can be released after a disruption of cell integrity. Compounds that belong in the previously mentioned group are still considered to be constitutive, as they are immediately derived from pre-existing constituents. A great amount of plant compounds have been reported to reveal antifungal potential. Well-studied examples include flavonoids (Ciric et al., 2011; Karioti et al., 2011; Weidenbörner and Jha 1997), lactones (Djeddi et al., 2007; Janackovic et al., 2002; Skaltsa et al., 2000a,b; Vajs et al., 1999, 2004), proteins (Giudici et al., 2000), sulfur compounds (Ilić et al., 2012), cyanogenic glycosides and glucosinolates (Osbourn, 1996), and essential oils (Daouk et al., 1995; Dzamic et al., 2010; Garg et al., 1992; Glamočlija et al., 2006, 2009; Marinković et al., 2002; Mishra and Dubey, 1994; Muller-Riebau et al., 1995; Rančić et al., 2005; Shimoni et al., 1993; Soković, 2001; Soković et al., 2002, 2008a,b, 2009a,b,c; Stojković et al., 2011; Thompson, 1989).

Essential oils

Essential oils of aromatic and medicinal plants origin have been well known since the ancient times for exhibiting a wide spectrum of biological activities. They form one of the most studied categories of secondary metabolites, with an expanding interest in their use in the pharmaceutical and agrochemical industries. Over the last hundred years, the antimicrobial potential of common spice oils has been thoroughly tested and a great number of studies can be found in worldwide literature concerning antifungal potential of essential oils (Daouk et al., 1995; Garg et al., 1992; Glamočlija, 2006; Glamočlija et al., 2009; Kalemba and Kunicka, 2003; Mishra and Dubey, 1994; Muller-Riebau et al., 1995; Shimoni et al., 1993; Soković, 2001; Soković et al., 2013; Thompson, 1989).

During the last years, scientific literature concerning studies on antifungal potential of natural products has mainly included aromatic plant families, demonstrating in many cases that different kinds of essential oils possess antifungal effects and therefore can be beneficial to humans (Bakkali et al., 2008; Reichling et al., 2009). However, the information concerning the potential of essential oils against specific fungal pathogens is limited and, thus, the scientific results and details that follow will review recent research literature in this particular field.

In vitro activity of essential oils

Maruzzella and Balter (1959) demonstrated that out of 119 spice oils tested, 100 oils exhibited an antagonistic effect to at least one of twelve fungi while 50 of them revealed a wide range of antifungal potential against all tested species. The essential oil of *Origanum majorana* found to possess considerable inhibitory potential concerning *Aspergillus flavus*, *A. niger*, *A. ochraceus*, *A. parasitcus*, and *Trichoderma viride* (Deans and Svoboda, 1990). A comparative evaluation of the antifungal potential of essential oils extracted from thyme, rosemary, eucalyptus, and mugwort was performed by a group of researchers against 39 fungi. From the results it can be seen that essential oil of thyme was the most effective one (Conner and Beuchat, 1984). Essential oils deriving from allspice and cloves exhibited a total inhibition of *Trichoderma viride*, *Alternaria alternata*, *Fusarium oxysporum*, *Mucor circinelloides*, *Rhizopus stolonifer*, *Cladosporium cladosporioides*, *Aspergillus versicolor*, and *Penicillium citrinum* in a concentration of 2% (Schmitz et al., 1993). Essential oils of other plants, including Wormwood *Artemisia afra*, Lavender tree, *Heteropyxis natalensis*, and Sweet gale *Myrica gale*, revealed high potency against a variety of fungal strains. *Origanum syriacum* essential oil exhibited very strong antifungal activity against *Penicillium*, *Aspergillus*, and *Fusarium* species (Daouk et al., 1995). Very effective against *Aspergillus niger* and *A. flavus* was also found to be the essential oil deriving from Soldier's herb *Piper angustifolium* (Trillini et al., 1996). Lemon mint *Monarda citriodora* and Tea tree *Melaleuca alternifolia* contain essential oils, which were tested in vitro for their activity against fifteen fungal strains. It was observed that these essential oils possess strong antifungal activity in both direct contact and the vapor phase. Lemon mint oil was found to be, in general, more potent than that one of Tea tree and specifically, concerning rapidly growing fungi (Bishop and Thornton, 1997). Essential oils from eight commercial plants (*Cinnamomum zeylanicu* (Cinnamon), *Cananga odorata* (Ylang ylang), *Ocimum basilicum* (Sweet basil), *Citrus limon* (Lemon), *Cymbopogon citratus* (Lemon grass), *Baswellia thurifera* (Boswellia), *Majorana hortensis* (Marjoram), and *Rosmarinus officinalis* (Rosemary) were tested by Baratta et al. (1998) and were almost all found able to inhibit the growth of *A. niger*, even at a concentration of 1 μL/mL broth, except lemon and rosemary oils, which achieved inhibitory effects in higher concentrations. Earlier results revealed weak antifungal potential for sage, while thyme, oregano, and savory were found to be potent against all tested strains

(Özcan and Boyraz, 2000). Considering the importance of essential oils as potential antifungals, 75 different oils were studied for their activity against *A. niger* finding that all of them possess strong potential (Pawar and Thaker, 2006). The antifungal activity of a variety of essential oils from several plant families (Compositae, Labiatae, Lauraceae, Apiaceae, Cupressaceae, Poaceae, Illiaceae, Myrtaceae, Verbenaceae) was evaluated. In particular, Essential oils of Wild marjoram *Origanum onites*, Thyme-leaved savory *Satureja thymbra*, Greek sage *Salvia fruticosa*, and *S. pomifera* subsp. *calycina* plants growing wild in Greece and their ingredients, **carvacrol**, **camphor**, and **1,8-cineole** (Fig. 4.10), were tested for their antifungal capacity against thirteen fungi and found to inhibit all of them. Oils containing carvacrol (Wild marjoram and Thyme-leaved savory) revealed the strongest and broadest potential, while the one deriving from Greek sage was found to be the least effective among them (Soković et al., 2002).

The oils from three *Micromeria* species revealed remarkable antifungal activity (MICs, MFCs): *M. dalmatica* (0.2–0.4 μL/mL), *M. thymifolia* (0.4–2 μL/mL), and particularly *M. albanica* (0.2–0.4 μL/mL) against *Aspergillus niger*, *A. ochraceus*, *Penicillium ochrochloron*, *Cladosporium cladosporioides*, *Fusarium tricinctum*, *Phomopsis helianthi*, and *Trichoderma viride* (Marinković et al., 2002). *Foeniculum vulgare* essential oil was evaluated for its antifungal activity against nine different fungal strains and found to be effective against *C. cladopsorioides* and *P. helianthi* (MICs, MFCs of 0.8–4.0 μL/mL), more potent than bifonazole (Mimica-Dukić et al., 2003). The antifungal capacity of oils from eight *Stachys* species was tested against *A. niger* and *Penicillium ochrochloron* revealing that the strongest activity possessed *S. scardica* due to the high percentage of sesquiterpene hydrocarbons (69.3%) in this oil (Skaltsa et al., 2003). Essential oil from *Juniperus excelsa* (Cupressaceae) was evaluated for its antifungal activity against twelve micromycetes from the genera *Alternaria*, *Aspergillus*, *Cladosporium*, *Fusarium*, *Penicillium*, *Phomopsis*, and *Trichoderma*, exhibiting moderate-to-high potency with MICs range of 8.0–40.0 μL/mL and MFCs of 10.0–50.0 μL/mL. From the obtained results it was elucidated that this oil exerted the strongest activity against *Phomopsis helianthi*, while *Trichoderma viride* was

Carvacrol **Camphor** **1,8-cineole**

FIG. 4.10 Chemical structures of **carvacrol**, **camphor**, and **1,8-cineole**.

found to be the most resistant strain. It would be interesting to point out that this essential oil possessed stronger antifungal activity than the commercial fungicide bifonazole (Soković et al., 2008a).

Several essential oils deriving from different species of Lauraceae family (*Aniba rosaeodora* Rosewood, *Laurus nobilis* Bay tree, *Sassafras albidum* Sassafras, and *Cinnamomum zeylanicum* Cinnamon) were tested against seventeen micromycetes. In the effort to elucidate fungistatic and fungicidal concentrations (MICs and MFCs) both diffusion and microdilution assays were used. Essential oil deriving from cinnamon was found to be the most potent as antifungal, followed by rosewood and sassafras oils, while laurel oil exerted the lowest activity of all (Simic et al., 2004). Minimal inhibitory and fungicidal concentrations of essential oils isolated from Sweet cicely *Myrrhis odorata*, St. John's wort *Hypericum perforatum*, and *Helichrysum arenarium* were determinate by using microdilution test and were found to cover the range of 0.5–60.0 μg/mL (MICs and MFCs). The best antifungal activity was obtained by sweet cicely oil, while *H. arenarium* revealed the lowest antifungal potency. In addition, sweet cicely oil exerted stronger activity than reference commercial compound and was very effective against all tested fungi (Rančić et al., 2005). Essential oils and their components from of *Salvia* species (Labiatae) were examined for their antifungal potential. The corolla, calyx, and leaf oils of *Salvia brachyodon* were assayed using microdilution method against *Aspergillus niger*. The oil from the calyx was found to exhibit the strongest antifungal capacity probably due to synergistic activity of all compounds present. The previously mentioned oil contains high concentrations of sesquiterpenes and diterpenes, which are lacking from the other tested samples (Sokovic et al., 2005). Clary sage (*Salvia sclarea*) of *Salvia* genus was also tested to determine its antifungal capacity. At 25.0 μL/mL revealed fungicidal activity against *Aspergillus*, *Penicillium*, *Trichoderma*, and *Fusarium* species (Dzamic et al., 2008). The essential oil of Sardinian sage *S. desoleana* and its main ingredients were tested for their antifungal potential against eleven micromycetes using macro- and microdilution assays, being diluted in ethanol and Tween 80 in both mentioned methods. Essential oil from Sardinian sage and its main components, namely, **linalyl acetate** (Fig. 4.11), **1,8-cineole** (Fig. 4.10) and **linalool** (Fig. 4.11), which together represent 52.71% of the total oil, exhibited strong antifungal potential, while 1,8-cineole exerted only moderate activity. From the obtained results should be mentioned that linalyl acetate possessed the lowest antifungal capacity. It should be noted that MIC values found to be lower with microdilution assay (Soković et al., 2009b).

Essential oils from *Cymbopogon winterianus* (Poaceae) and *Carum carvi* (Apiaceae) were tested against seventeen fungi in order to reveal their antifungal capacity. In particular, oil of Karawya at concentrations of 0.25–2.5 μL/mL inhibited the growth of all used micromycetes except *T. viride*. Concerning this resistant strain, the concentration of Karawya oil had to be increased to 10 μL/mL in order to be effective. On the other hand, citronella's essential oil was found

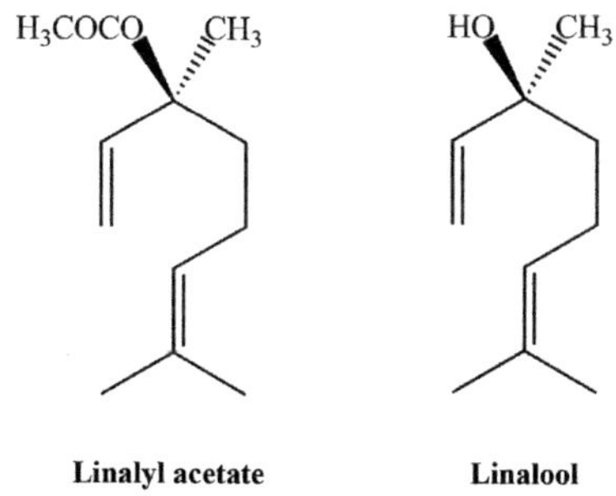

FIG. 4.11 Chemical structures of **linalyl acetate** and **linalool**.

to be less potent but still more effective than bifonazole. In particular, MIC and MFC values of this oil were in the range of 1–20 μL/mL (Simic et al., 2008). A collection of essential oils isolated from the aerial parts during the flowering and vegetative phases, roots and seeds of the plant *Portenschlagiella ramosissima* were evaluated and found to be potent as antifungal agents against several micromycetes (*Aspergillus flavus*, *A. niger*, *A. versicolor*, *Penicillium ochrachloron*, *P. funiculosum*, and *Trichoderma viride*). Among them the most resistant fungi were *Aspergillus flavus* and *A. versicolor* (Soković et al., 2008b). Essential oils deriving from Star anise *Illicium verum* (Illiaceae) and Clavos *Eugenia caryophyllata* (Myrtaceae) were tested as potential antifungals and from the results it was clear that the most resistant fungal species were *Trichoderma viride*, *Penicillium*, and *Aspergillus* ones. It is well established that, in general, the antimicrobial activity of star anise is mainly due to **anethole**, with **eugenol** (Fig. 4.12) being responsible particularly for the antifungal potency of cloves' oil.

It was proposed by the authors that interactive effects of other compounds that are present in smaller quantities could possibly contribute in the activity observed. Both oils, and especially the one deriving from cloves, exerted powerful antifungal capacity (Dzamic et al., 2009). The antifungal potential from different essential oils of wild carrot *Daucus carota* L. (Apiaceae), which were collected in Serbia, was evaluated. Moreover, the inhibitory activity of oils from the ripe fruits, unripe fruits, flowers, root, leaves, and stem was studied against eight fungal species (*Fusarium sporotrichoides*, *Fulvia fulvum*, *Trichoderma viride*, *Penicillium ochrochloron*, *P. funiculosum*, *Aspergillus*

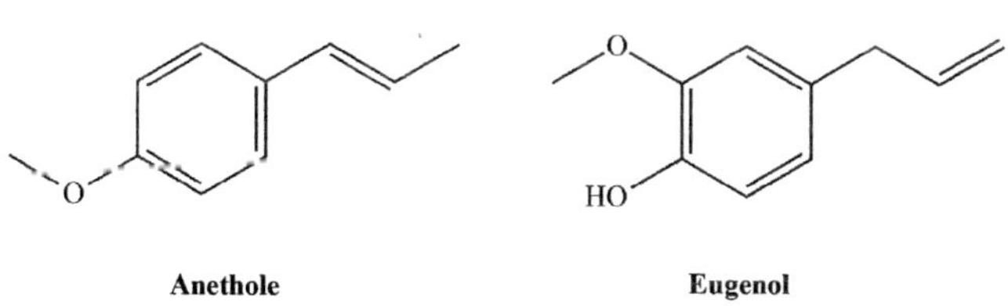

FIG. 4.12 Chemical structures of **anethole** and **eugenol**.

ochraceus, *A. flavus*, and *A. fumigatus*) using the microdilution method. The mentioned oils demonstrated higher potency than the commercial drug bifonazole and were found to be even more efficient as far as ketoconazole was concerned. It should be noted that the essential oils, which derive from ripe and unripe fruits of wild carrot oil, revealed the most prominent antifungal activity (Soković et al., 2009c). The chemical composition and effectiveness of the essential oil obtained from *Echinophora spinosa* (Apiaceae) was studied concerning its inhibitory activity against several fungi. The most resistant microorganisms were found to be *Penicillium ochrochloron* and *P. funiculosum*, while *Trichoderma viride* was proved to be the most sensitive one. It should be pointed out that this essential oil exhibited stronger antifungal activity against *T. viride* than bifonazole and ketoconazole (Glamočlija et al., 2011a). *Lippia alba* Bushy lippia's (Verbenaceae) essential oil revealed antifungal inhibitory effect against human pathogenic strains. In particular, it was found to be potent against *Aspergillus ochraceus*, *A. niger*, *A. versicolor*, *A. fumigatus*, *Penicillium ochrochloron*, *P. funiculosum*, and *Trichoderma viride*. Moreover it can be classified as citral type and taking into consideration the obtained results, it could be proposed as a potential alternative to synthetic antifungal drugs (Glamočlija et al., 2011b). In another study, the chemical ingredients showed lower antifungal capacity than the essential oils isolated from *Satureja thymbra*. As far as the fungi are concerned, *A. versicolor* and *A. fumigatus* were found to be the most sensitive species, while *P. ochrochloron* was the most resistant concerning these particular oils. Both oils demonstrated much stronger antifungal potential than reference drug, ketoconazole (Marković et al., 2011). The essential oil of *Seseli montanum* subsp. *tommasinii* was studied for its antifungal capacity on four specific fungi, namely: *Aspergillus ochraceus*, *A. fumigatus*, *Penicillium ochrochloron*, and *Trichoderma viride*. The activity against all the tested fungi could be characterized as moderate, although stronger than that of bifonazole against *A. fumigatus* and *T. viride*. This can be important especially for *A. fumigatus*, which is considered as a common and invasive fungus and can cause serious problems in cases of resistance to antifungal agents (Siljegovic et al., 2011b). From the results concerning antifungal activity of *Seseli* species oils, it can be observed that all samples exhibited good potential (aerial parts of *S. anuum* (Milosavljević et al., 2007), *S. globiferum* fruits (Stojković et al., 2008a), *S. globiferum* aerial parts (Janaćković et al., 2011), and flowers of *S. rigidum* (Stojkovic et al., 2009)). The essential oil from the aerial parts of *S. montanum* subsp. *tommasinii* revealed stronger activity compared to that deriving from the aerial parts of *S. anuum* and specifically two-fold against *P. ochrochloron* and *T. viride*. On the other hand, *S. rigidum*'s essential oil demonstrated better antifungal capacity than the one of *S. montanum* subsp. *tommasinii* against *A. fumigatus* and *P. ochrochloron* (Siljegovic et al., 2011b). The essential oils of sixteen aromatic and medical plants and their components (Bitter orange *Citrus aurantium*, Lemon *C. limon*, *Hyssopus officinalis*, Lavender *Lavandula angustifolia*, Wild lavender *L. stoechas*, Chamomile *Matricaria chamomilla*, Melissa

Melissa officinalis, Peppermint *Mentha piperita*, Spearmint *M. spicata*, Sweet basil, Rosemary *Rosmarinus officinalis*, Sage *Salvia officinalis*, Sardinian sage *S. desoleana*, Clary *S. sclarea*, Thyme *Thymus vulgaris*, and *T. tosevii*) were studied for their antifungal properties using the microdilution method. From the obtained results it should be mentioned that *Thymus* essential oils were the most effective as far as in vitro tests were concerned, while those from *Citrus* species and *S. officinalis* had the lowest antifungal potential (Fig. 4.13) (Soković, 2001; Soković et al., 2009a).

Essential oils with phenolic components, **thymol** (Fig. 4.14), and **carvacrol** (Fig. 4.10), (Greek oregano *O. heracleoticum*, Pink savory *S. thymbra*, thyme and *T. tosevii* with MIC values ranging between 0.001 and 0.7 μL/disc and MFC values between 0.1 and 1.0 μL/disc) exerted significantly better antifungal capacity compared to those containing high concentrations of alcohols (Spearmint and Peppermint with MIC 1.0 μL/disc and MFC 2.5 μL/disc) and those with significant amounts of ketones (*S. pomifera* with MIC 1.0 μL/disc and MFC 15.0 μL/disc and *H. officinalis* with MIC of 5.0 μL/disc and MFC of 25.0 μL/disc) and especially compared to oils having monoterpenic hydrocarbons as dominant molecules (Lemon with MIC of 4.0 μL/disc and MFC of 5.0 μL/disc, Bitter orange with MIC of 0.7 μL/disc and MFC of 1.0 μL/disc and

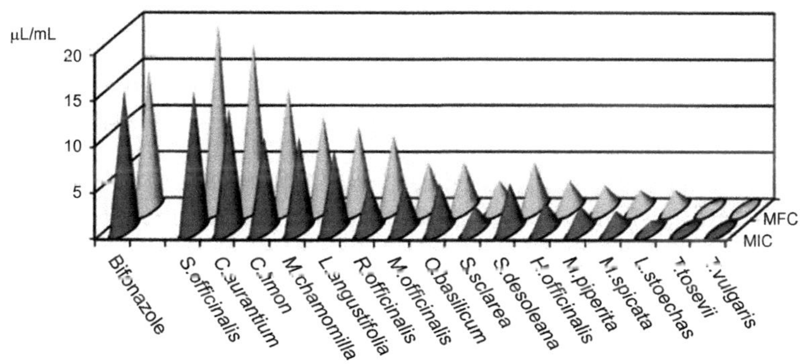

FIG. 4.13 Antifungal activity of 16 essential oils (MICs and MFCs), microdilution method.

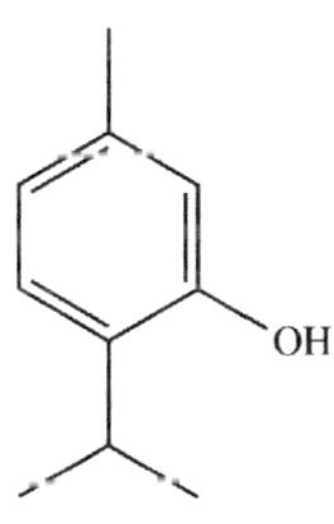

FIG. 4.14 Chemical structure of **thymol**.

Lavender with MIC of 1.5 μL/disc and MFC of 2.5 μL/disc). The commercial fungicide prochloraz demonstrated much lower antifungal effect than all the studied oils, with MIC of 5.0 μL/disc and MFC of 50.0 μL/disc (Glamočlija, 2006; Glamočlija et al., 2009).

In order to elucidate their antifungal potential, a series of aromatic plants from Lamiaceae family and in particular the oils from *Satureja montana*, *Lavandula angustifolia.*, *L. hybrida*, *Origanum vulgare Rosmarinus officinalis* have been studied against several different fungi and six chemotypes of *Thymus vulgaris* against *Candida albicans* (Giordani et al., 2004). It should be pointed out that thymol chemotype of the *T. vulgaris* essential oil achieved the strongest activity (MIC, MFC of 0.016 μg/mL). The antifungal potential of the essential oil of *L. angustifolia* and its main ingredients (i.e., **linalool** and **linalyl acetate**, Fig. 4.11) were tested against 50 clinical isolates of *C. albicans* (oropharyngeal and vaginal strains). As it was shown in the results, linalool alone was more effective than the essential oil, while linalyl acetate was found to possess almost no effect (D'Auria et al., 2005). It was observed that the essential oil of *O. gratissimum* exerted in vitro fungicidal activity against all of the *Candida* spp. tested (i.e., *C. albicans*, *C. krusei*, *C. parapsilosis*, and *C. tropicalis*). Moreover, further studies concerning the ultrastructure of the yeast cells revealed changes in the cell wall and in the morphology of some subcellular organelles (Nakamura et al., 2004).

Nepeta crispa's essential oil demonstrated antifungal capacity against four fungal strains, including *C. albicans* and *Microsporum gypseum* having as main constituents **1,8-cineole** (47.9%, Fig. 4.10) and **4aα,7α, 7aβ-nepetalactone** (20.3%, Fig. 4.15) (Sonboli et al., 2004).

As far as genus *Thymus* is concerned, essential oils revealed antifungal potency in anti-*Candida* activity screening assays. Specifically, the antifungal properties of Thymus pulegioides (thymol/carvacrol chemotype) oil and its main ingredients (**thymol**, **carvacrol**, **p-cymene**, and **γ-terpinene**, Figs. 4.10, 4.14 and 4.16) were examined against several clinical strains of *Candida* spp. that were isolated from recurrent cases of vulvovaginal candidiasis (*C. albicans*, *C. krusei*, *C. tropicalis*, *C.guillermondii*, and *C. glabrata*), four *Candida* spp. ATCC strains and dermatomycetes clinical strains that were isolated from nails and skin (*Microsporum canis*, *M. gypseum*, *Trichophyton rubrum*, *T. mentagrophytes*, and *Epidermophyton floccosum*).

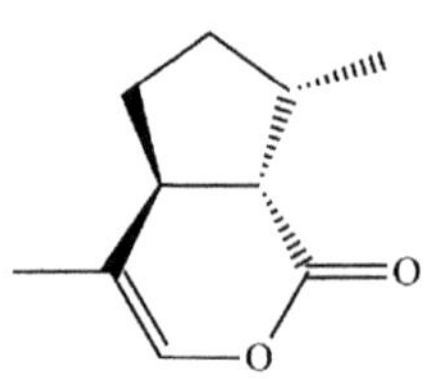

4aα,7α,7aβ-nepetalactone

FIG. 4.15 Chemical structure of **4aα,7α, 7aβ-nepetalactone**.

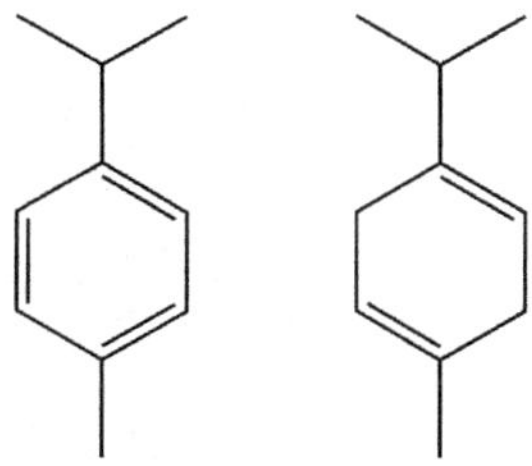

FIG. 4.16 Chemical structures of **p-cymene** and **γ-terpinene**.

The MIC and MLC values of the oil against all of the *Candida* spp. ranged between 0.32 and 0.64 μg/mL, respectively, while concerning the dermatomycetes were found to be lower and ranging from 0.16 to 0.32 μg/mL, respectively. When thymol was tested alone, a 0.16 μg/mL concentration was found to be lethal for almost all studied strains. On the other hand, when thymol was tested in combination with carvacrol, against *Candida* spp. a lethal dose up to 0.32 μg/mL was needed, while against dermatomycetes a lower concentration of 0.16 μg/mL was required. From the obtained results it was concluded that the essential oil from *T. pulegioides* is effective as a topical antifungal against the studied fungi that are pathogenic for humans and it was suggested that the mechanism of action of the oil was connected to a cytoplasmic membrane lesion that is created. It was proposed that further toxicity studies with emphasis on improving formulation and determination of optimal concentrations for clinical applications are in need. Moreover, it was suggested by researchers to proceed to comparative studies with commercial drugs in order to test the therapeutic effectiveness of essential oils concerning the control of mucocutaneous infections (Pinto et al., 2006). More recent studies with essential oils of three thyme species (*T. serpyllum*, *T. algeriensis*, and *T. vulgaris*) revealed that the growth of *Candida* spp. was inhibited in concentrations between 1 and 10 μg/mL, confirming that way the strong anticandidal capacity of these oils (Fig. 4.17) (Nikolić et al., 2014).

Concerning the essential oils of the plant species *Thymbra capitata*, which contain high amounts of carvacrol (60.0%–65.8%), they were found to possess antifungal capacity against all clinical isolates and ATCC strains of the *Candida* spp. that were used in the assay with all clinical dermatomycetes strains having IC_{50} values ranging between 0.08 and 0.32 μg/mL. It was also concluded that the fungicidal activity was mainly due to a direct lesion at the membrane (Salgueiro et al., 2004).

Regarding the in vitro antifungal potential studies that were performed for the essential oils and the main constituents of *Origanum vulgare* subsp. *hirtum*, *Mentha spicata*, *Lavandula angustifolia*, and *Salvia fruticosa*, could be suggested their potential use as effective antifungals against human pathogens *Trichophyton rubrum* and *Trichosporon beigelii*.

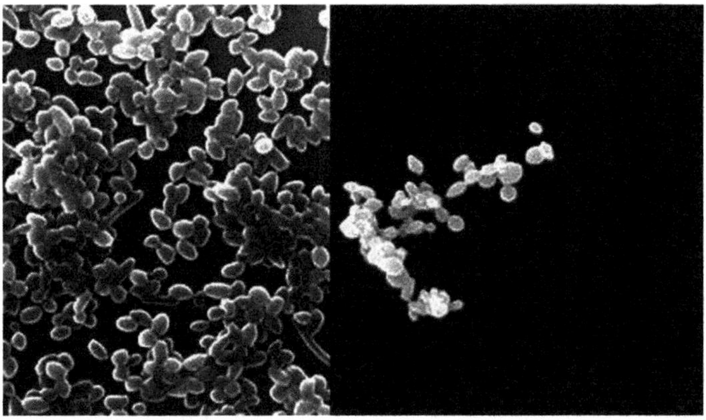

FIG. 4.17 Scanning electron microscopy (SEM) of *Candida albicans* biofilm treated with oil of *T. vulgaris* 6 h (E) and 24 h (F) 1000×.

It is well established that there is a strong correlation between the high antifungal potency of oils rich in phenolic constituents (particularly thymol and carvacrol), because these molecules appear to be responsible for the entire effect, although the rest of the oil ingredients may enhance the antifungal effect observed (Marković et al., 2011).

The oils from other five Lamiaceae species, including *M. piperita*, *M. pulegium*, *L. angustifolia*, *S. montana*, and *S. lavandulifolia*, were evaluated against fifty-right oral clinical isolates of *Candida* spp. and from the obtained results can be observed that they demonstrated strong inhibitory activity with the oil of *S. montana* as being the most potent one (MICs of 0.9–1.0 μg/mL and MFCs of 1.0–3.0 μg/mL) (Goncalves et al., 2010).

Four chemotypes (geraniol, linalool, thymol, and carvacrol) of *Thymus zygis* subsp. *sylvestris* from Portugal were evaluated for their antifungal activity against two dermatomycetes isolated from nails and skin (*Trichophyton mentagrophytes* and *Microsporum canis*), two standard dermatomycetes (*T. rubrum* and *M. gypseum*), two *Candida* strains isolated from recurrent cases of vulvovaginal candidiasis (*C. krusei* and *C. guillermondii*). Potent against the dermatomycetes strains was found to be the carvacrol chemotype of the oil with MIC of 0.16 μL/mL, lacking also cytotoxicity in this concentration, in mouse skin dendritic cells. It was proposed that further studies should follow in order to evaluate the possibility of introducing this oil in clinical practice for the treatment of fungal diseases that involve mucosal and cutaneous infections as well as an alternative to synthetic antifungals (Goncalves et al., 2010) (Fig. 4.18).

From the in vitro antifungal evaluation of the leaf oils of five endemic *Psiadia* species from Mauritius was concluded that only *Psiadia lithospermifolia* oil, with major constituent of (E)-isoasarone (51.5%), significantly inhibited the growth of *Candida pseudotropicalis* (Govinden-Soulange et al., 2004).

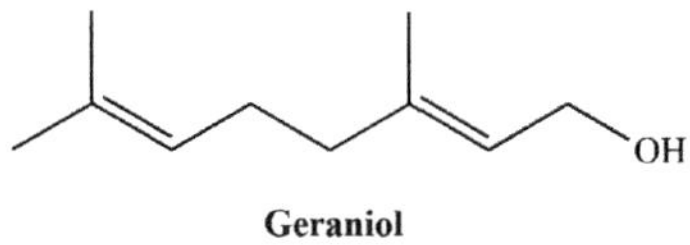

FIG. 4.18 Chemical structure of **geraniol**.

The essential oil of *Ageratum houstonianum* leaf revealed antifungal potency concerning the control of human nails infestation, *Tinea unguium*, by *Candida albicans* and *Trichophyton rubrum* that are considered the main causes of this type of infections. The MIC values for both species were 400 ppm, while at 500 ppm the oil was lethal with minimum killing times of 30 s and 40 s respectively (Kumar, 2014).

The antifungal activity of *Matricaria recutita* flower's essential oil (rich in **chamazulene**, Fig. 4.19, at 61.3%) was studied against dermatomycetes and opportunistic saprophytes that are considered important in terms of medicine. Percent of growth inhibition of the dermatomycetes exposed to serial two-fold concentrations of the oil (2.5–80 µg/mL) was measured. It was suggested that the oil of *M. recutita* could be considered as a potential candidate for the design of effective formulations against dermatomycoses (Jamalian et al., 2012).

The *Lippia graveolens* and *Lantana achyranthifolia* aerial parts derived essential oil (both oils rich in carvacrol at 37.8% and 30.6%, respectively) was evaluated in vitro against *Trichophyton mentagrophytes* revealing that both oils are able to exhibit antifungal effect (20–110 µg/mL). The results of this study are in agreement with the dermatological use of the oils of previously mentioned species in the traditional medicine (Hernandez et al., 2008).

Two chemotypes, **citral** (geranial-neral mixture) and **carvone** (Fig. 4.20), of the essential oil of *Lippia alba* were evaluated in vitro. The MICs of the citral chemotype against *Trichophyton rubrum* and *T. mentagrophytes* were found to be 31.3 and 125 µg/mL, respectively, while MICs for the carvone chemotype were proved to be higher (Tangarife-Castaño et al., 2012). From the obtained results concerning *Lippia gracilis* essential oil (thymol and two carvacrol chemotypes) against *Trichophyton rubrum* was elucidated that this strain is highly susceptible to the oil, which demonstrated antifungal activities similar these of

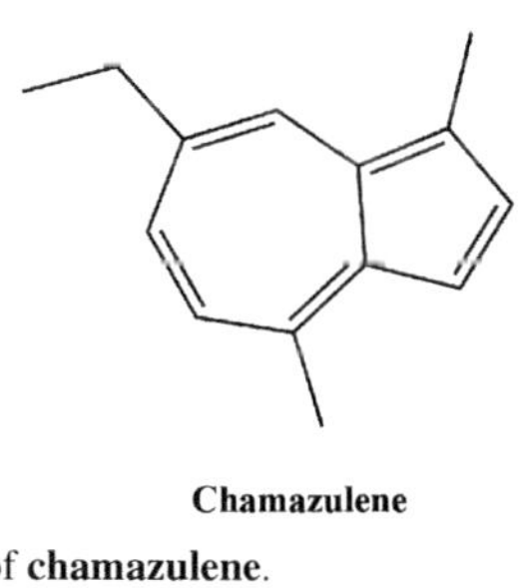

FIG. 4.19 Chemical structure of **chamazulene**.

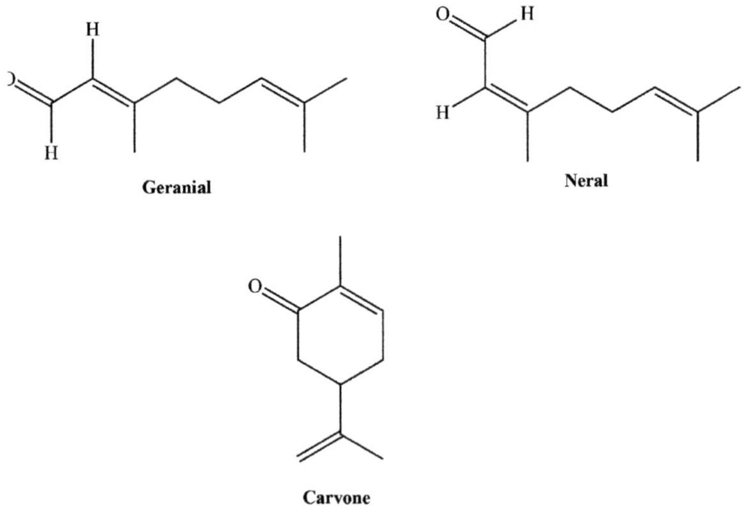

FIG. 4.20 Chemical structures of **citral** (geranial-neral mixture) and **carvone**.

fluconazole. The thymol chemotype of the oil (61.8%) revealed stronger potential compared to the carvacrol chemotypes of the oil (54.5% and 48.9%) and thus, it was proposed that thymol contributes substantially to the antifungal capacity observed (de Melo et al., 2013).

The sensitivity of common *Candida* spp. (ATCC strains and clinical isolates) recovered from human oral cavities (*C. albicans*, *C. krusei*, and *C. glabrata*) was evaluated against four commercial essential oils from *Boswelia carterii*, *Canarium luzonicum*, and *Commiphora myrrha*. The *B. carterii* oil, with a high **α-pinene** (Fig. 4.21) content (31.8%), demonstrated the strongest potential. It should be noted that the average MIC and MFC values ranged between 1.25 and 1.34 mg/mL and 2.50–3.75 mg/mL, respectively, depending on the fungal strain. The results of this study support the possibility of using essential oils from the *Bursecaceae* family in the treatment of *Candida* spp. infections in human populations (Nikolić et al., 2014).

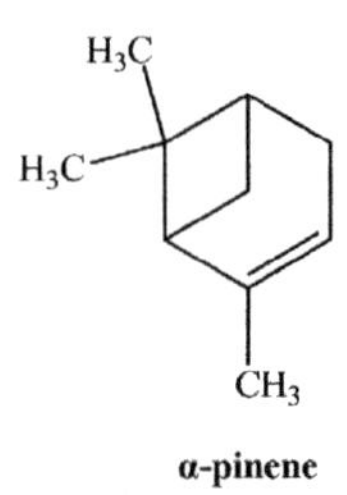

FIG. 4.21 Chemical structure of **α-pinene**.

Three *Juniperus* spp. oils were tested for their antifungal potential. One rich in **α-pinene** (65.5%, Fig. 4.21) and **delta-3-carene** (5.7%, Fig. 4.22), was found to be the most potent against *E. floccosum*, *T mentagrophytes*, *M. canis*, and *M. gypseum* (MICs and MLCs between 0.08 and 0.32 μL/mL) and the *Candida* strains of *C. albicans*, *C. glabrata*, *C. tropicalis*, *C. krusei*, and *C. parapsilosis* (MICs and MLCs ranging from 0.32 to 2.5 μL/mL). Researchers came to the conclusion that the oil could be an effective alternative antifungal against the tested strains of dermatomycetes, having delta-3-carene as crucial ingredient for the desired pharmacological result (Cavaleiro et al., 2006).

From the obtained results regarding the antifungal capacity of *Metasequoia glyptostroboides* essential oil against a variety of dermatomycetes, including *Trichophyton rubrum*, *T. mentagrophytes*, and *Microsporum canis*, it was observed that the most resistant fungi were *T. rubrum* and *M. canis* with a MIC value of 500 μg/mL. Nevertheless, the range of MIC values for the remaining strains of the dermatomycetes was between 62.5 and 125 μg/mL. It was concluded that the oil might be a source of natural and novel dermatological antimycotics, being able in the same time to restrain superficial human fungal infections (Bajpai et al., 2009).

The essential oils of eight *Cinnamomum* spp. (*C. pubescens*, *C. impressicostatum*, *C. microphyllum*, *C. scortechinii*, *C. rhyncophyllum*, *C. cordatum*, *C. zeylanicum*, and *C. mollissimum*) and their main ingredients were tested for their antifungal capacity against six dermatomycetes (*Trichophyton rubrum*, *T. mentagrophytes*, *T. tonsurans*, *Microsporum canis*, *M. gypseum*, and *M. audouini*) and four yeasts (*Candida albicans*, *C. glabrata*, *C. tropicalis*, and *C. parapsilosis*). The majority of the essential oils exhibited moderate-to-high potency against all tested fungi: the leaf and bark oils of *C. zeylanicum* revealed the strongest activity (MICs of 0.04–0.63 μg/μL), while the leaf oil of *C. cordatum* and the bark and twig oils of *C. pubescens* and *C. impressicostatum* were also found to be potent inhibitors. **Cinnamaldehyde**, which is the main constituent of bark oil from *C. zeylanicum*, demonstrated the highest potency against all the fungi that were used in the assay. The lowest MIC values (<0.04 μg/μL) were observed for the following ingredients: **benzyl salicylate** against *T. rubrum*, *T. mentagrophytes*, *M. gypseum*, and *M. audouini*; **benzyl benzoate** against *T. rubrum* and *T. mentagrophytes*; and **geraniol** against *T. rubrum* and *M. audouini*. On the basis of these results it was concluded that the high levels of **cinnamaldehyde** (Fig. 4.23), **eugenol** (Fig. 4.12), **geraniol** (Fig. 4.18), **benzyl benzoate** (Fig. 4.23), and **methyl cinnamate** (Fig. 4.23) in the essential oils

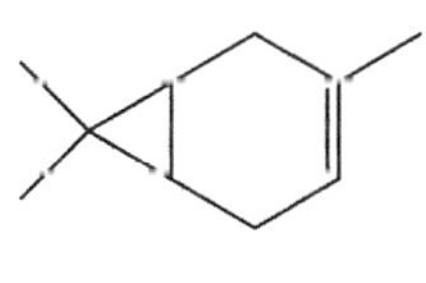

Delta-3-carene

FIG. 4.22 Chemical structure of **delta-3-carene**.

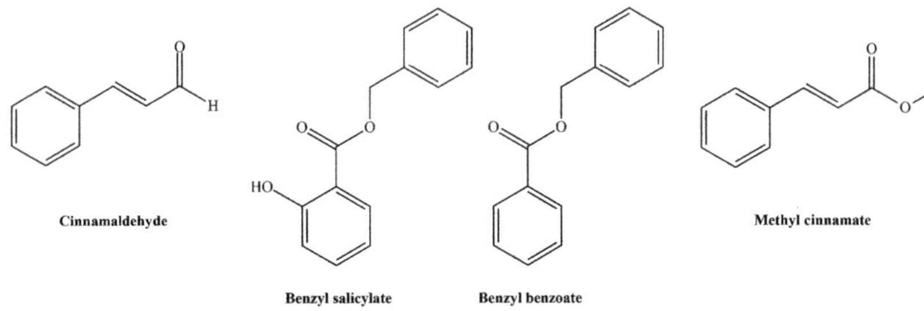

FIG. 4.23 Chemical structures of **cinnamaldehyde**, **benzyl salicylate**, **benzyl benzoate,** and **methyl cinnamate**.

and combined with the rest of the oil components, are responsible for the strong antifungal potential of these natural products (Jantan et al., 2008).

Biological evaluation of the essential oil of *G. procumbens* revealed inhibition of planktonic cell culture of *C. albicans* and anti-*Candida* activity against a formed biofilm (Nikolić et al., 2014). Knowing that biofilm structures are often resistant and act as a reservoir for infections and cross-contaminations, the prevention of *Candida* adhesion and biofilm formation essential oils that derive from plants is noteworthy.

One of the most studied essential oils for activity against *Candida* spp. is the one that can be obtained from *Melaleuca alternifolia* and has been widely used for its biological properties. There is a great amount of studies that justify its efficacy against *Candida* (Carson et al., 2006; Nikolić et al., 2012) and thus its use is recommended due to its potency and satisfactory tolerability.

Two other essential oils deriving from plants of the same family, *Leptospermum petersonii* and *Syzygium aromaticum*, were also found to possess inhibitory activity against *Microsporum canis*, *Trichophyton mentagrophytes*, *T. rubrum*, *Epidermophyton floccosum*, and *Microsporum gypseum*. Potential of *S. aromaticum* oil at 0.2 mg/mL exceeded 60%, while eugenol was the most effective constituent against *T. mentagrophytes* and *M. canis*. Moreover, the oil of *L. petersonii* applied at 0.2 mg/mL was found to be even more potent against all tested fungal strains (above 90%), except *T. rubrum*, having geranial as the strongest antifungal ingredient (Park et al., 2007).

The antifungal capacity of an essential oil isolated from *Cymbopogon winterianus* was screened against sixteen strains of *T. rubrum*. As it was observed, the essential oil in all concentrations effectively inhibited the mycelium growth. A series of morphological changes that were caused by the use of the essential oil included loss of conidia, alterations in the form, and pigmentation of hyphae. It was concluded that *C. winterianus* oil might be potentially used as antifungal agent, especially for preventing infections by *T. rubrum* (Pereira de Oliveira et al., 2011).

The essential oil deriving from *Cymbopogon martini*, which is rich in ***trans*-geraniol** (60.9%, Fig. 4.18), was found to be more potent than the oil

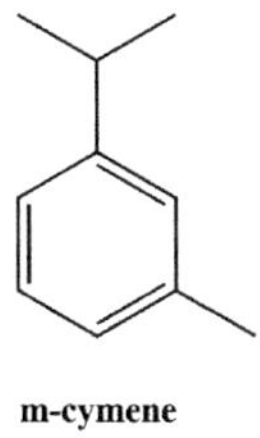

FIG. 4.24 Chemical structure of **m-cymene**.

of *Chenopodium ambrosioides*, which is rich in **m-cymene** (43.9%, Fig. 4.24) against a series of dermatomycetes. The authors came to the conclusion that these two oils and their combinatory use could be potentially employed in the future for the treatment of *Tinea corporis* (ringworm) among other superficial mycoses in human populations (Prasad et al., 2010).

In the continuous effort to introduce more stable and natural antifungal agents in clinical practice, the activity of essential oils from the *Allium sativum* var. *pekinense*, *A. fistulosum*, and *A. cepa*, were evaluated for their activity against *T. rubrum*, *T. erinacei*, and *T. soudanense*, which are the cause of a series of serious mycoses in humans. The essential oil of *A. sativum* was found to possess the highest growth inhibition for all three *Trichophyton* spp., with MIC at 64 μg/mL, exhibiting 12.5%–25% of the activity of ketoconazole. These results led the authors to consider this oil as a remarkable natural product, especially taking into account the side effects and toxicity of ketoconazole. Moreover, the oil also revealed noteworthy synergistic antifungal potency in combination with ketoconazole (Pyun and Shin, 2006).

From a variety of essential oils, the leaf oil of *Pelargonium graveolens* and its main ingredients, citronellol and geraniol, alone and in combination with ketoconazole, were evaluated and demonstrated, in general, strong potential against all tested *Trichophyton* spp. strains (*T. erinacei*, *T. mentagrophytes*, *T. rubrum*, *T. schoenleinii*, *T. soudanense*, and *T. tonsurans*). The combinatorial use of this oil and its components with ketoconazole revealed very strong synergistic activity against the lesser-known African dematomycetes *T. schoenleini* and *T. soudanense* (Shin and Lim, 2004).

The essential oil deriving from the seeds of *Nigella sativa* and its major constituent thymoquinone (42.4%) were tested against *T. mentagrophytes*, *M. canis*, and *M. gypseum*. This study presents a first step in the search for new drugs against dermatomycetes, supporting the use of *N. sativa* seeds in traditional medicine for dermatophytic infections (Mahmoudvand et al., 2012). The chemical constituents and antifungal activity of the oil samples from Iberian endemic plant species, traditionally used in the treatment of contact dermatitis and skin infections, were studied (Tavares et al., 2010). In particular, the oil of *Distichoselinum tenuifolium* (rich in **myrcene** at 47.7%–84.6%, Fig. 4.25) was assayed against dermatomycetes (*M. canis*, *T. rubrum*, *M. gypseum*, and *E. floccosum*) and *Candida* spp. (*C. albicans*, *C. tropicalis*, *C. krusei*, *C. guillermondii*,

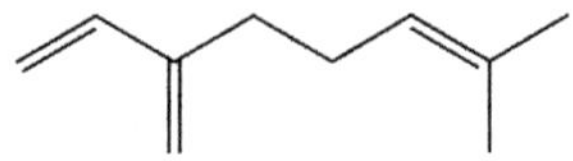

Myrcene

FIG. 4.25 Chemical structure of **myrcene**.

and *C. parapsilosis*). It was concluded that the greater antifungal capacity of the *D. tenuifolium* oil compared to that of myrcene alone is explained by the synergistic effect of the diverse molecules that are present in the oil. *D. tenuifolium* essential oil demonstrated stronger antifungal potential against the dermatomycetes (MICs and MLCs ranging between 0.64 and 1.25 μL/mL) compared to their activity against *Candida* spp. (MICs of 2.5–5.0 μL/mL and MLCs of 2.5–10 μL/mL).

The essential oil from *Piper betle* L. showed antifungal effect against thirteen fungi, including *Alternaria*, *Aspergillus*, *Botryodiplodia* (syn. Lasiodiplodia), *Penicillium*, and *Rhizopus* species in concentration of 0.20–0.50 μL/mL (Guha and Nandi, 2019). Basak and Guha (2017) showed that microemulsion of this essential oil was toxic to *Aspergillus flavus* and *Penicillium expansum* (0.40 μL/mL), while sporicidal effect was detected at a higher concentration (15 μL/mL). Prakash et al. (2010) observed that essential oil of betel leaf was effective against 1651 fungal strains, from fourteen species, among which six species of *Aspergillus*, with MIC at 70 μL/mL. Aflatoxin production was achieved at 0.60 μL/mL of this oil.

The composition of *Heracleun pyrenaicum* subsp. *pollinianum* (Bertol.) F. Pedrotti & Pignatti (HPP), and *H. orphanidis* Boiss. (HO) essential oils was analyzed by GC and GC-MS. The most abundant compounds were β-pinene (35.1%) in HPP root oil, (Z)-falcarinol (80.0%) in HO root oil, while (E)-nerolidol (28.5%) was dominant constituent in HPP leaf oil. Antimicrobial effect was tested on eight fungal strains. The best antifungal activity was exhibited by HPP fruit oil (MICs, MFCs of 0.30–1.25 mg/mL) and HO leaf oil (MICs, MFCs of 0.15–1.25 mg/mL) (Usjak et al., 2016a).

The other species from this genera, *Heracleum verticillatum* Pančić and *H. ternatum* Velen, were also tested for the antimicrobial activity of their oils. The GC and GC/MS analysis showed that the root oils were rich in monoterpenes, mostly β-pinene (23.5% and 47.3%, respectively), while in leaf oil monoterpenes were dominant, mainly limonene (20.3%) and in lower content were present sesquiterpenes, mostly (E)-caryophyllene (19.1%). In leaf oil of *H. ternatum* phenylpropanoids were the dominant constituents, with (Z)-isoelemicin (35.1%). Aliphatic esters were dominant in both fruit oils, mostly octyl acetate (42.3% in *H. verticillatum* oil and 49.0% in *H. ternatum* oil). *H. verticillatum* root oil showed the strongest effect against *Staphylococcus aureus*, *Salmonella typhimurium*, and *Trichoderma viride* (MICs, MFCs of 0.14–0.28 mg/mL) (Usjak et al., 2016b).

In vivo activity of essential oils

Regarding the in vivo antifungal activity of essential oils, only limited information can be found in international literature, while the corresponding experiments are related with a series of problems as far as the application of the oils is concerned. Some of these problems are related to volatility issues as well as their low water solubility that must be resolved prior to the in vivo studies. It should be noted that the abidance of volatile oil of that kind on a plant that is under treatment and thus its protection against pathogens has limited duration (Soković, 2001; Soković et al., 2010, 2013). This issue brings questions related to environmental pollution risk and even phytotoxicity and is the reason for an emerging scientific interest in performing experimental studies in order to overcome such complications.

The use of essential oils that can be isolated from spices alongside with other aromatic herbs in dental preparations makes their potential general use particularly advisable. As far as the antimicrobial activity is concerned, the essential oils of oregano and thyme, as well as their major constituents (carvacrol and thymol, respectively) we found to be the most potent as antimicrobials (Soković et al., 2010). It should be mentioned that several antimicrobials of plant origin, due to synergistic effect, exhibit higher antimicrobial/antifungal activity when applied in combination compared to their potency when applied individually.

The analysis of chemical composition and examination of antimicrobial activity of the essential oils and their main ingredients deriving from *Vitex agnus-castus* L. revealed that they could be potentially used in vitro as well as in vivo (Stojković et al., 2011). A series of essential oils have been used as proactive therapeutic agents as well as reactive care solutions and mostly the ones deriving from red thyme, eucalyptus, peppermint, cinnamon bark, and lavender (Nikolić et al., 2015). The constituents of the essential oils have been extensively studied and were found to be able to act in distinct and also synergistic ways: antimicrobial, tissue conditioning, connective tissue rebuilding, and antihalitosis.

The study of twenty essential oils against a series of fungi, including *Epidermophyton floccosum, Microsporum gypseum, Trichophyton rubrum*, and *Candida albicans*, revealed that the oils of *Ocimum gratissimum* and *Trachyspermum ammi* were the most potent with fungicidal activity at 50 ppm. The ointments of two oils, prepared in PEG (polyethylene glycol), managed to intensify skin penetration and inhibit dermatomycoses induced by *E. floccosum* and *M. gypseum* in 5- and 8-month-old guinea pigs, suggesting that they could potentially be used efficiently as herbal medicines. Moreover, the ointment made by using *O. gratissimum* oil cured the disease in the guinea pigs in 11 and 9 days, respectively, while the ointment with *T. ammi* oil had the same results in 13 and 11 days, respectively (Tiwari et al., 2003). From the in vivo assays came up promising results regarding the therapeutic efficacy of oregano oil on rats, which were infected with *T. rubrum*. From the study of the main oil constituents, carvacrol and thymol exhibited the stronger antifungal capacity (Adam et al., 1998).

Several Lamiaceae essential oils alongside with their main ingredients were evaluated in vivo for their antifungal potential against experimentally induced dermatomycoses in 2-month old male Wistar rats, caused by some of the most usual dermatomycetes (*Trichophyton mentagrophytes*, *T. rubrum*, and *T. tonsurans*). The results revealed that *Lavandula angustifolia* oil, rich in linalool (27.2%) and linalyl acetate (27.5%), exerted therapeutic activity within 13 days following the treatment; on the other hand, rats treated with *Ocimum basilicum* oil, with linalool as main constituent (69.3%) were cured in 25 days. The shortest period of achieving complete treatment (12 days) was observed for *Salvia officinalis* oil, while the longest cure period (45 days) was observed when *Citrus aurantium* and *Citrus limon* oils, both rich in limonene (71.8% and 59.7%, respectively), were applied. As far as the major essential oil components were concerned, camphor exhibited therapeutic antifungal capacity in 14 days and linalool in 32 days, while the rats that were treated with 1,8-cineole and limonene were cured in 40 and 50 days, respectively (Soković, 2001).

The essential oil deriving from *Spilanthes acmella* was tested for its antifungal activity against *Trichophyton mentagrophytes* infections that were induced on the skin of *Mus musculus*. The oil was applied topically on a series of mice with skin lesions at a concentration of 1 μL/mL (v/v, oil/emulsifier) and terbinafine (5 mg/mL, *W/V*) was used as control drug. The *S. acmella* essential oil (1 μL) exhibited therapeutic activity comparable to that of terbinafine against the tested strain. Moreover, terbinafine reached absolute growth inhibition on day 8, while the oil achieved a 75% inhibition on day 11. Consequently, it was concluded that this oil, could be potentially used topically as a natural antifungal agent in patients with hypersensitivity issues where application of terbinafine would not be recommended (Padhan and Pattnaik, 2014).

From the in vivo studies regarding the essential oil of *Mentha* x *piperita* and its major ingredient **menthol** (Fig. 4.26) can be observed that the oil completely cured the Wistar rats that were infected with *T. metagrophytes* in 15 days, *T. tonsurans* in 29 days, and *T. rubrum* in 30 days. In addition, menthol demonstrated strong therapeutic and antifungal potentials, curing the animals in 10 days, being more effective compared to marketed drug bifonazole (Soković et al., 2006).

The assay results concerning antifungal capacity of the essential oil deriving from *Lavandula angustifolia* revealed therapeutic and antifungal activities

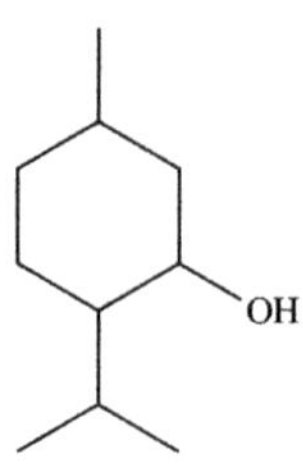

Menthol

FIG. 4.26 Chemical structures of **menthol**.

during the 13-day observation period, completely curing the rats during the same time period (Soković et al., 2007). As far as the essential oil of *Thymus vulgaris* and its major component thymol are concerned, they were also studied for their in vivo therapeutic potential. Two-month-old male Wistar rats infected with *T. mentagrophytes*, *T. rubrum*, and *T. Tonsurans* were completely cured in 24, 37, and 32 days respectively, while, on the other hand, the rats which were treated with the antifungal drug bifonazole were cured in 14–15 days. In addition, thymol exhibited a stronger therapeutic and antifungal potential than the whole oil curing the rats in 14 days (Soković et al., 2008c).

Studies on the in vivo antifungal capacity of essential oils from the Lamiaceae family and their components revealed that the most potent oil was that of *S. officinalis*, followed by the oil of *L. angustifolia* and menthol as the single oil component with good therapeutic and antifungal potentials (Soković et al., 2012). Thus, it was concluded that these oils and menthol could contribute in the field of medical mycology as effective choices in controlling fungal skin infections.

The susceptibility of the turpentine oil deriving from Cluster pine *Pinus pinaster* oleoresin was assayed in vitro against three Sudanese clinical isolates of *Actinomadura madurae*, which is considered to be the main cause of actinomy cetoma in humans and animals. The minimum inhibitory concentrations (MICs) of the oil were found to be within the range of 100.3–124.8 μL/mL, while the minimum microbicidal concentrations (MMCs) were between 100.3 μL/mL and 150.0 μL/mL. The major constituent of the oil, **α-pinene** (Fig. 4.21), demonstrated prominent biological activity with MICs between 3.3 and 5.0 μL/mL and MMC value of 10.0 μL/mL against the same strains. From these results it can be suggested that Cluster pine turpentine oil and α-pinene could be used alternatively as antifungals for the treatment of mycetoma caused by *A. madurae* (Stojković et al., 2008b).

Different modes of action of essential oils and their constituents

From the obtained results was observed that each essential oil demonstrated lower antifungal potential compared to their constituents and thus it can be concluded that the individual components of the oils are the ones that are responsible, in general, for the activity. Thus, it is possible that there can be interactions between the ingredients of essential oils that can lead to blockage of the activity of individual components when the whole essential oil is applied. Moreover, antagonistic effects can be observed (Davidson and Parish, 1989), revealing that the role of individual components of essential oils concerning the expression of antifungal potential can be significant. Menthol, camphor, and thymol, which demonstrated the strongest in vivo antifungal activity among all studied components, are the main constituents of the essential oils of Spearmint and Thyme and that fact can justify the high in vivo antifungal capacity of these oils. The essential oil obtained from Sweet basil that is well known for its beneficial activity on the skin, wounds, etc. revealed good antifungal potential, better than the oils of lemon and orange. The major constituent of mentioned oil is linalool and it proved to possess good antifungal effect as a single compound. Limonene, the other tested component, is one of the weaker fungicides in this case. Consequently, Lemon and Bitter orange oil showed the lowest antifungal potential among tested oils, containing limonene in high levels among other components. That certainly influenced the decrease in the efficiency of these oils, which have a high proportion of limonene. Essential oils of *S. officinalis* and *L. angustifolia* were found to be the most effective in the treatment of experimentally induced dermatomycoses. Comparison of the data that were obtained from the in vitro investigation of the antifungal capacity of essential oils with the ones from in vivo experiments showed that the essential oils of Sage and Lavender achieved the lowest potential in vitro, while concerning they in vivo assays appeared to be the most effective ones (Fig. 4.27) (Soković, 2001). Moreover, it is known that Sage, Lavender, and others have been used to treat various skin diseases and can be found as ingredients in various cosmetic products for skin care (Bremnes, 1994).

Lavender essential oil has as main constituents linalool, linalyl acetate, limonene, cineole, and camphor. The potency of the essential oil can be justified by the interactions between individual ingredients and especially the ones between linalyl acetate and linalool (Lis-Balchin et al., 1998). The mentioned essential oils and their components could represent potential alternative solutions for the treatment of infections by dermatomycetes in humans and animals. Nevertheless, there are only a few studies regarding the in vivo antifungal activity of essential oils against human fungal pathogens (Soković, 2001). The antifungal capacity of limonene was tested against *C. albicans* using the bioautographic method on TLC plates, revealing stronger activity compared to bifonazole. When the activity of limonene toward five dermatomycetes was evaluated using the microdilution and agar diffusion methods, it was found to be more potent against the same pathogens in humans and animals compared to commercial drug bifonazole (Rančić et al., 2003).

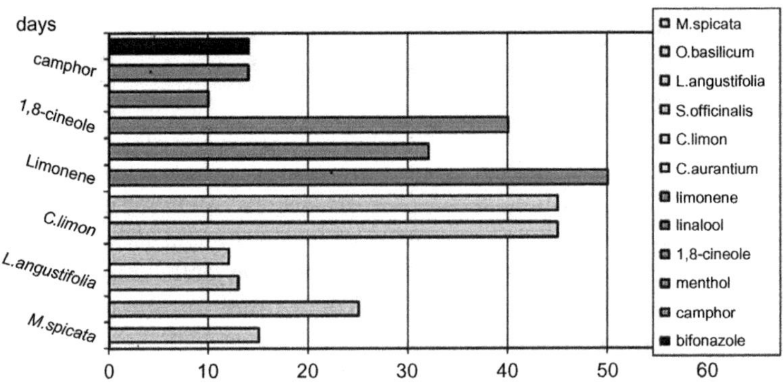

FIG. 4.27 Antifungal activity of essential oils and their constituents, in vivo.

Flavonoids

Flavonoids are widely distributed in plants (more than 6000 types) and represent a diverse group of metabolites found in almost all fruits and vegetables. They are responsible for the colors in vegetables and fruits, alongside with carotenoids.

Flavonoids are known as very potent antioxidants and possess many other biological activities, with anti-inflammatory and immune system benefits. Historically they have been used in traditional medicinal practice.

Structure

Chemically, flavonoids have the general structure of a 15-carbon skeleton, which consists of two phenyl rings (A and B) and heterocyclic ring (C). This carbon structure can be abbreviated as C6-C3-C6.

Most flavonoids occur in edible plants and foods as β-glycosides, bound to one or more sugar molecules (Williamson et al., 2005) with exceptions of **catechins** and **proanthocyanidins** and fermented soy-based products. The rest are present in the form of flavonoid aglycones (they are not bound to a sugar molecule).

Flavonoids are divided into several significant groups: **anthocyanidins**, **flavanols**, **flavones**, **flavonols**, **flavonones**, and **isoflavones** (Fig. 4.28) (Manach et al., 2004; Xiao et al., 2013a).

- Flavones

  They can be found in various herbs like celery, parsley, peppers, and many others. Flavones are famous for their strong antioxidant potential and they are connected to metabolism of drugs. The most prevalent in nature are **luteolin** and **apigenin** (Fig. 4.29).

- Anthocyanidins

  Rich source for flavonoids of this group are pomegranates, red, purple and blue berries, plums, red wine, grapes, and some others. This group of flavonoids is connected to antioxidant potential, heart health, obesity, and diabetes prevention. Some famous anthocyanidins are **cyanidin**, **peonidin**, **pelargonidin**, and **malvidin** (Fig. 4.30).

- Flavonones

  Flavonones are usually found in Citrus fruits. They are also associated with antioxidant and anti-inflammatory activities and also cardiovascular health. **Hesperetin**, **eriodictyol**, and **naringenin** (Fig. 4.31) are some of the most abundant flavonones.

- Isoflavones

  Isoflavones are very often found in soybeans and their products, but also in legumes. They can act as phytoestrogens. This characteristic marked them as a beneficial in lowering the risk of breast, endometrial, and prostate cancers and for the treatment of menopausal symptoms. **Genistein**, **glycitein**, and **daidzein** (Fig. 4.32) are members of this group of flavonoids.

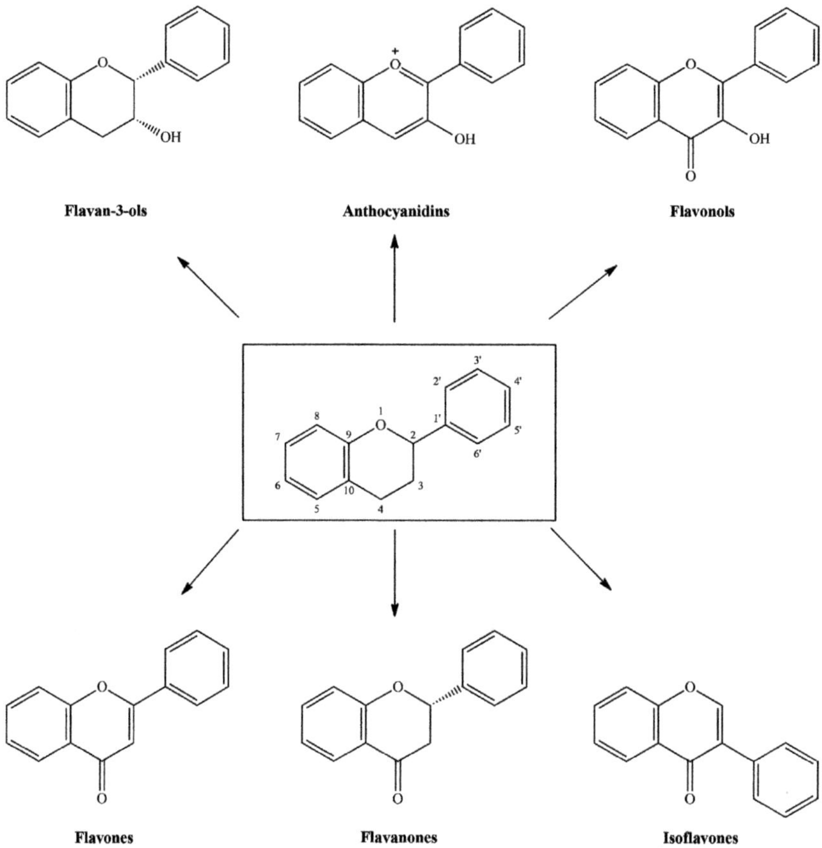

FIG. 4.28 Flavonoid subclasses.

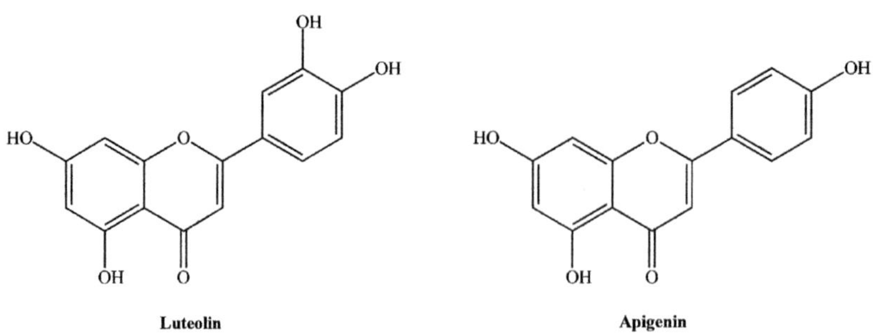

FIG. 4.29 Chemical structures of **luteolin** and **apigenin**.

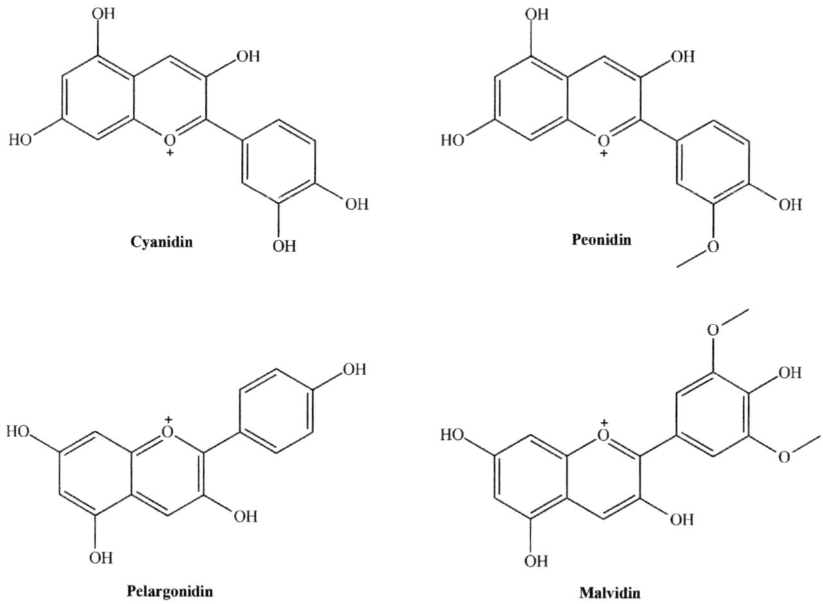

FIG. 4.30 Chemical structures of **cyanidin**, **peonidin**, **pelargonidin**, and **malvidin**.

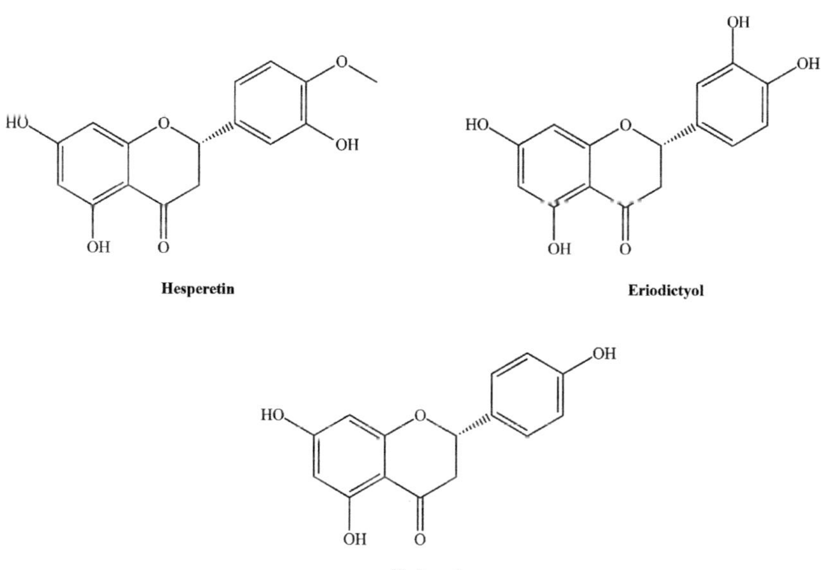

FIG. 4.31 Chemical structures of **hesperetin**, **eriodictyol**, and **naringenin**.

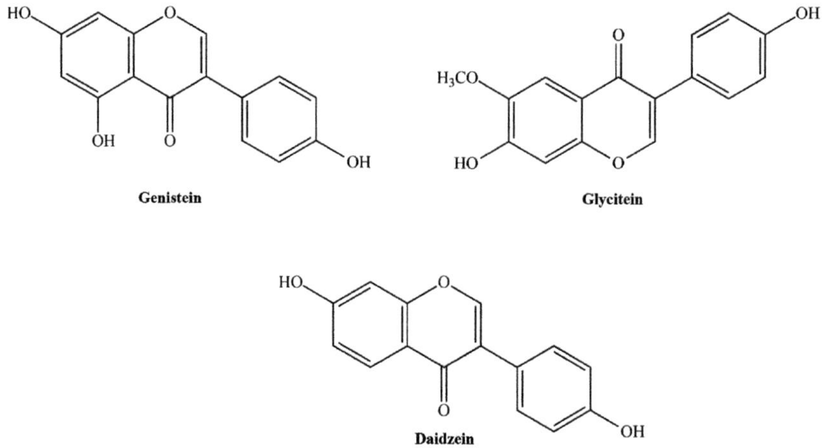

FIG. 4.32 Chemical structures of **genistein**, **glycitein**, and **daidzein**.

- Flavonols

  Apples, onions, Brussels sprouts, broccoli, leeks, kale, tea, berries, and beans are rich in flavonols. They are powerful anti-inflammatory and antioxidant agents and very valuable in chronic diseases prevention. The most famous flavonols are **quercetin** and **kaempferol** (Fig. 4.33).
- Flavanols

  Flavanols are further divided into monomers (catechins), dimers, and polymers. They can be found in apples, berries, teas, cocoa, grapes, and red wine. **Catechins** (Fig. 4.34) are found in white and green teas, while dimers are especially common in black tea, which are associated with lowering cholesterol. Catechins are also helpful in the prevention and treatment of cardiovascular and neurological conditions.

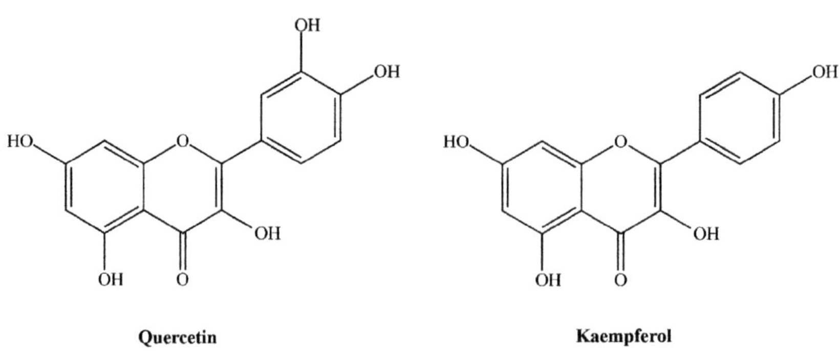

FIG. 4.33 Chemical structures of **quercetin** and **kaempferol**.

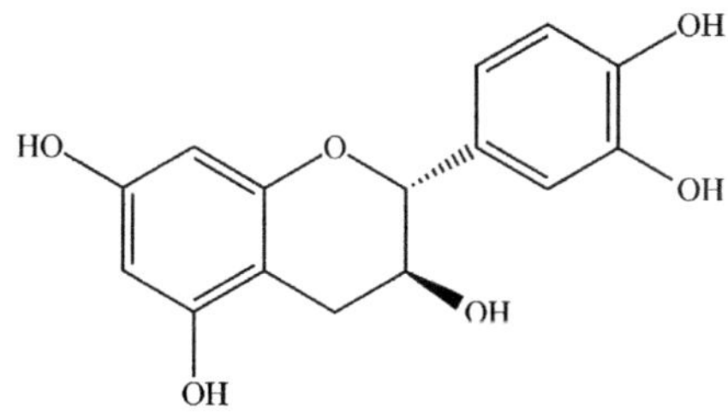

Catechin

FIG. 4.34 Chemical structure of **catechin**.

Biosynthesis

Flavonoids are synthesized by the phenylpropanoid pathway, where the amino acid phenylalanine produces 4-hydroxycinnamoyl-CoA (Ververidis et al., 2007). 4-hydroxycinnamoyl-CoA combined with malonyl-CoA leads to the synthesis of flavonoids with two phenyl rings (chalcones). Conjugate ring closure of chalcones leads to the three-ringed structure of a flavone. Series of modifications gives flavanones, dihydroflavonols, and anthocyanins. Alongside with the steps of this this pathway, the other products can be formed; flavonols, flavan-3-ols, proanthocyanidins (tannins), and other polyphenoles (Fig. 4.35).

Antifungal activity of flavonoids

Flavonoids consist of a group of secondary metabolites that demonstrate great antifungal potential. It has been established that at least in a number of cases their presence can serve as a chemical barrier against pathogenic microorganisms. Because they are compounds of natural origin that exhibit highly specific antifungal and in general antimicrobial activity, flavonoids could be employed as alternatives to marketed fungicides in order to control plant diseases that are caused by fungi.

Twenty-five flavonoids were evaluated for their activity on the mycelial growth of a crop pathogen, *Verticillim albo-atrum*. A number of flavonoids inhibited hyphal growth and several compounds were found to be ineffective at the highest used concentrations. Active compounds did not share a common pattern concerning their substitutions. In particular, the unsubstituted flavonoids were found to be more potent growth inhibitors and in most of the cases, when the number of substitutions (hydroxylation, methoxylation, and glycosylation) was increased, this had as result the loss of antifungal effect (Picman et al., 1995). Moreover, the fungicidal capacity of two isoflavones, one isoflavanone, and seven isoflavans was studied against *Aspergillus repens*, *A. amstelodami*, *A. chevalieri*, *A. flavus*, and *A. petrakii* (Weidenbörner et al., 1989). While the isoflavones demonstrated low activity, the two isoflavans were found to be highly potent inhibitors (Weidenbörner and Jha, 1997). Two isoflavones that can be found in nature, **genistein** (Fig. 4.32) and **biochanin A** (Fig. 4.36), and

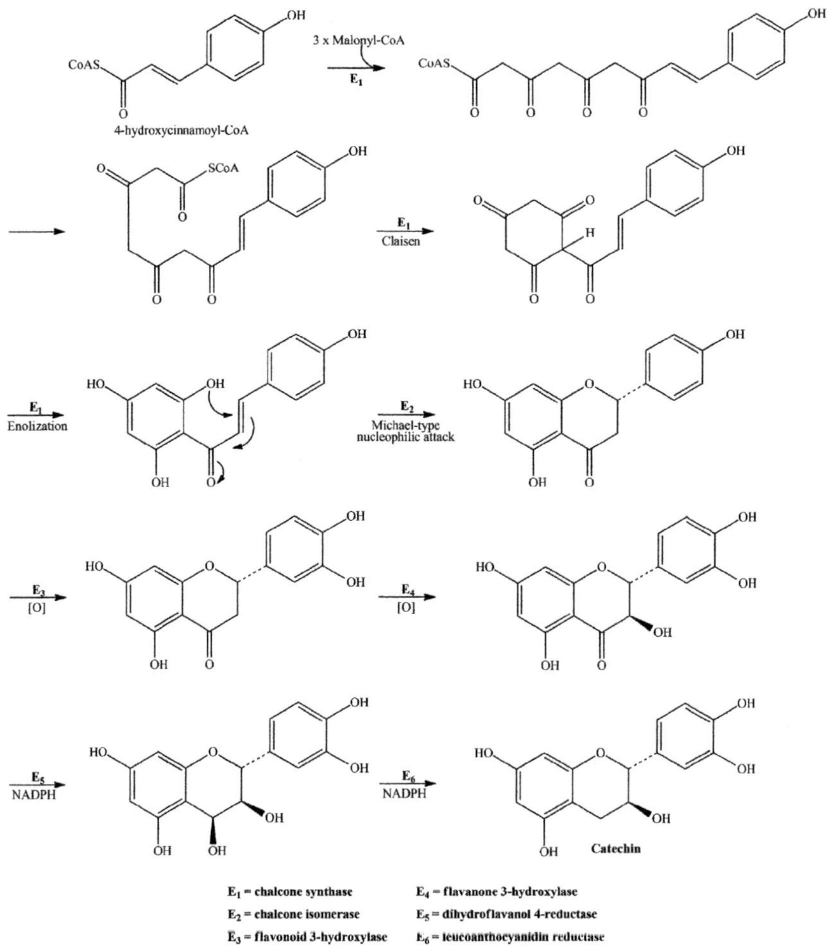

FIG. 4.35 Biosynthesis of flavonoids.

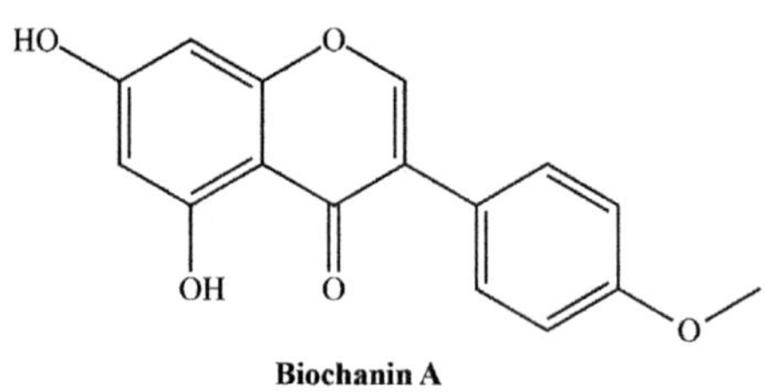

FIG. 4.36 Chemical structure of **biochanin A**.

their dihydroderivates (isoflavanones) alongside with nine perhydrogenated isoflavones (isoflavans) were studied for their antifungal potential against *Rhizoctonia solani* and *Sclerotium rolfsii.* All the isoflavonoids of the biochanin A series revealed strong antifungal activity. Genistein and the other isoflavans that possess two hydroxyl groups and one methoxy group were proved to be fungitoxic, while isoflavans with two or three methoxy groups were found to be almost inactive (Weidenbörner and Jha, 1997), although earlier results revealed that isoflavans exhibit, in general, greater potency than the corresponding isoflavones and isoflavanones.

Since the individual unsubstituted flavonoids exhibited strong antifungal capacity, a series of different mixtures have been examined against fungal strains of grain targeting to enhance the fungicidal activity of each substance by synergistic effects. As a general observation, it was observed that the best potential was exhibited with a combination of flavones and flavanones in different proportions. Nevertheless, it should be mentioned that a mixture containing flavonoids substituted with a single methoxy group and several hydroxyl groups demonstrated, in general terms, stronger activity compared to a mixture containing only hydroxylated flavonoids. Thus, it is obvious that combinatorial use of several flavonoids with preferable substitutions might lead to even greater increase in antifungal capacity. In that way, lower concentrations of several flavonoids could be applied for the desired effect, proving them more attractive as natural protective agents (Silva et al., 1998; Weidenbörner and Jha, 1997). Regarding the antifungal potential of flavonoid glycosides, it should be mentioned that there wasn't observed any substantial effect (Weidenbörner and Jha, 1997).

Krauze-Baranowska et al. (1999) studied **cupressuflavone** and **4′-*O*-methylcupressuflavone** (Fig. 4.37), isolated from leaves of *Cupressocyparis leylandrii.* From the obtained results can be observed that compounds exerted antifungal activity against *Alternaria alternata, Cladosporium oxysporum, Fusarium culmorum,* and *F. avenaceum.*

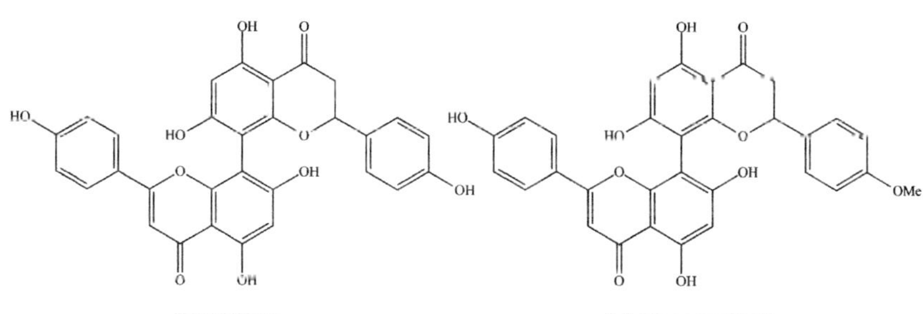

FIG. 4.37 Chemical structures of **cupressuflavone** and **4′-*O*-methylcupressuflavone**.

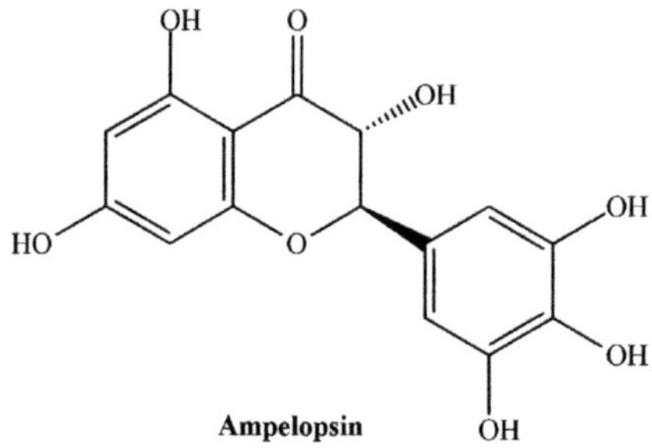

FIG. 4.38 Chemical structure of **ampelopsin**.

Matsumoto and Tahara (2001) isolated the flavonol **ampelopsin** (Fig. 4.38) from *Salix sachalinensis* leaves and reported its antifungal potential against *Cladosporium herbarum.*

In another study, ethanolic extract of *Phlomis fruticosa* (Jerusalem sage) (Labiatae) was evaluated by diffusion method inhibiting *Aspergillus niger, Penicillium ochrochloron, Trichoderma viride, Fusarium tricinctum,* and *Phomopsis helianthi.* In addition, this extract revealed fungicidal activity against *Cladosporium cladosporioides* and *Aspergillus ochraceus* at a very low concentration range (10–20 μg/mL), (Ristić et al., 2000). This study showed that when this extract was treated with HCL and μ-glucosidases, which have the ability to remove sugars from flavonoids, it resulted in stronger antifungal potential when compared to the original ethanol extract. Moreover, it is generally established that flavonoid glycosides exhibit lower antimicrobial activity than aglycones.

The antifungal capacity of a series of phenol derivatives that were isolated from Holm Oak *Quercus ilex* leaves and can be categorized as flavonoids, proanthocyanidins, and phenolic acids, have been tested against fourteen fungi (Karioti et al., 2009). Two coumarins, **scopoletin** and **isoscopoletin,** two simple phenolic acids, **protocatechuic acid** and **isovanillic acid** (Fig. 4.39), as well as one flavonoid, **eriodictyol** (Fig. 4.31), which were separated from the aerial parts of *Centaurea spruneri,* revealed fungistatic effect at 0.259–2.38 μmoL/mL and fungicidal at 0.69–2.6 μmoL/mL, respectively, against all studied strains. The flavonoid, eriodictyol, exhibited the strongest antifungal activity with MICs and MFCs at 0.694–1.388 μmoL/mL, while the observed effect of protocatechuic acid was found to be weaker (0.65–1.3 μmoL/mL for MICs and 1.3–2.6 μmoL/mL for MFCs). It should be noted that from the simple phenolic acids, protocatechuic acid demonstrated the lowest antifungal potential (Ćirić et al., 2012).

From the obtained results can be concluded that the inhibitory activity of flavonoids generally diminishes when the number of substitutions on the compounds increases. Accordingly, the most potent inhibitors were found to be the unsubstituted derivatives.

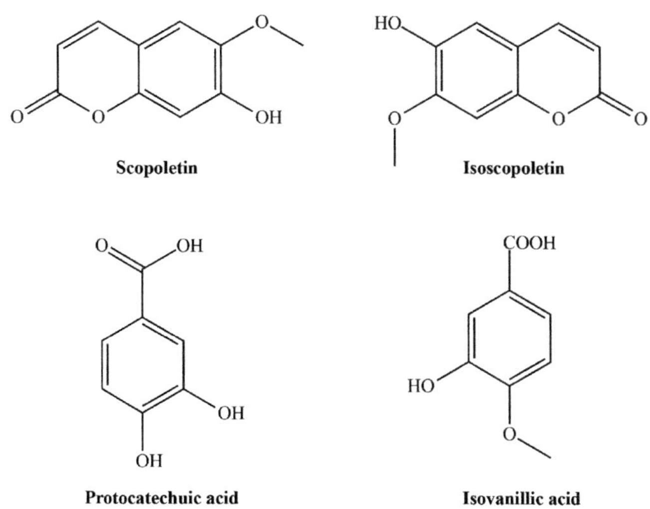

FIG. 4.39 Chemical structures of **scopoletin**, **isoscopoletin**, **protocatechuic acid**, and **isovanillic acid**.

Apigenin (Fig. 4.29) is a nontoxic flavon, present in fruits and vegetables (cardoon, celery, artichoke, parsley, etc.), some of which are widely marketed as dietary and herbal supplements (Sharma et al., 2010). It has been thoroughly studied mainly due to its notable anticancer, antiviral, antibacterial, antioxidant, proapoptotic, and anti-inflammatory activities (Kanazawa et al., 2006; Cai et al., 2011).

Apigenin-7-*O*-glucoside (AP7Glu, Fig. 4.40) is a stable flavonoid of natural origin, demonstrating better solubility than many other flavonoids, such as apigenin. It would be interesting to point out that both compounds also exhibit similar anti-inflammatory activity (Kowalski et al., 2005). AP7Glu demonstrates a wide range of biological activities and is employed for the treatment of inflammatory diseases such as upper respiratory infections (Bhaskaran et al., 2010). It has been recently established that AP7Glu exhibits anxiolytic

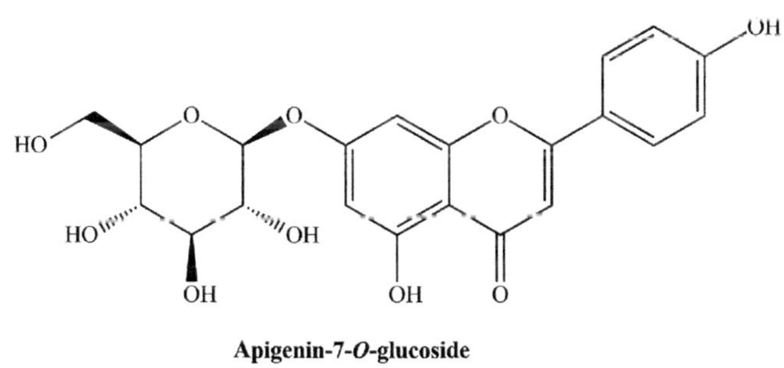

FIG. 4.40 Chemical structure of **apigenin-7-*O*-glucoside**.

activity in rats, comparable to apigenin and diazepam (Kumar and Bhat, 2012). Moreover, new data came to sight concerning the antifungal and cytotoxic activity of apigenin-7-*O*-glucoside (inn colon cancer cells) in comparison with the corresponding potential of apigenin. Besides comparing the activity of the two compounds, the target was to collect more data regarding potential dual effect that would be important for treating colon cancer (Smiljkovic et al., 2017).

Testing of different flavonoids showed that flavonoid aglycones (**luteolin**, **myricetin**, and **quercetin**, Fig. 4.41) exhibited higher antifungal activities than other tested flavonoid glycosides, with luteolin possessing the best antifungal activity.

It was noticed that the most active flavonoids were the polyhydroxylated ones (myricetin, datiscetin, quercetin, luteolin, and kaempferol), with the exception of flavones lacking hydroxyl groups. Flavonoids that showed good activity possess the obligatory C-4 keto group and hydroxyl group substitutions at C-3, C-5, and C-7 and at least one hydroxyl group at ring B. This indicates that the most hydrophilic compounds exhibited stronger antifungal potential than the least hydrophilic ones, with indication that hydroxyl group at C-3 is necessary for the activity.

Among several flavonoids that were tested for their potential antifungal activity, luteolin showed the best capacity. **Chlorogenic acid**, **cynarine**, **luteolin 7-rutinoside**, and **cynaroside** (Fig. 4.42) compounds isolated from *C. scolymus* leaves revealed high activity against *Aspergillus*, *Penicillium*, and some other species (Zhu et al., 2005).

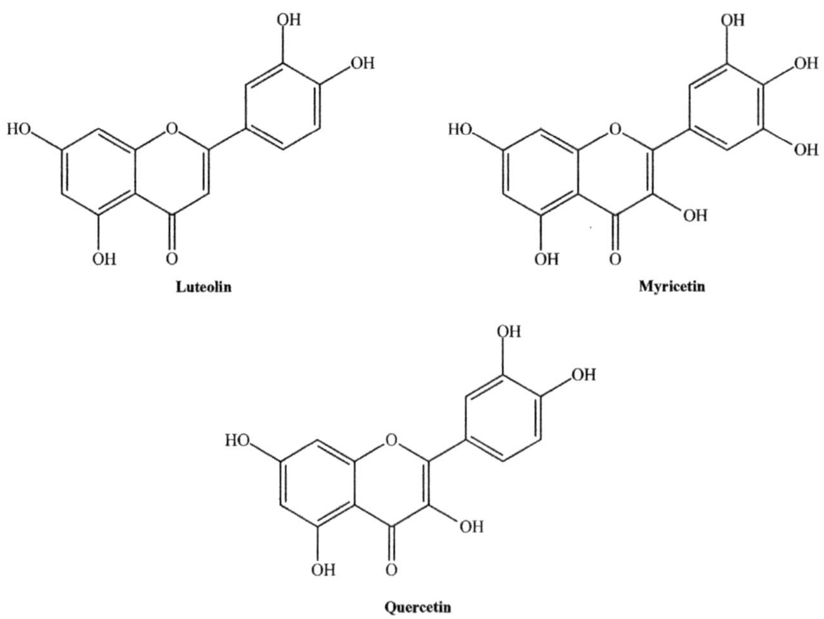

FIG. 4.41 Chemical structures of **luteolin**, **myricetin**, and **quercetin**.

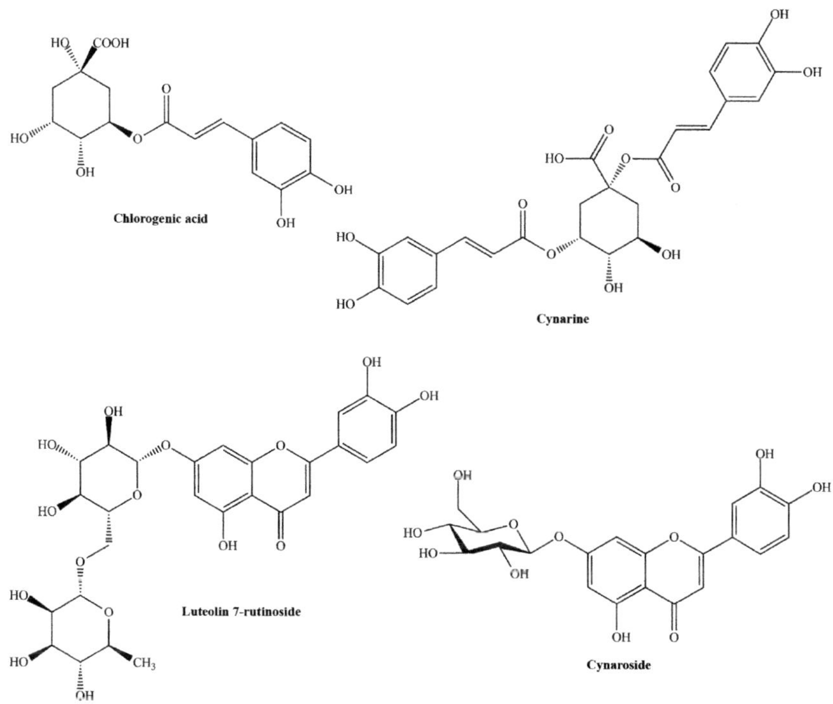

FIG. 4.42 Chemical structures of **chlorogenic acid**, **cynarine**, **luteolin 7-rutinoside**, and **cynaroside**.

Antifungal activities of apigenin, apigenin 7-glucoside, and luteolin were previously reported (Aljancic et al., 1999; Tshikalange et al., 2005). It has been reported that a mixture of **ellagic acid**, **gallic acid**, and **punicalins** (Fig. 4.43), which had been isolated from pomegranate juice byproducts, revealed antimicrobial activity against pathogenic microorganisms such as *Clostridia species*, *Candida albicans*, *Cryptococcus neoformans*, *Staphylococcus aureus*, *Aspergillus fumigatus*, and *Mycobacterium* spp.

Ballota macrodonta Boiss. et Balansa. is an endemic medicinal plant traditionally used for the treatment of a variety of diseases in Turkey. The presence of nineteen components (e.g., hydroxybenzoic acid, catechin hydrate, vanillic acid, caffeic acid, syringic acid, epicatechin, coumaric acid, ferulic acid, luteolin-7-O-glucoside, rutin, and quercetin) was confirmed. The water and methanolic extracts were rich in high total phenolic and flavonoid contents. Activity against eight fungi (*Aspergillus*, *Penicllium*, *Trichoderma*, *Candida*) (MICs of 0.0125–0.20 mg/mL) was observed (Uysal et al., 2018).

There are still ongoing studies regarding the impact of flavonoids' glycosylation on different bioactivities. Due to the complexity of the matter and the intense scientific interest about it, more data are being collected in order to enlighten the general influence of sugar moieties in biological activities. This

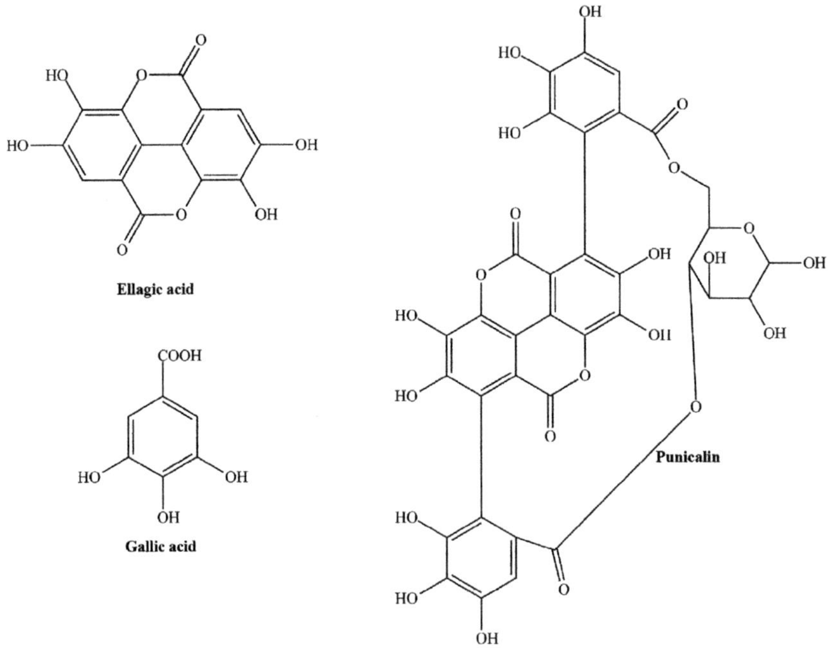

FIG. 4.43 Chemical structures of **ellagic acid**, **gallic acid**, and **punicalin**.

situation can be explained also partially by the lack of clinical and in vivo studies that could support the results from the in vitro assays. It should be mentioned, that in a small portion of clinical trials, the in vivo data differ from the ones obtained in vitro. Moreover, some results revealed that glycosylated flavonoids, when applied in vivo, exhibited similar or even stronger biological activities (antidiabetic, antistress, antiallergic, antidegranulating, and anti-inflammatory) than their corresponding aglycones (Xiao et al., 2013b). In some of the more recent studies was reported that apigenin glucosides can penetrate into cancer cells followed by effective hydrolyzation (Srivastava and Gupta, 2009).

The most active flavonoids, flavones, and flavanones exert excellent potential as novel antifungal agents of natural origin (Picman et al., 1995).

Lactones

As a result of the increasing interest in ethnopharmacology, various biologically active compounds have been discovered and studied, such as **sesquiterpene lactones** (SL).

Sesquiterpenes is a class of terpenoids, which are commonly found in nature. They can be produced by plants, microorganisms, and inhabitants of oceans such as invertebrates and sponges. Sesquiterpene lactones (SL) are colorless phytochemicals of bitter taste and lipophilic properties. They can be mainly detected and isolated from flowering heads and leaves of Asteraceae family species with over 3000 reported different structures (Chaturvedi, 2011), but also from plants of other families, such as Acanthaceae, Amaranthaceae, Anacardiaceae, Apiaceae (=Umbelliferae), Aristolochiaceae, Burseraceae, Cortiariaceae, Cannellaceae, Euphorbiaceae, Hepaticae, Illiciaceae, Lauraceae, Lamiaceae, Menispermaceae, Magnoliaceae, Rutaceae, and Winteraceae. It should be noted that the yield of plant-isolated SLs (dry weight) is usually higher than 1% (Jennings, 1977).

In an effort to gain a deeper understanding of this particular group of plant-derived secondary metabolites, herein are presented sesquiterpene lactones of known chemical structure, isolated from various plant species that have been qualitatively and quantitatively analyzed for different kinds of biological activities.

Structure

An important and recognizable and usual characteristic of the sesquiterpene lactones is the presence of a γ-lactone ring (closed toward C-6 or C-8), in general containing an α-methylene group. Among other modifications, the incorporation of hydroxyls or esterified hydroxyls and epoxide ring is common. Some of SLs are present in glycoside form and could contain halogen or sulfur atoms (Chen et al., 1995). These compounds are built from head-to-tail structure with three isoprene units and oxidative transformation, subsequent cyclization, being able to produce *cis* or *trans* forms of lactones (Dewick, 2002; Chaturvedi and Dwivedi, 2016).

Synthesis

Biosynthesis in plants may produce a variety of products and further modifications, oxidation and glycosylation could result in a great number of varied structures. Regulation of the biosynthetic pathways is strictly controlled in species and sesquiterpenes are produced as a response in different kinds of stress.

The mechanism of SLs production in species is presented in Fig. 4.44. Biosynthesis of SLs in plants is carried out with the three isoprene units, via farnesyl pyrophosphate (FPP) and it occurs in the endoplasmic reticulum. Their basic structure consists of a C15 sesquiterpene backbone and a lactone moiety and the sesquiterpene backbones are mainly germacranolide, eudesmanolide, and guaianolide skeletons, derived from germacrene A (GA) (Fig. 4.44) (de Kraker et al., 1998). Besides the backbone, for their diversity some other conformation are of great importance; C_6 or C_8 positions and stereoselective formation of the lactone ring (α or β).

Synthetic pathway of SLs starts with the cyclization of farnesyl diphosphate (FDP) by germacrene A synthase (GAS) to germacrene A (GA), which is subsequently oxidized at its isopropenyl side chain by cytochrome P450 enzyme, germacrene A oxidase (GAO), to form germacra-1(10),4,11(13)-trien-12-ol (GOH). Next step is oxidation to germacra-1(10),4,11(13)-trien-12-al (GAL) and germacra-1(10),4,11(13)-trien-12-oic acid (GAA) (Liu et al., 2011).

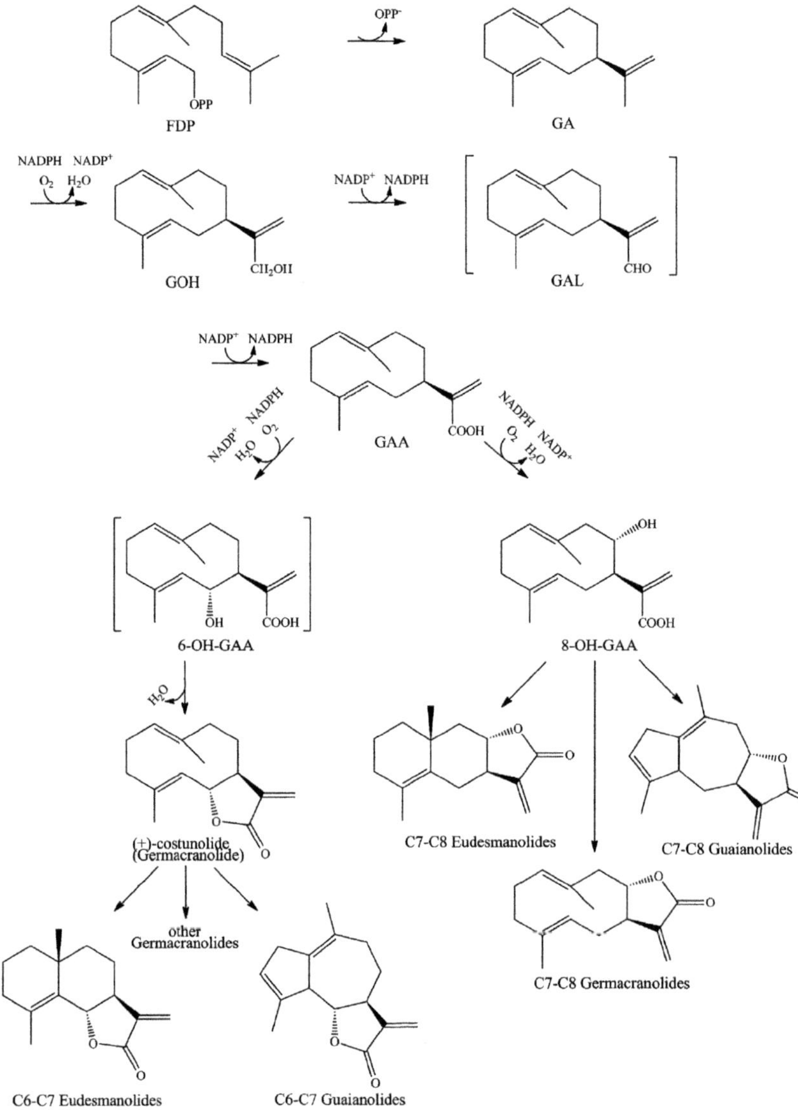

FIG. 4.44 Proposed biosynthetic mechanisms of sesquiterpene lactones.

Starting from GAA there are two possible biosynthetic pathways that can be followed for the synthesis of C6-C7 and C7-C8 types of STLs, respectively. Thus, hydroxylation of GAA at the C6-α position by costunolide synthase (COS) leads to an unstable intermediate (6-OH-GAA), which is cyclized to form costunolide, the precursor of the C6-C7 types of STLs, while hydroxylation at the C8 position of GAA and subsequent spontaneous lactonization results to the formation of the C7-C8 types of SLs (de Kraker et al., 2002).

The majority of the more than 4000 known different structures of SLs are germacranolides, guaianolides, and eudesmanolides (de Kraker et al., 2002). Germacranolides are recognized, biogenetically, as a primitive class of SLs and probably all the other sesquiterpene lactones evolved from this type. The basic and simplest form of germacranolides, **(+)-costunolide** (Fig. 4.44), is accepted as the basic compound and common intermediate of all of the three groups of SLs. Germacranolides are also the largest group of naturally occurring SLs and most significant type of SLs concerning to their function in humans. They are presented with a 10-member ring (Fig. 4.45A); Guaianolides have 7- and 5-membered rings and a methyl group at C-4 position (Fig. 4.45B); Eudesmanolides (Fig. 4.45C) are presented with two fused 6-member rings. All three types contain a fused 5-membered lactone group (γ lactone) with a carbonyl moiety at the alpha position (Chadwick et al., 2013).

Antifungal activity of sesquiterpene lactones

Sesquiterpene lactones display a wide spectrum of biological activities, considering as one of the most important of these their antifungal activity.

As it is presented in literature, most SLs exhibit a variety of biological activities according to plant origin, including hypoglycemic, antibacterial (Bach et al. 2011), antifungal (Vajs et al., 1999), antiplasmodial (Medjroubi et al., 2005), antinociceptive, antipyretic (Akkol et al., 2009), and anti-inflammatory (Al-Saghir et al., 2009), antifeedant, antifungal, antitumor (Kolli et al., 2012), etc. They possess a specific range of pharmacological properties, while a few of them, mainly guaianolides, can also potentially present a toxic profile.

A group of sesquiterpene lactones was tested in vitro and analyzed for potential antifungal activity, against *Aspergillus niger*, *A. ochraceus*, *Penicillium*

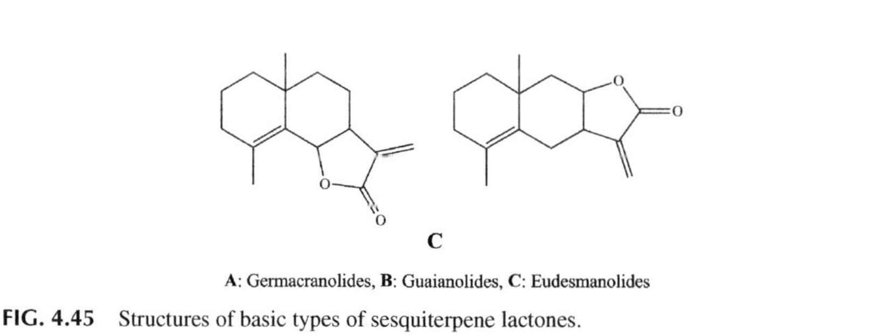

A: Germacranolides, **B**: Guaianolides, **C**: Eudesmanolides

FIG. 4.45 Structures of basic types of sesquiterpene lactones.

ochrochloron, *Trichoderma viride*, *Fusarium tricinctum*, *Phomopsis helianthi*, and *Cladosporium cladosporioides*. Guaianolides from *Centaurea nicolai* were found to be highly potent (0.8–25 μg/mL) (Vajs et al., 1999), while sesquiterpene lactones isolated from *Centaurea achaia*, *C. thessala*, and *C. attica* also exhibited excellent antifungal activity against all tested fungi mentioned in the previous study, with germacranolides possessing the greatest antifungal potency (Skaltsa et al., 2000a,b).

The fungicidal activities of thirty-six natural and synthetic sesquiterpene lactones with guaianolide, trans-germacranolide, cis-germacranolide, medampolide, and eudesmanolide carbon skeletons were evaluated against the fungi *Colletotrichum acutatum*, *C. fragarie*, *C. gloeosporioides*, *Fusarium oxysporum*, *Botrytis cinerea*, and *Phomopsis* sp. by Wedge et al. (2000). It was observed that dehydrozaluzanin showed the highest antifungal potency due to the presence of an α, ß-unsaturated carbonyl group in the cyclopentananone ring. Additional to previously isolated sesquiterpene lactones, 11,13-dihydrocnicin and 11,13-dihydro-19-desoxycnicin, the aerial parts of *Centaurea pullata*, afforded three minor sesquiterpene lactones, namely, a new germacranolide, 8R-*O*-(4-acetoxy-5-hydroxyangeloyl)-11,13-dihydrocnicin, and two new eudesmanolides, 8R-*O*-(4-hydroxy-2-methylenebutanoyloxy)-11,13-dihydrosonchucarpolide and 8R-*O*-(4-hydroxy-2-methylenebutanoyloxy)-11,13-dihydro-4-*epi*-sonchucarpolide. The antifungal activity of all previously mentioned compounds and some more recently isolated sesquiterpene lactones (a novel elemanolide with an α-methyl-γ-lactone moiety, 8α-O-(4-hydroxy-2-methylenebutanoyloxy)melitensine, in addition to other four known sesquiterpene lactones containing the same ring, melitensine, 11β-dihydrosalonitenolide, 8α- hydroxy-11β-13-dixidroxy-4-epi-sonchucarpolide and 8α-hydroxy-11β-13-dihydroxyonopordaldehyde) from *Centaurea pullata* were tested against eight fungal species, using microdilution method. All the tested compounds exhibited greater antifungal activity compared to positive controls. Additionally, the pharmacokinetic profile of tested compounds was investigated using computational methods and the results were proved to be in agreement with the in vitro data (Djeddi et al., 2007, 2008).

Comparing antifungal results of three linear sesquiterpene lactones, namely, anthecotulide, hydroxyanthecotulide, and acetoxyanthecotulide, isolated from the aerial parts of *Anthemis auriculata*, with previously published data (Konstantinopoulou et al., 2003) sesquiterpene lactones of *A. altissima*, it was observed that linear lactones were generally the most potent. Nine sesquiterpene lactones, anthemin A, 1α-hydroxydeacetylirinol-4 α,5β-epoxide, anthemin C, tatridin A, 1-epi-tatridin B, anthemin B, 6-deacetyl-β-cyclopyrethrosin, elegalactone A, and 1β,4 α,6trihydroxyeudesm-11-en-8-12-olide, were isolated from the aerial parts of *Anthemis melanolepis*. All tested compounds exhibited inhibitory effect against almost all the studied fungi (Saroglou et al., 2005), supporting results from previous studies on related compounds isolated from other *Anthemis* species (Konstantinopoulou et al., 2003). The antifungal activity of sesquiterpene lactones can be generally justified by the alkylating properties of α-methylene-γ-lactone moiety. Additionally, the presence of other alkylating sites such as epoxides and conjugated carbonyls may enhance their potency as antifungals.

The general mechanism of SLs action contains alkylation of biological nucleophiles such as cysteine (cys) and glutathione or other sulfhydryl-containing systems, phosphofructokinase and glycogen synthase by α, ß-unsaturated carbonyl structures in a Michael-type addition (Koukoulitsa et al., 2005).

In corresponding scientific studies, the structures (Fig. 4.46A–D) and the biological activities of chlorine-containing sesquiterpene lactones isolated from *Centaurea* species have been also thoroughly tested. Halogen-bearing sesquiterpenes have recently started to be subjects for comparative studies. Presently, there is access to a plethora of scientific data showing that halogenated sesquiterpenes, unlike the halomonoterpenes and haloditerpenes, which are almost exclusively produced by marine organisms, they can be found also among metabolites of higher plants (Dembitsky and Tolstikov, 2002). The first mentioned chlorine-containing sesquiterpenoids were eupachlorin (**1**), its acetate (**2**) and eupachloroxin (**3**), which were isolated from *Eupatorium rotundifolium* in 1968. All these compounds exhibited

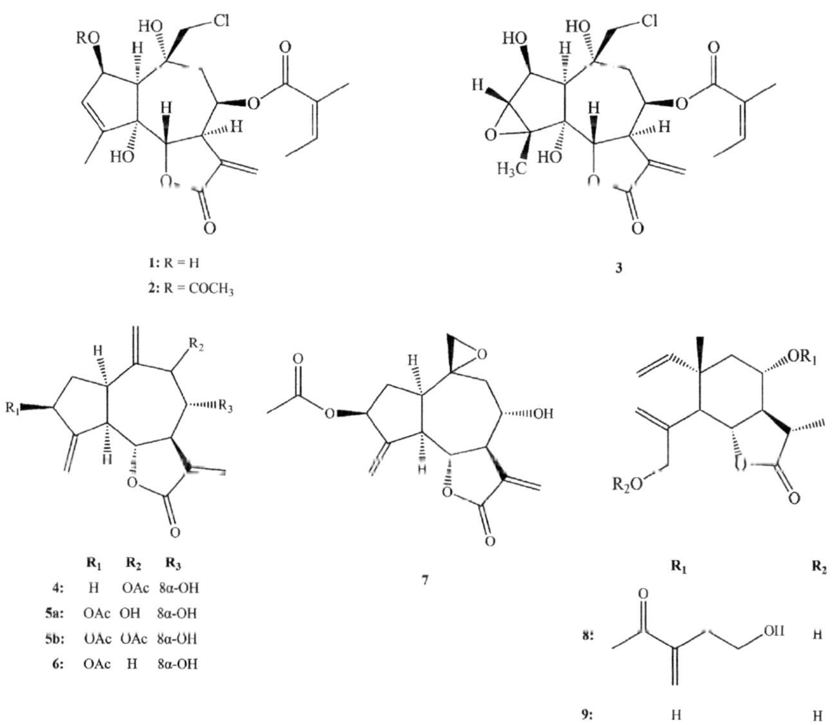

FIG. 4.46 (A) Chemical structures of sesquiterpene lactones **1–9**.

(Continued)

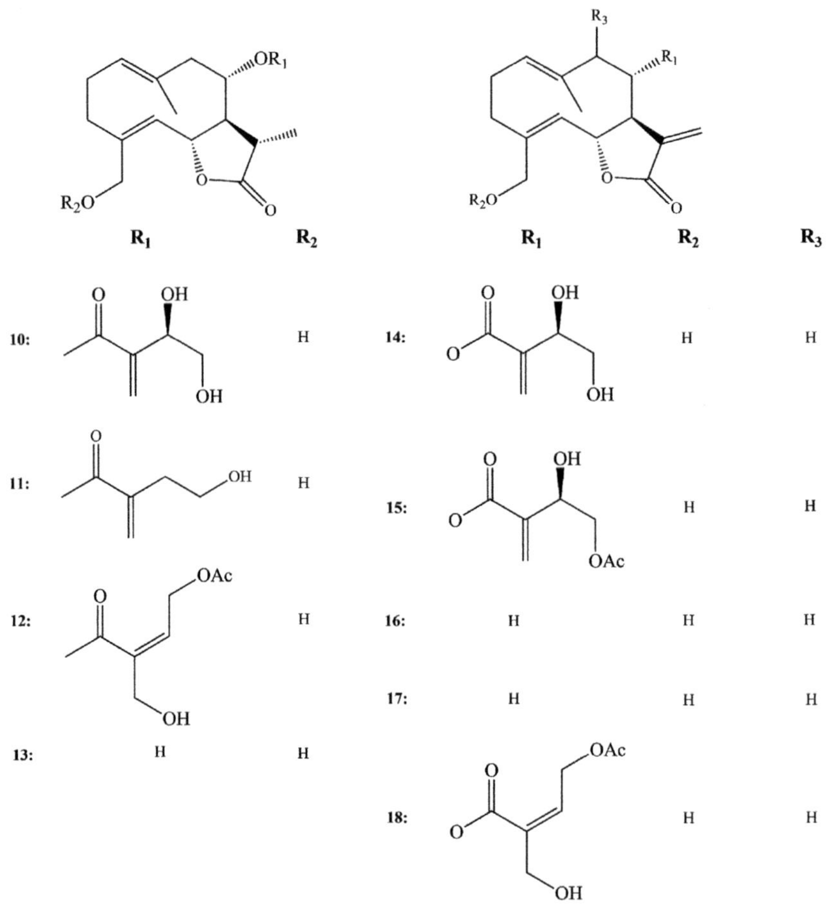

FIG. 4.46, CONT'D (B) Chemical structures of sesquiterpene lactones **10–18**.

notable anticancer potential. Thorough studying of SLs of different species has turned *Centaurea* genus into a producer of a large group of chlorine-containing sesquiterpene lactones, with the majority of them belonging to guaianolides. Depending on the position of the chlorine atom, *Centaurea* spp. sesquiterpene lactones can be further divided into three categories: lactones of category 1 have chlorine atom positioned in the guaiane skeleton; concerning lactones of category 2, chlorine atom is allocated in the acid substituent; lastly, lactones of category 3 are esters of chlorine-containing guaianolides and chlorine-containing oxyacids (Dembitsky and Tolstikov, 2002; Milošević-Ifantis et al., 2013).

The guaianolide, 3-deacetyl-9-*O*-acetylsalograviolide A (**4**), and four related lactones, salograviolide A (**5a**), 9-*O*-acetylsalograviolide A (**5b**), kandavanolide (**6**), and salograviolide B (**7**), have been isolated from aerial parts of *C. nicolai.*

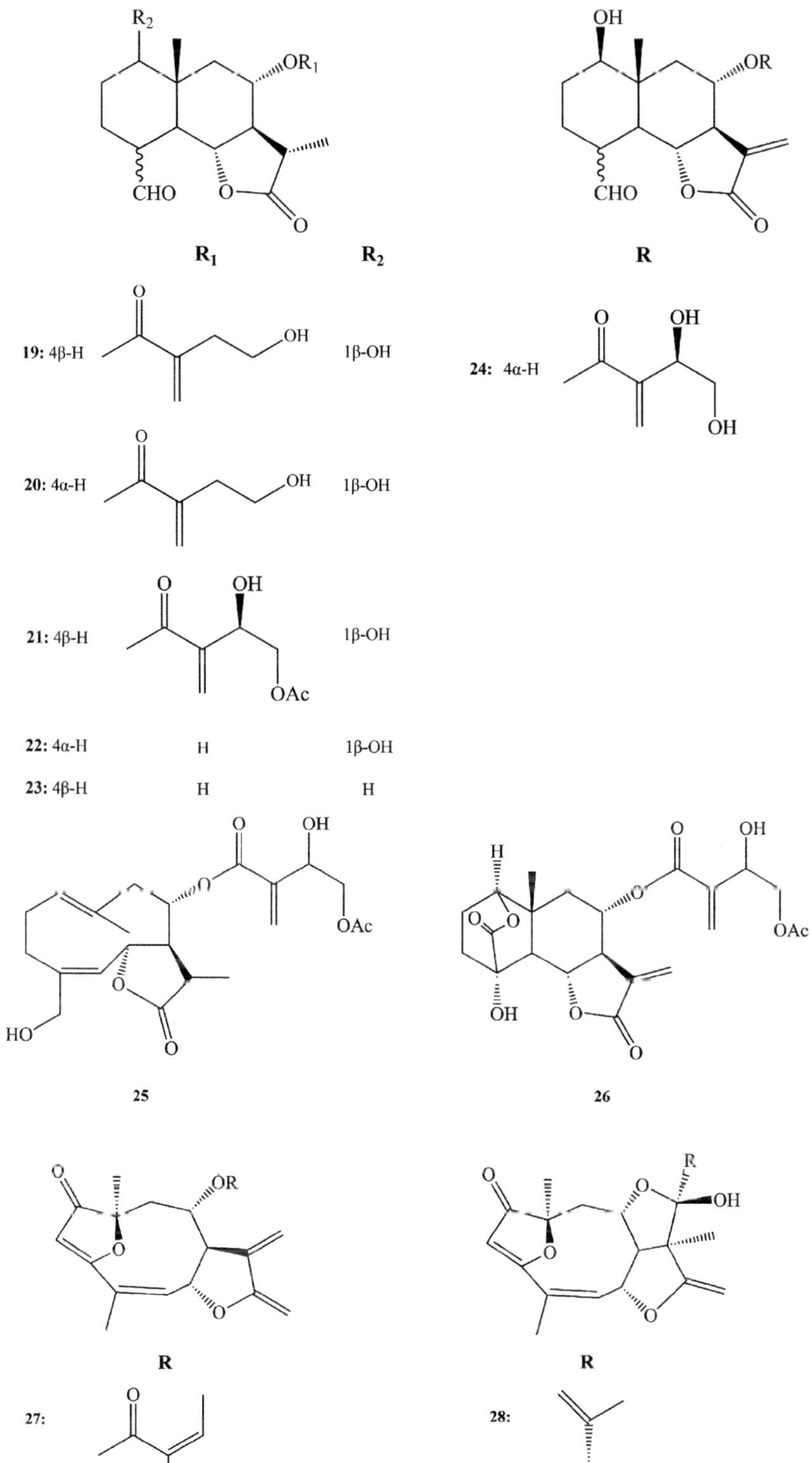

FIG. 4.46, CONT'D (C) Chemical structures of sesquiterpene lactones **19–24**. (D) Chemical structures of sesquiterpene lactones **25–28**.

Testing salograviolide A and its 9-*O*-acetyl-derivative showed antifungal activity against *Phomopsis helianthi*, *Cladosporium cladosporoides Aspergillus niger*, *A. ochraceus*, *Penicillium ochrochloron*, and *Fusarium tricinctum*, while *Trichoderma viride* was found to be totally resistant against the studied molecules (Vajs et al., 1999).

Sesquiterpene lactones originating from *Centaurea* spp., growing wild in Greece, containing α-methylene-γ-lactone ring were tested for their antifungal potential and results showed that in general terms, compounds exhibited strong activity (Skaltsa et al., 2000a,b; Karioti et al., 2002). SLs derived from mentioned species (Fig. 4.46A–D, Table 4.1) were evaluated for their antifungal activity on eight particular fungi. All tested lactones showed higher antimicrobial potential than miconazole.

The strong antimicrobial potential of these lactones led to further studies concerning the antimicrobial properties of corresponding lactones of similar structure isolated from *C. pullata* (Djeddi et al., 2007, 2008), a taxon growing wild in Algeria. In particular, 11β,13-dihydrocnicin (**10**) and 11β,13-dihydro-19-desoxycnicin (**11**), 8α-O-(4′-acetoxy-5′-hydroxyangeloyl)-11β,13-dihydrocnicin (**12**), 8α-O-(4′-hydroxy-2′-methylenebutanoyloxy)-11β,13-dihydrosonchucarpolide (**19**) 8α-O-(4′-hydroxy-2′-methylenebutanoyloxy)-11β,13-dihydro-4-*epi*-sonchucarpolide (**20**), 8α-O-(4-hydroxy-2-methylene-butanoyloxy)melitensine (**8**), melitensin (**9**), 11β, 13- dihydrosalonitenolide (**13**), 8α-O-hydroxy-11β,13-dihydro-4-*epi*-sonchucarpolide (**22**) and 8α-O-hydroxy-11β,13-dihydro-onopordaldehyde (**23**) were tested for their antimicrobial activity on eight fungal species. Although these SLs do not bear an α-methylene-γ-lactone ring, the results of their activity were similar to the previously obtained results for other *Centaurea* spp. (Karioti et al., 2002; Koukoulitsa et al., 2002; Saroglou et al., 2005; Skaltsa et al., 2000a,b). All the studied molecules showed higher antifungal potency than positive control drugs (Djeddi et al., 2007, Milošević-Ifantis et al., 2013).

Thirteen sesquiterpene lactones were isolated from the aerial parts of *C. zuccariniana* growing wild in Greece. Three of them were new isolated and identified lactones: one heliangolide, (1E,4Z)-15-hydroxy-8a-O-(4′-acetoxy-3′-hydroxy-2′-methylidenebutanoyl)-6βH,7αH-germacra-1,4,11(13)-trien-6,12-olide (**25**) and two eudesmanolides, 8a-(4′,5′-diacetoxyangeloyl)-sonchucarpolide (**21**) and an unusual eudesmanolide containing an oxygenated bridge linking C(1) and C(4), named zuccarinin (**26**). These were isolated for the first time and present new structures for science. Malacitanolide (**24**) was the most isolated SL from this species, followed by cnicin (**14**), and 4′-*O*-acetylcnicin (**15**) (Ćirić et al., 2012). Phytochemical analysis results were in agreement with those obtained from the previously evaluated *Centaurea* sp. from Greece (section Acrolophus) (Cass.) DC. Furthermore, the antimicrobial potential of these novel sesquiterpene lactones toward eight fungal species was tested and according to the results they were found to possess strong antifungal activity. Among the most sensitive species were found to be the *Aspergillus* ones (with *A. versicolor* as the most

TABLE 4.1 Antifungal activity of some sesquiterpene lactones.

Centaurea species	Sesquiterpene lactones	Antimicrobial activity assay/Results	Microorganisms used
C. achaia Boiss. & Heldr	salonitenolide, onopordopicrin, 8α-O-[4,5-dihydroxytigloyloxy]-salonitenolide, 8α-O-(4-hydroxy-3-methyl-buten-2-oyloxyl)-salonitenolide, 8α-O-(methacryloyl)-salonitenolide, 8α-O-(4-hydroxy-3-methyl-butanoyloxy)-salonitenolide, llβ,13-dihydrosalonitenolide, 8α-O-(4-hydroxy-3-methylene-butanoyloxyl)-llß,13-dihydrosalonitenolide, 8α-O-(4,5-dihydroxytigloyloxy)-llβ,13-dehydromelitensin, 8α-O-(4-hydroxymethylacryloyl)-llβ,13-dehydromelitensin and its elemane derivative, elemacarmanin 4-*epi*-carmanin 8α-O-(4-hydroxymethylacryloyl)-sonchucarpolide, 8α-O-(4-hydroxymethylacryloyl)-4-*epi*- sonchucarpolide	Microdilution method MIC 0.1–6.00 nmoL/mL MFC 0.75–12.00 nmoL/mL	*A. alternata, A. flavus, A. niger, A. ochraceus, A. versicolor, C. cladosporoides, P. funiculosum, P. ochrachloron, T. viride*

Continued

TABLE 4.1 Antifungal activity of some sesquiterpene lactones—cont'd

Centaurea species	Sesquiterpene lactones	Antimicrobial activity assay/Results	Microorganisms used
C. attica Nyman ssp. *Attica*	cnicin, 4′-acetylcnicin, 8α-O-[(4-acetoxy-5-hydroxy)-angelate]salonitenolide, 8α-O-(3,4-dihydroxy-2-methylene-butanoyloxy)- llβ,13-dehydromelitensin, malacitanolide, 8α-O-(3-hydroxy-4-acetoxy-2-methylene-butanoyloxy)-llβ,13-dehydromelitensin, methyl 8α-O-(3,4-dihydroxy-2-methylene-butanoyloxy)-6α,15-dihydroxyelema-1,3,11 (13)-trien-12-oate, 8α-O-hydroxy-4-*epi*-sonchucarpolide, 4′-acetoxymalacitanolide, Atticin	Microdilution method MIC 0.03–2.00 μg/mL MFC 0.125–4.00 μg/mL	*Alternaria alternata, Aspergillus flavus, A. niger, A.ochraceus, A. versicolor, Cladosporium cladosporoides, Penicillium funiculosum, P. ochrachloron, Trichoderma viride*
C. deusta Ten. [= *C. alba* subsp. *deusta* (Ten.) Nyman	8α-O-(4-acetoxy-3-hydroxy-2-methylene-butanoyloxy, 8α-O-(3,4-dihydroxy-2-methylene-butanoyloxy)-6α,15-dihydroxy-elema-1,3,11(13)-trien-12-oate, 8α-O-(4-acetyl-2-hydroxymethyl-buten-2-oyloxy)-salonitenolide, 8α-O-(4-acetoxy-3-hydroxy-2-methylene-butanoyloxy)-sonchucarpolide	Microdilution method MIC 0.06–5.00 nmoL/mL MFC 0.06–15.00 nmoL/mL	*A. alternata, A. flavus, A. niger, A. ochraceus, A. versicolor, C. cladosporoides, P. funiculosum, P. ochrachloron, T. viride*
C. nicolai Bald.	kandavanolide, salograviolide A, 9-*O*-acetyl- salograviolide A 3-*O*-deacetyl-9-*O*-acetyl-salograviolide A, salograviolide B	Microdilution method MIC 0.78–25.0 μg/mL	*A. niger, A. ochraceus, P. ochrochloron, C. cladosporoides, Fusarium tricinctum, Phomopsis helianthi*

C. pullata L.	8α-O-(4-hydroxy-2-methylenebutanoyloxy)melitensine, melitensin, 11β, 13 dihydrosalonitenolide, 8α-O-hydroxy-11β,13-dihydro-4-*epi*-sonchucarpolide, 8α-O-hydroxy-11β,13-dihydro-onopordaldehyde, 11β,13-dihydrocnicin, 11β,13-dihydro-19-desoxycnicin, 8α-O-(4-acetoxy-5-hydroxyangeloyl)- 11β,13-dihydrocnicin, 8α-O-(4-hydroxy-2- methylenebutanoyloxy)-11β,13-dihydrosonchucarpolide, 8α-O-(4-hydroxy-2-methylenebutanoyloxy)- 11β,13-dihydro-4-*epi*-sonchucarpolide	Microdilution method MIC 0.1–0.4 μg/mL MFC 0.2–0.5 μg/mL	*A. alternata*, *A. flavus*, *A. niger*, *A. ochraceus*, *Fusarium tricinctum*, *P. funiculosum*, *P. ochrachloron*, *T. viride*
C. raphanina Sm. ssp. *mixta* (DC)	Cnicin	Microdilution method MIC 0.06–0.25 μg/mL MFC 0.25–0.5 μg/mL	*A. alternata*, *A. flavus*, *A. niger*, *A.ochraceus*, *A. versicolor*, *C. cladosporoides*, *P. funiculosum*, *P. ochrachloron*, *T. viride*
C. solstitialis L. ssp. *solstitialis*	Centaurepensin = chlorohyssopifolin A, Chlorojanerin, 13-acetyl solstitialin A	Microdilution method MIC 16–256 μg/mL	*C. albicans*, *C. parapsilosis*
C. thessala Hausskn. spp. *drakiensis* (Freyn and Sint.) Georg	cnicin, 4′-acetylcnicin, 8α-O-[(4′-acetoxy-5′-hydroxy)-angelate]salonitenolide, 8α-O-(3′,4′-dihydroxy-2-methylene-butanoyloxy)- dehydromelitensin, 8α-O-(3′-hydroxy-4′-acetoxy-2′-methylene- butanoyloxy)-dehydromelitensin, malacitenolide, 8α-O-hydroxy-4-*epi*-sonchucarpolide	Microdilution method MIC 0.03–1.00 μg/mL MFC 0.125–1.00 μg/mL	*A. alternata*, *A. flavus*, *A. niger*, *A.ochraceus*, *A. versicolor*, *C. cladosporoides*, *P. funiculosum*, *P. ochrachloron*, *T. viride*

Source: Sokovic M., Ciric A., Glamocija J., Skaltsa H., 2017. Biological activities of sesquiterpene lactones isolated from the genus Centaurea L. (Asteraceae). Cur. Pharm. Des. 23 (19), 2767–2786.

sensitive), followed by *Penicillium* species and *Trichoderma viride*. Chemical analysis of *C. raphanina* ssp. *mixta*, an edible species, justified the presence in traces of only one sesquiterpene lactone, namely cnicin, while the other isolated compounds were found to be flavonoids and lignans (Panagouleas et al., 2003). As far as the isolated compounds are concerned, only cnicin exhibited antifungal activity better than the control drugs.

Cnicin is the most frequently isolated germacranolide and can be found in significant amounts in several *Centaurea* species (Bissett, 1994; Panagouleas et al., 2003; Tešević et al., 2014). Historically, cnicin was isolated for the first time in 1837, mainly from *Cnicus benedictus* (blessed thistle) and can be found in eighty-three *Centaurea* species (Bruno et al., 2013). It is one of the first evaluated compounds that were found to exhibit antibacterial and antifungal activities (Skaltsa et al., 2000a; Cavallito and Bailey, 1949; Kelsey and Locken, 1987).

Six *Centaurea* species from Morocco and Spain were subjected to chemical analysis. Several sesquiterpene lactones were detected, isolated, and characterized (Barrero et al., 2000), namely: cnicin, salonitenolide (**16**), costunolide (**17**), lychnopholide (**27**), and eremantholide C (**28**). Subsequently, they were evaluated for their antifungal activity against micromycete *Cunninghamella echinulata*.

Although microdilution test is the most common available screening method for fast and rapid determination of antimicrobial activity of natural products, in some cases, other methods, such as bioautographic TLC, were used to test the biological activity of isolated SLs as well. Zuccarinin (**26**) and eudesmanolide exhibit the best activity among tested sesquiterpene lactones isolated from *Centaurea* spp. from Greece (Skaltsa et al., 2000a) (Table 4.1), while presented results are in agreement with previously reported data (Djeddi et al., 2007, 2008; Koukoulitsa et al., 2002, 2005; Saroglou et al., 2005; Skaltsa et al., 1999, 2000a,b), proving that this series of compounds possess the strongest antifungal potency.

The strong antifungal activity of studied SLs led, as it was expected, to further tests, such as investigation of the pharmacokinetic profile of twenty-one compounds of this class, which were isolated from the following *Centaurea* taxa, growing wild in Greece: *C. achaia* (Skaltsa et al., 2000a), *C. attica* ssp. *attica* and *C. thessala* ssp. *drakiensis* (Skaltsa et al., 2000b) and *C. deusta* ssp. *deusta* (Skaltsa et al., 1999), all containing α-methylene-γ-lactone moiety (Table 4.1).

The in silico study was performed by using the VolSurf computational procedure, which was designed to provide descriptors linked to pharmacokinetic properties, starting from 3D molecular maps (Koukoulitsa et al., 2005). All compounds were projected on the following precalculated ADME models: Plasma protein affinity, blood-brain barrier permeation, Caco-2 cell permeability, and thermodynamic solubility.

In silico study for the selected compounds presented a nonoptimal pharmacokinetic profile. Low solubility and limited permeability could suggest that in most of the cases topical application might be preferable. Low brain blood barrier (BBB) penetration in BBB suggests a corresponding low level of toxicity for central nervous system (CNS). Tested compounds are expected to show diverse pharmacokinetic behavior compared to miconazole. Moreover, this commercial drug showed higher affinity for the plasma proteins, higher Caco-2 cell permeability and higher BBB permeability than the tested compounds. These SLs were predicted to achieve a low BBB penetration potential and, as expected, lower toxicity. Among tested lactones, the most interesting was found to be 8α-O-[(4′-acetoxy-5′-hydroxy)-angelate]-salonitenolide (**18**), which exhibited the best activity, with lower minimal inhibitory concentrations (MICs) on six of the nine tested fungi (*Aspergillus niger*, *A. versicolor*, *A. ochraceus*, *Penicillium ochrochloron*, *P. funiculosum*, and *Trichoderma viride*) and best minimal fungicidal concentrations (MFCs) for two fungal species (*Aspergillus niger*, *A. ochraceus*) with low Caco-2 permeability predicted. Accordingly, the optimization of the pharmacokinetic profile of selected molecules as far as absorption, distribution, metabolism, and excretion (ADME) are concerned, was screened in silico by using a semisynthetic compounds virtual library. Prediction revealed that many derivatives would achieve higher Caco-2 cell permeability, preserving the pharmacokinetic behavior as far as BBB penetration, solubility, and protein binding are concerned. Several of the sesquiterpene lactones were presented as good candidates, according to their antifungal potential, for further synthetic work and to be subjects for their pharmacokinetic properties experimental evaluation (Koukoulitsa et al., 2005). Moreover, sesquiterpene lactones that contain an α-methyl-γ-lactone moiety (**10–20**), (**8–23**) and were isolated from *C. pullata* L. have also been studied by using precalculated models. According to prediction results, tested compounds would not be able to penetrate the intestinal epithelium, they would have low affinity to the plasma protein and finally would not be able to pass BBB. They acquire medium-to-low solubility in water, as it is stated in previous results on prediction of the pharmacokinetic profile and antifungal potential of twenty-one sesquiterpene lactones possessing α-methylene-γ-lactone group. Concluding, it would be interesting to point out that the pharmacokinetics does not depend on the presence of an exocyclic double bond conjugated to the γ-lactone. This fact could be useful for the design of semisynthetic antifungal agents. Furthermore, one of the sites of particular interest could be the position of the esterification at C-8. Accordingly, introducing a modification on the side chain that contains the conjugated carbonyl could lead to pharmacokinetic properties optimization.

In all tested cases concerning sesquiterpene lactones, it should be noted that lipophilicity seems to be of great importance in the observed antifungal activity. It is widely known that chemical composition of fungal cell walls can be described as highly lipophilic. Consequently, they can act as biological barriers against penetration and transport of hydrophilic molecules through the outer lipid layer. Moreover, studies have shown that a general inverse relationship exists between polarity and antifungal activity for sesquiterpene lactones

(Skaltsa et al., 2000a). Accordingly, their polarity decreases following the order eudesmanolides > elemanolides > germacranolides.

It would be interesting to point out that in all cases, despite the presence of the exocyclic double bond in the lactone ring, the investigated compounds exhibited higher antifungal activity than reference drug miconazole. This could be because of differences in scaffold types and functional groups of the molecules. Specifically the high potency of germacranolides can be related to their lipophilic character that is important for sesquiterpenoids in order to pass through the cell wall of fungi. Structurally, the cell wall of fungi is very lipophilic and consequently acts as a significant barrier for the permeability of hydrophilic compounds. Taking these facts into consideration, there is a hypothesis for an inverse correlation between polarity and antifungal activity for sesquiterpene lactones. Therefore, it seems that a relatively low polarity is also responsible for high antifungal activity (Inoue et al., 1995; Barrero et al., 2000).

If we take a look on the obtained MICs and MFCs, germacranolides are clearly the most active molecules among the tested compounds. The biological activities of lactones are related in chemical terms with the existence of α,β-unsaturated carbonyl moieties such as α,β-unsaturated cyclopentenone, α-methylene γ-lactone, or a conjugated ester. These chemical groups can react through a Michael-type addition with nucleophiles and especially cysteine sulfhydryl group (Picman and Schneider, 1993; Kupchan et al., 1971; Sokovic et al., 2017).

The diversity in antimicrobial activity among SLs could be explained by various structural factors (Heilmann et al., 2001; Karabunarliev et al., 1996; Kupchan et al., 1970). Lipophilicity, chemical environment, and molecular geometry, or the target sulfhydryl group can also contribute to the observed biological activity (Heilmann et al., 2001; Kupchan et al., 1971). In particular, besides the lipophilicity, molecular accessibility play an important role in antifungal potential of SLs (Robles et al., 1995), while differences in the morphology and physiology of the fungal species are also of high importance and should always be taken into consideration.

It is obvious that sesquiterpene lactones are very potent as antifungal agents and could be further used for preclinical testing toward human pathogenic fungi. According to literature, it seems that in order to obtain desirable antifungal potency, the presence of α-methylene-γ-lactone and α,β-unsaturated cyclopentenone ring, alongside with a generally low polarity of the designed compound, are key factors that need to be considered by researchers working in the field.

4.3 Natural compounds from sponges

In the course of ongoing research and screening of antimicrobial activity of natural products, one of the invertebrate organisms, the freshwater sponge *Ochridaspongia rotunda,* was investigated (Pejin et al. 2014a). From the estimated 15,000 sponges, about 150 species live in freshwater habitats, lakes, and rivers, while the rest populate marine environments (Pejin et al., 2014b). The

FIG. 4.47 Freshwater sponage, *Hyalinella punctate*, collected from Sava river, Belgrade, Serbia.

diversity of secondary metabolites present in fresh water sponges is low, with lipid molecules being notable exceptions. More than 100 novel and rare fatty acids, unusual ones, especially phosphonolipids and sterols, are isolated from the freshwater sponges (Fig. 4.47) (Dembitsky et al. 2003).

The search for novel and valuable compounds of pharmaceutical interest in sponges started in the early 1950s, when some nucleosides, spongothymidine, and spongouridine, from marine sponge *Cryptotethia crypta*, were detected. That was the basis for the synthesis of the new **compound (Ara-C**, Fig. 4.48) that became the first marine-derived anticancer drug (Pejin et al., 2014a). The majority of the investigations are focused, so far, on marine representatives and just a few papers were conducted on freshwater sponges. Compounds isolated from sponges showed antitumor, anti-inflammatory, antiviral, immunosuppressive or neurosuppressive, antimalarial, antibiotic, and antifouling activities (Sipkema et al., 2005).

The sponge *Ochridaspongia rotunda*, endemic species of the oldest lake in Europe, Lake Ohrid, was tested for its potential antifungal potential. In vitro

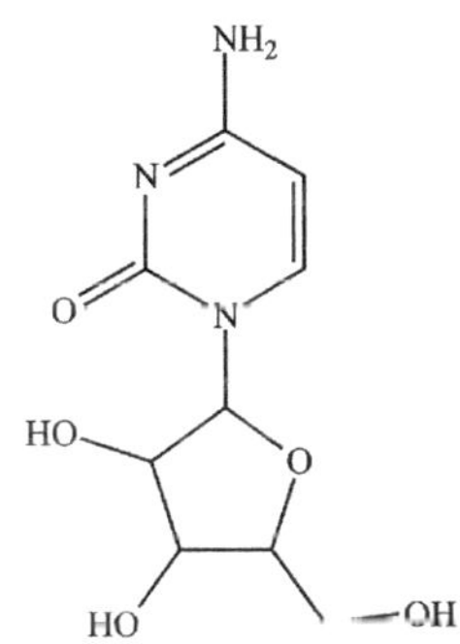

Compound (Ara-C) or cytarabine

FIG. 4.48 Chemical structure of **compound (Ara-C)**.

antifungal activity of five extracts (aqueous, methanol, ethyl acetate, acetone, and methylene chloride) against eight fungal species was tested using a microdilution method (CLSI, 2009; Espinel-Ingroff, 2001; Tsukatani et al., 2012).

All the tested extracts were found to be active against all fungal strains. The antifungal potency followed the order: acetone, methanol, ethyl acetate, methylene chloride, aqueous. The acetone extract possessed the highest activity (MICs of 7.5–45.0 mg/mL and MFCs of 15–60 mg/mL), while the aqueous extract showed the lowest antifungal effect (MICs of 15–120 mg/mL and MFCs of 30–240 mg/mL). All tested extracts revealed higher antifungal potential than the positive controls, bifonazole and ketoconazole. The most sensitive fungus was *Trichoderma viride*, while *Candida albicans* was the most resistant species concerning all tested extracts.

Chemical analysis of *O. rotunda* extracts (methanol and acetone, respectively) showed similarity in chemical composition. Alcohols, carbonyl compounds, sterols, and fatty acids can be found in both extracts. Antimicrobial activity of marine sponges has been tested for a long time and very intensively (Ravichandran et al. 2011), while there are no similar reports until now for their freshwater species. Testing of the sponge *O. rotunda* indicates that methanol and acetone extracts are promising sources of novel antifungal compounds of natural origin.

The increasing interest in testing compounds that derive from sponges can be attributed mainly to their unique organic compounds. A sesquiterpene hydroquinone, **avarol** (Fig. 4.49), isolated from the sponge *Dysidea avara* was evaluated for its antifungal potential. The structure of this compound was established by chemical degradation and stereocontrolled synthesis, characterized by 4,9-friedodrimane rearranged bicyclic sesquiterpene skeleton attached to a hydroquinone moiety. Avarol was found to be active against a variety of fungal species.

This marine-isolated compound demonstrated considerable potential against all fungi tested: the MIC and MFC values were at 0.004–0.015 mg/mL and 0.008–0.030 mg/mL, respectively. The most susceptible fungi were *A. versicolor* and *P. verrucosum* var. *cyclopium*, while avarol was more potent than both commercial amtimycotics, ketoconazole and bifonazole (Pejin et al., 2014a,b).

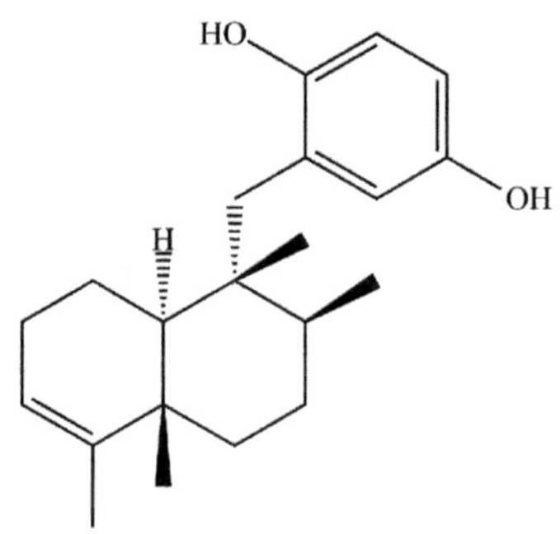

Avarol

FIG. 4.49 Chemical structure of **avarol**.

4.4 Other tested compounds

Triterpenoid saponins, with **hederagenin** or **oleanolic acid** (Fig. 4.50) as aglycone, were tested for their antifungal effects against *C. glabrata*, *C. albicans*, *Trichosporon beigelii*, *Penicillium avelaneum* UC-4376, *Pyricularia oryzae*, *Cryptococcus neoformans*, *Coccidioides immitis*, *Saccharomyces cerevisiae*, *Microsporum canis*, and *Trichophyton mentagrophytes*. All studied saponins showed great antifungal activity (Du et al., 2003).

The antifungal potential of **allicin** (Fig. 4.51) and similar organo-sulfur products derived by microwave-assisted transformation of allicin in ethanol was tested against eight fungi: *Aspergillus flavus*, *A. niger*, *A. fumigatus*, *Penicillium funiculosum*, *P. ochrochloron*, *Trichoderma viride*, *Candida albicans*, and *C. krusei*. Allicin exhibited powerful antifungal effect with MICs at 0.001 to 0.008 mg/mL and MFCs at 0.004 to 0.03 mg/mL. The main mechanism involved in the antimicrobial effect of allicin could be the inhibition of thiol-containing enzymes in fungi by the reaction with thiosulfinates and thiol groups. Furthermore, molecular accessibility and lipophilicity of these compounds are important factors concerning their antifungal potential (Yamada and Azuma, 1997). Allicin transformation products also revealed very strong antifungal effect, but lesser than the one that was demonstrated by allicin itself (Ilić et al., 2012).

The antifungal activity of garlic bulb powder, allicin, and the lozenge with 15% of garlic powder was evaluated using broth microdilution method against *C. albicans*. The tested garlic powders, as well as the lozenge, have

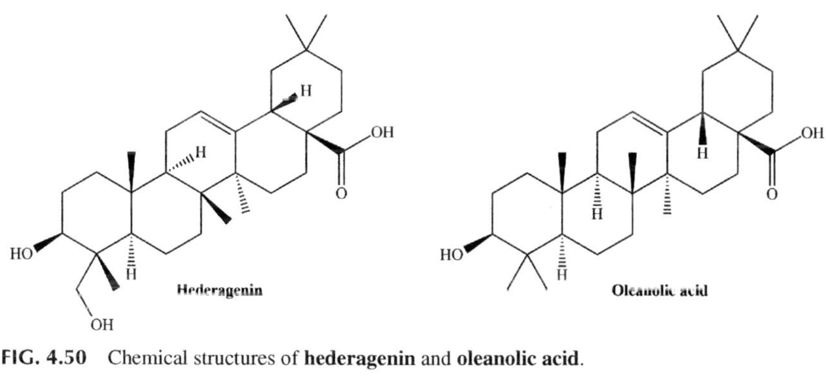

FIG. 4.50 Chemical structures of **hederagenin** and **oleanolic acid**.

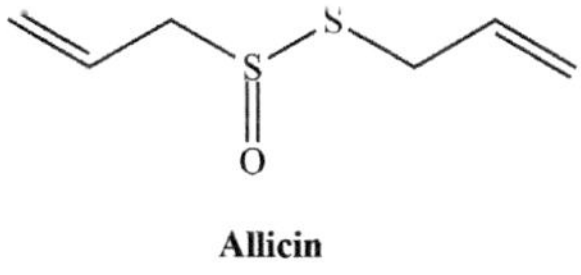

FIG. 4.51 Chemical structure of **allicin**.

shown activity with MICs of 1.25–7.50 mg/mL. The major compound, allicin, was highly active at a very low concentration with MIC = 0.4 μg/mL. It would be interesting to point out that these concentrations are lower than the ones of commercially available fungicides (Kundaković et al., 2011). These antifungal agents could possibly offer safer drug interaction profiles and good tolerance.

Insect secretions that are being tested for their antifungal effects

The antimicrobial activity of gland secretions isolated from adults from beetle species *Carabus ullrichii*, C. coriaceus, and Abax parallelepipedus was tested. Secretion of *C. ullrichii* possessed the highest antimicrobial potential toward all fungi. *Penicillium funiculosum* was the most sensitive fungal species, while *Penicillium* species were the most resistant micromycetes. The samples of C. coriaceus were active against *Aspergillus fumigatus*, *A. versicolor*, *A. ochraceus*, and *P. ochrochloron*. Pygidial gland secretion of *A. parallelepipedus* showed antifungal potential against five out of eight tested micromycetes (A. fumigatus, *A. versicolor*, *A. ochraceus*, *Trichoderma viride*, and P. verrucosum var. *cyclopium*) (Nenadić et al., 2016a).

Gland secretion of ground beetle Laemostenus (pristonychus) *punctatus* was tested for its antimicrobial effect. *Aspergillus fumigatus* was the most resistant fungus, while *Penicillium* species were the most sensitive micromycetes (Nenadić et al., 2016b).

Gland secretion isolated from *Calosoma sycophanta*, carabid species, showed significant antifungal effect against eight strains of treated micromycetes (*Aspergillus*, *Penicillium*, *Trichoderma*, *Candida*). *Aspergillus versicolor* was the most sensitive fungus. Secretion exhibited much higher antifungal potential against all fungi compared to commercial agents ketoconazole and bifonazole (Nenadic et al., 2017a).

The antifungal activity of gland secretion from *Laemostenus punctatus* has been investigated. Fungal species of genera *Aspergillus*, *Penicillium*, *Alternaria*, *Cladosporium*, *Rhizopus*, *Trichoderma*, *Arthrinium*, *Aureobasidium*, *Epicoccum*, *Talaromyces*, and *Fusarium* were tested for their sensitivity on isolated secretion. The most sensitive species was *Penicillium* sp. This results gave opportunities for research of insects secretions against pathogens, with possible application in medicine (Nenadic et al., 2017b).

4.5 Morphophysiological changes in fungi due to inhibition activity by natural compounds

The studied extracts of a series of plants were found to be potent inhibitors of the growth of micromycetes (1–20 μg/mL). On the other hand, all tested extracts that derived from mosses demonstrated effect against all used fungi, with the ethanolic extracts of silver moss exhibiting the strongest activity (MICs, MFCs of 0.29–0.52 μg/mL). All studied oils revealed antifungal capacity ranging from 0.1–1250 μL/mL (using different kinds of assays). The essential oil of cinnamon was found to be the most potent antifungal agent in concentration range of 0.1–1 μL/mL according to microdilution method results. Moreover, essential oils' constituents such as carvacrol, thymol, menthol exhibited very strong antifungal potential (0.02–300 μL/mL), with carvacrol showing the strongest activity (0.1–0.5 μg/mL). Among all tested and analyzed compounds, sesquiterpene lactones, etc., lactones from *Centaurea pullata* showed the highest antifungal potential (0.0001–0.0007 μmoL/mL). Consequently, comparing the results of all the studied agents (extracts, essential oils and components, pure compounds, etc.), it can be concluded that the sesquiterpene lactones were the most potent as antifungals, followed by essential oils and their constituents.

Many teams investigated plant pathogens, but only a small portion of them work on animal pathogens. Essential oils are the most studied natural products in the present studies concerning antifungal activity and should be noted that they demonstrated diverse effects on plant and human pathogens. In particular cases (Soković, 2001), essential oils in general were found to possess stronger antifungal activity against plant pathogen species (*Phomopsis helianthi*, *Cladospoirium cladosporioides*, *Alternaria alternata*, *Fusarium* species) compared to human ones (*Trichophyton* species, *Microsporum cannis*, *Epidermophyton flocosum*) (Fig. 4.52).

The results revealed different ranges of Minimal Fungicidal Concentrations (MFCs), with human pathogens being more resistant than the ones of plant origin: *Mentha spicata* 1–2 μL/mL for plant pathogens (pps) and 2 μL/mL for human pathogens (hps), *M. piperita* 1.5–2.5 μL/mL pps and 2.5 μL/mL for hps, Thyme 0.125–0.25 μL/mL pps and 0.25 μL/mL for hps., *Salvia* species for pps 2–15 μL/mL and 3–20 μL/mL for hps, *Lavandula* species 0.5–9 μL/mL for pps and 0.5–10 μL/mL hps, *Citrus aurantium* 10 μL/mL for pps and 7–15 μL/mL for hps. A portion of the oils exhibited the same antifungal capacity against

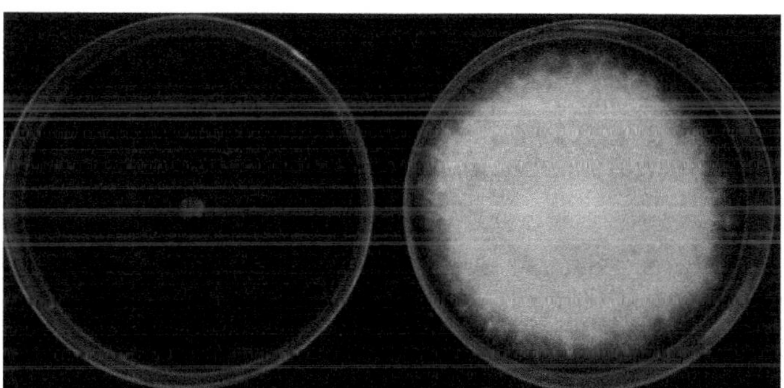

FIG. 4.52 Antifungal activity of the essential oil from *Thymus vulgaris* against *Trichophyton mentagrophytes (left)* vs. a control colony *(right)*.

both pps and hps: *Melissa officinalis* (5–6 μL/mL), *Rosmarinus officinalis* (6–8 μL/mL), *Citrus limon* (7–10 μL/mL), while *Matricaria chamommilla* demonstrated stronger antifungal activity against hps (6–9 μL/mL) than for pps (7–9 μL/mL) and Sweet basil (4–5 μL/mL for pps) and 3–5 μL/mL for hps. In some other studies, similar results were obtained and it was established that dermatomycetes are the most resistant group, although numerous essential oils can be very potent against them. Among twenty-two essential oil samples from eleven species of *Cinnamomum*, the one from *C. suvabenium* was the most potent against *Microsporum canis*, *Trichophyton mentagrophytes*, and *T. rubrum* alongside with some *Candida* strains (*C. albicans* and *C. glabrata*) (Kalemba and Kunicka, 2003). It can be concluded that different essential oils can exhibit diverse effectiveness. Moreover, the mechanism of action of essential oils is not always the same against different fungal species, depending on the type of each fungus and being related mainly to their cell wall structure and the outer membrane arrangement.

Series of different essential oil constituents exhibited very strong antifungal activity against several microfungi. In particular, eugenol demonstrated very high antifungal potential against *Aspergillus nidulans* and *A. niger*, while caryophyllene inhibited *A. glauca*. Cineole, on the other hand, demonstrated favorable inhibitory potential against *Alternaria alternata* and good-to-moderate effect against the remaining studied fungi. The essential oil of *Foeniculum vulgare* and its ingredients (**anethole**, **fenchone**, and **camphor**, Figs. 4.10, 4.12, and 4.53) were assayed in order to elucidate their antifungal activity against nine plant strains. Anethole demonstrated the best activity (1.3–2.8 μL/mL), followed by fenchone (3.7–6.0 μL/mL), while camphor possessed the lowest antifungal potential (2.8–9.7 μL/mL) (Mimica-Dukić et al., 2003).

As far as the essential oil constituents were concerned (**β-caryophyllene**, **β-caryophyllene oxide**, **α-pinene**, **δ-cadinene**, and **linalool**, Figs. 4.11, 4.21, and 4.54) from eight *Stachys* species, were evaluated for their antifungal activity. From the obtained results it is clear that linalool revealed the strongest effect (0.03 mg/mL), while the lowest activity was obtained by β-caryophyllene (0.3 mg/mL) (Skaltsa et al., 2003).

The antifungal activity of limonene against fourteen fungi was evaluated using microdilution and agar diffusion assays. According to the results, this compound demonstrated inhibitory effect against all tested strains, in concentrations of 8.0–13.0 μL/mL regarding diffusion method and 6.0–10.0 μL/mL in

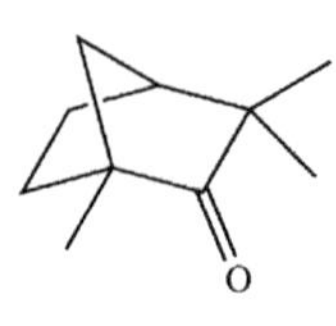

FIG. 4.53 Chemical structure of **fenchone**.

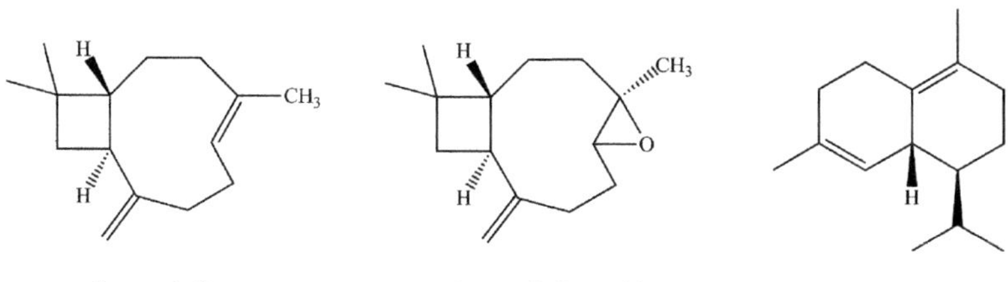

FIG. 4.54 Chemical structures of **β-caryophyllene**, **β-caryophyllene oxide,** and **δ-cadinene**.

microdilution assay, respectively. The observed diversities in MICs as far as these two methods are concerned could be attributed to the low solubility of limonene in the agar medium, which was used in the diffusion assay. Limonene exhibited higher antifungal capacity than that of the commercial drug bifonazole, especially in the microdilution method, when the component was dissolved in Tween (Rančić et al., 2003). Phenolic molecules were found to be the most potent among the studied compounds. Limonene and linalyl acetate were less active against micromycetes. The essential oils and their constituents were investigated against twelve fungi: *Aspergillus niger*, *A. ochraceus*, *A. versicolor*, *A. flavus*, *A. terreus*, *Alternaria alternata*, *Penicillium ochrochloron*, *P. funiculosum*, *Cladosporium cladosporioides*, *Trichoderma viride*, *Fusarium tricinctum*, and *Phomopsis helianthi* (Soković, 2001; Soković et al., 2009a,b,c) (Fig. 4.55). The ingredients linalyl acetate, linalool, limonene, α-pinene, β-pinene, 1,8-cineole, camphor, carvacrol, thymol, and menthol were tested for their antifungal activity against all the mentioned fungi and also *Verticillium* and *Trichoderma* spp. The strongest and broadest potential was demonstrated by carvacrol and thymol with very low MICs and MFCs (0.02–1.5 μL/mL), while linalyl acetate and limonene exhibited the lowest capacity (MICs/MFCs of 5.0–11.0 μL/mL) (Soković and van Griensven, 2006).

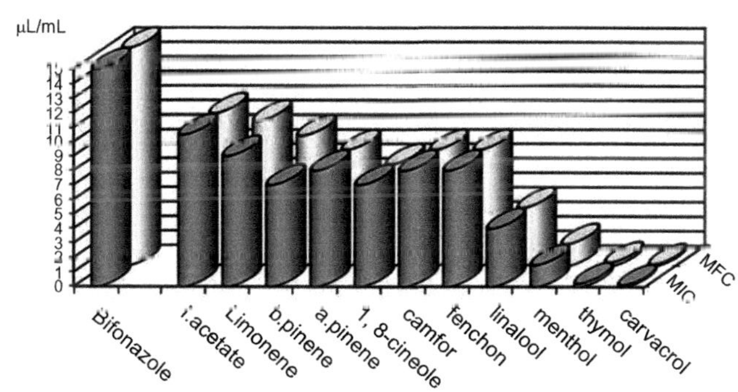

FIG. 4.55 Antifungal activity of essential oils components (MIC and MFC).

Carvacrol, eugenol, and thymol demonstrated antifungal activity, with carvacrol obtaining the strongest effect against *C. albicans* biofilms. It is interesting to point out that carvacrol interfered in the disruption of both Ca^{2+} and H^{+} homeostasis in yeast, leading to disarrangements and loss of cell viability.

On the other hand, thymol demonstrated synergistic effect with azoles, such as fluconazole, that could be attributed to either disruption of the cell wall/membrane integrity or to the creation of lesions in the plasma membrane in addition to the inhibition of ergosterol biosynthetic pathway by fluconazole. Thus, it can be concluded that phenolic compounds can not only act alone, but they are also able to enhance the antifungal effects of widely used agents (Guo et al., 2009; Faria et al., 2012).

From the previously mentioned studies is clear that the growth of the evaluated fungi responded in different ways to the essential oils and their constituents, indicating that different components might have different modes of action or that the metabolism of some fungal species have the ability to overcome the activity of the oil or adapt to it. Terpenic compounds are able to inhibit electron transport, proton translocation, phosphorylation steps, and other enzyme-dependent reactions or even act directly on the fungal cell membrane. Subsequently, their antifungal capacity depends on the chemical composition of the cell wall, as well as on the chemical structure of the terpenoids.

Terpenic hydrocarbons are not soluble in water, exhibiting in the same time poor activity, while among the water soluble molecules, **vanillin**, **piperonal** (Fig. 4.56), and **camphor** (Fig. 4.10), were not particularly potent, whereas the nonaromatic ester **borneol acetate** (Fig. 4.56) was found to possess antiseptic properties. Moreover, aliphatic alcohols, such as linalool or citronellol and ketones like **piperitone** (Fig. 4.56) or carvone, demonstrated antifungal activity.

Phenolic oil constituents revealed very strong antifungal capacity despite their relatively low water solubility (Knobloch et al., 1988; Soković et al., 2006, 2007, 2010, 2013). Phenols were the most potent terpenoids, followed by aldehydes and ketones, alcohols and hydrocarbons. It should be mentioned that thymol and carvacrol were the most active compounds, achieving complete inhibition of oxidative phosphorylation. The ability of terpenoids to inhibit the reactions described here is related to their lipophilic properties that enhance their

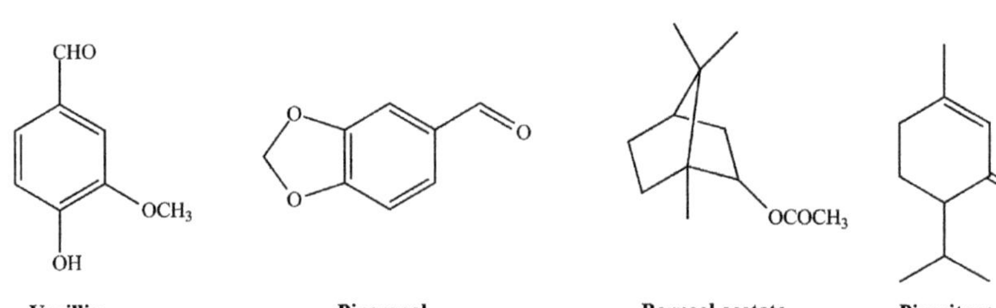

FIG. 4.56 Chemical structures of **vanillin**, **piperonal**, **borneol acetate**, and **piperitone**.

ability to dissolve in the cell membrane, as well as to their functional groups, which can interfere with enzymes (Griffin et al., 1999; Knobloch et al., 1988; Soković, 2001). From studies regarding the antimicrobial activity of essential oils and their ingredients it was concluded that terpene acetates and hydrocarbons tended to be relatively inactive, no matter their chemical structure. Their inactivity seems to be closely connected to their low hydrogen bonds-forming ability and water solubility in general. Ketones, aldehydes, and alcohols possessed antifungal effect but with diverse specificities that were not always related to their functional groups but rather associated with hydrogen-bonding properties (Griffin et al., 1999). Obtained results about the antifungal capacity of several different essential oils and their constituents indicate diverse potential. Moreover, the modes of action of essential oils can differ among tested fungal strains. It should be mentioned that the strong antifungal potential of some oils (*Mentha* species, Thyme, Oregano) can be explained by the high percentage of their active ingredients such as menthol, thymol, and carvacrol. For the rest of the essential oils, there is no significant connection between antifungal activity and the quantity of their main components and thus, the compounds that are present in larger amounts are not necessarily more responsible for the observed effect. The diverse antifungal potential that is exerted by an essential oil, in comparison with the activity of its main constituents can be justified by either the synergistic effect of the different components of the oil and/or by the existence of other ingredients that can potentially be active, even in small concentrations (Soković and van Griensven, 2006).

To evaluate further the issue of the lack of universal criteria, more studies regarding the antifungal potential of essential oils should be taken into account. Janssen et al. (1983) studied the characteristics of complex mixtures alongside with the methods that were used for their evaluation and came to the conclusion that a high percentage of the results are difficult to be compared due to the wide diversity of the testing methods. It was suggested that in future studies, the strain number of the evaluated fungi, the composition of the essential oil, and the specific conditions under which this was obtained, should be included as an integral part of the report. Kalemba and Kunicka (2003) reviewed the classical assays that are usually used to study the antibacterial and antifungal capacity of essential oils, such as the agar diffusion method (paper disc and well), the dilution method (agar and liquid broth), and turbidimetric and impedimetric monitoring of the growth of microorganisms when these oils are applied. Except from coming into conclusions about factors that can influence the in vitro antifungal activity of essential oils and their modes of action, they presented an overview of the susceptibility of human and foodborne fungi against several essential oils and their components. The most potent ones, which included oils from Thyme, Oregano, Mint, Cinnamon, Sage and Clove, have antimicrobial properties. Some of the other criteria that were pointed out were the evaluation of the plants that were used for their antifungal properties, as well as the study of the use of spices as antifungals. The criteria that were mentioned seem to be sufficient to justify the studies, but Rios and Recio (2005) suggested that research efforts should be focused on establishing definitive knowledge facts about each plant and its properties. Taking into consideration that the chemical diversity of natural products is increasingly accepted and also suited to provide core scaffolds for novel antifungals, there will arise further developments in the use of new products of natural origin and chemical libraries that will be focused on natural products (Harvey, 2008). The techniques that were employed should also be taken into deep consideration. As far as nonpolar extracts are concerned, the use of diffusion methods is probably not sufficient, although there are numerous published reports referring to the use of these techniques. Solid dilution assays can be sufficiently used for examining plant extracts or nonpolar molecules. It should be noted that only when the amount of the sample is small, then the use of diffusion methods are possibly more suitable (Rios and Recio, 2005).

Comparative studies of the obtained results from different kinds of assays revealed that MICs of essential oils and their constituents are in general higher in disc-diffusion and macrodilution techniques compared to microdilution assay. Subsequently, the lower potential concerning the two first methods can be attributed to the low water solubility of the oil and its ingredients that limits diffusion through the agar medium. It should be pointed out that only the more water-soluble molecules, such as 1,8-cineole, are able to be diffused into the agar. Consequently, the hydrocarbon compounds either remain on the surface of the medium or evaporate. This could justify the better results that are obtained by using the microdilution assay. Moreover, the essential oils and their constituents demonstrated greater antifungal activity when diluted in Tween 80. Both MICs and MFCs were found to be lower in microdilution compared to diffusion method, especially when Tween 80 was used. Nonionic emulsifiers, such as Tween 20 or 80, are relatively inactive when studied alone and have been widely used as emulsifying agents (Soković, 2001; Soković et al., 2002; Soković and van Griensven, 2006).

It was observed that some oils and their constituents not only exerted fungicidal effect, but also inhibited effectively sporulation of a series of different fungal strains (Glamočlija et al., 2006; Soković and van Griensven, 2006). The antifungal activity was not only revealed in solution or by contact, but even with a vapor treatment, enhancing inhibition of fungal growth by a smaller amount of essential oil while also inhibiting effectively sporulation. It should be noted that in this case vapor concentration and exposure duration are crucial. Subsequently, the gaseous contact effect was exhibited primarily at the maximum vapor concentration at an early incubation stage. Nevertheless, maintaining a high vapor concentration for long periods of time seemed to be unnecessary. Moreover, essential oil vapors could be applied to control proliferation of molds that are now dealt with a series of other sanitizing agents. It should be mentioned that essential oils and their constituents establish high vapor pressures and are considered to be relatively volatile. Therefore, solutions and emulsions that are applied in the form of sprays, with or without a carrier, represent the preferable formulation in which this type of agents should be used when large areas are targeted on casing soil surface. Evaporation by heating should also be taken into account as a possibility. One of the advantages of the volatility of essential oils is that no or only little residue will be ultimately left on the product after the treatment (Glamočlija et al., 2006).

Scientists suggest as optimal the microdilution method, carried out in microtiter trays, which involves a larger number of samples and the use of small volumes of test substance and growth medium (Soković, 2001; Soković et al., 2006, 2009a,b,c). Nevertheless, it should be noted for the study of some plants, fungal extracts, separation of fractions and compounds of high purity, bioautographic methods on TLC plates were the choice of preference. This is a qualitative assay that can only provide a general aspect of the presence or absence of compounds (in very small amounts), which possess antifungal activity. This technique was found to be handy for screening plants and fungal strains concerning antifungal potential, alongside with the bioassay-guided isolation of antifungal compounds of natural origin. Moreover, bioautography provides easy localization of activity, even for a matrix as complex as that of natural products. With this method, comparative chromatograms can be developed under identical conditions and are visualized with the use of appropriate chromogen reagents, providing informative data regarding the nature of the biologically active compounds (Fig. 4.6) (Rancic et al., 2006; Ćirić et al., 2012).

Due to significant regulations in legislation and increasing consumer trends concerning the use of natural alternatives to chemical fungicides (Brul and Coote, 1999), the potential therapeutic use of essential oils came to the surface as they are considered to be common plant derivatives. Among the natural antifungals that were evaluated in a series of studies, the essential oils of Oregano and Thyme and their components, carvacrol and thymol were found to be the most promising. It would be interesting to point out that adding a variety of plant-derived antifungal agents can improve both the spectrum of the total activity and the extent of inhibition due to their synergistic effects. Therefore, the combinatorial use of such compounds might have as a result even stronger antifungal effect.

Currently, the use of essential oils can be characterized as limited, having as possible cause their strong smell and taste when applied at doses that can achieve a desired effect (Skandamis and Nychas, 2000). Even knowing that most of the essential oils are classified as Generally Recognized As Safe (GRAS) (Kabara, 1991), their use in foods as preservatives is often limited due to flavor preferences and choices, when effective antimicrobial doses could exceed the organoleptically acceptable levels. Thus, there is a continuous demand for more precise reports regarding the minimum inhibitory (effective) concentrations (MICs) of essential oils in order to achieve a balance between sensory acceptability and antifungal efficacy (Lambert et al., 2001).

Taking into account previously presented in vitro and in vivo data, it is clear that a great amount of essential oils and their selected components do possess antifungal potential. Reviewing the in vivo antifungal activity of essential oils and their constituents, the composition of these oils and the proportion of each individual component could elucidate the diversities between their effects. For example, menthol and thymol exhibited stronger antifungal activity than the essential oils of *Mentha spicata* and *Thymus vulgaris*, respectively, which have these compounds as their main components. Therefore, (Davidson and Parish, 1989) it can be concluded that the observed effect can be attributed to the individual compounds of these oils. Despite the well-studied synergistic effects of the essential oils constituents that are widely reported in the literature, antagonistic effects between several oil ingredients can also take place. It would be interesting to point out that some oil constituents can diminish or even block each other's activities. Therefore, the efficacy of the evaluated oils and their constituents should be analyzed and interpreted with great attention. For instance, the essential oil of Sweet basil exhibited stronger in vivo antifungal activity than lemon and bitter orange oils. All the oils have main constituent linalool. When studied alone, linalool was found to be a good antifungal agent, although less potent than limonene. Lemon and bitter orange oils, on the other hand, demonstrated lower antifungal capacity. Limonene, although present in a high percentage in the oils, is suspected for the diminishing the potency of the entire oils (Davidson and Parish, 1989).

The essential oils of *S. officinalis* and *L. angustifolia* exhibited the strongest potential in the treatment of experimentally induced dermatomycoses. Comparing the results obtained from the study of the in vitro antifungal activity of the essential oils with these of the in vivo assays, it can be stated that the oil of *S. officinalis* and *L. angustifolia* demonstrated lower in vitro potential. Regarding the in vivo results, these two oils proved to be very potent, possessing better therapeutic activities than most of the other studied oils. Moreover, it is widely known that sage and lavender have traditionally been used for the treatment of several skin disorders and can be found in numerous cosmetic products for skin care (Bremnes, 1994). Knowing that the potency of an essential oil can be explained by interactions of the individual components, it should be pointed out that in the case of the lavender oil, this effect is attributed mainly to linalyl acetate and linalool (Lis-Balchin et al., 1998).

Some of the extracts and components could provoke demelanizing activity on different structure of microfungi, mainly spores. This effect was significant when extract of *Phlomis fruticosa,* but also some other compounds, was applied on *Aspergillus fumigatus*, *A. niger*, *Penicillium funiculosum*, and *P. verrucosum* var. *cyclopium* (Fig. 4.57) (Soković, 2001).

The tested extract provoked demelanizing effect of *A. fumigatus* and influenced lower amount of conidia and nude vesicle without conidia (Fig. 4.58B). Lower concentration of this extract reduced the number of spores and vesicles, while higher amount completely reduced spores and induced thinny hyphae (Fig. 4.58B). This effect is usually observed when extract or compound is applied in lower amount than MICs. This represents Minimum Demelanizing

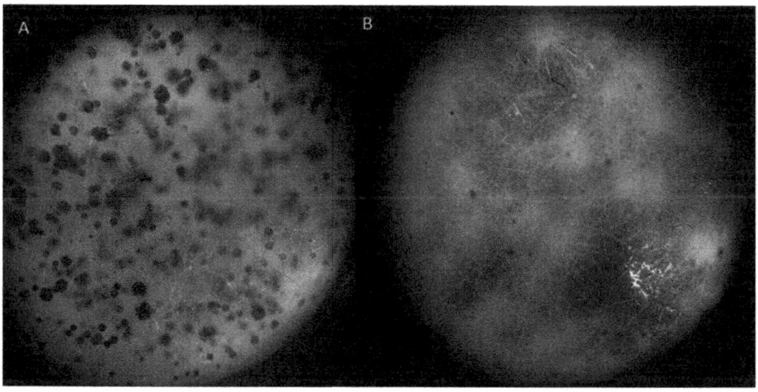

FIG. 4.57 Demelanizing activity of Phlomis fruticose extracts on Aspergillus niger (A-control, B-treated).

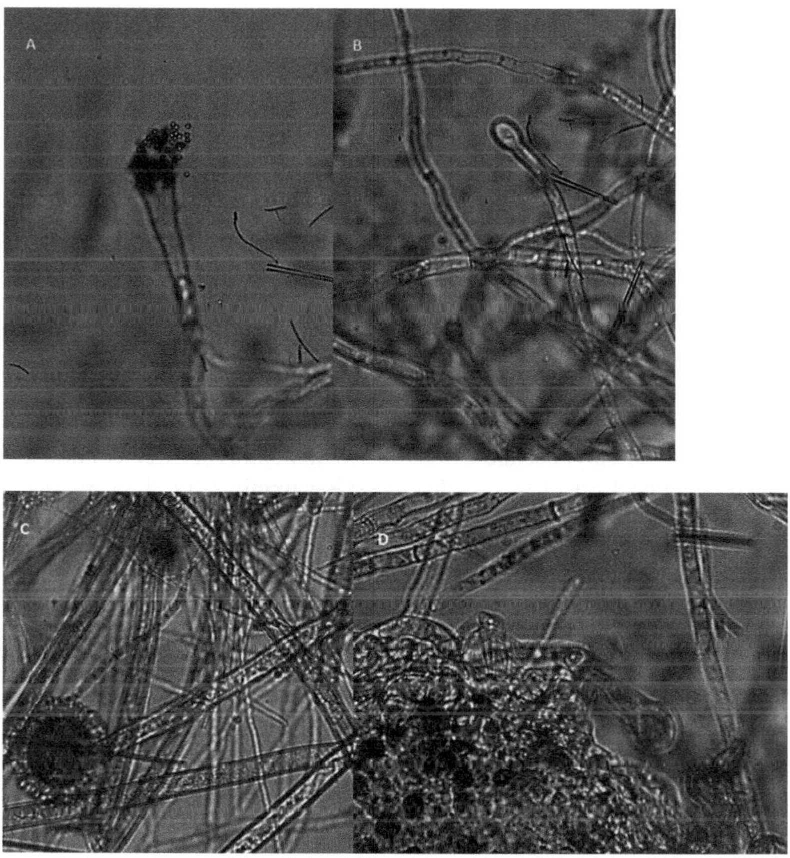

FIG. 4.58 Demelanizing activity of *Phlomis fruticosa* extract against fungi: (A) *A. fumigatus-control*, (B) *A. fumigatus-treated*, (C) *Aspergillus niger-control*, (D) *Aspergillus niger-treated.*

Concentrations (MDCs). Morphological changes and demelanization of *A. niger* are presented in Fig. 4.58C and D. These changes in *Aspergillus* and *Penicillium* species could be seen as depigmentation, morphological changes of conidiphores, seen as unusually small number of heads, different shape of heads and nude vesicles and phialides in comparison to those in untreated culture (Fig. 4.58A–D). *Aspergillus* and *Penicillium* species have colored conidiophores and contain pigments belonging to the group of melanins. This pigment could be green, chromoprotein, as in *A. flavus*, *A. fumigatus*, etc., or black insoluble pigment, which could be seen in *A. niger* (Eisenman and Casadevall, 2012). Melanin production by fungi is very important in their metabolism, contributing to the virulence of pathogens (Rosa et al., 2010). The same effect was observed in a few microfungi and in previous studies of antifungal activity of some mushroom extracts (*Morchella esculenta* and *Ganoderma lucidum*) (Heleno et al., 2013).

Changes provoked by natural compounds toward *Candida albicans*

C. albicans is a dimorphic fungus found as a commensal inside the human body. In the immune-deficient organisms it could show its pathogenic effect due to its dimorphic nature. Nonpathogenic form is presented in circular shape, yeast form, and when it is transforming into a pathogen, elongated structures, i.e., hyphal form, could cause infections. This form is better adapted and could easily penetrate the epithelium and provoke the immune response (Prasad et al., 2010). Since the dimorphic characteristic could be a key of pathogenicity and virulency of this pathogen inside the host, this could be a target in order to evade the pathogenic effect of this fungus.

Some new phenolic compounds isolated from *Baseonema acuminatum* showed antifungal activity against *C. albicans*. Different phenolic acid derivatives isolated from ethyl acetate extract of *Lycium chinense* (Lee et al., 2004) exhibited antifungal effect on *C. albicans*. **Curcurmin** (Fig. 4.59) produced by *Curcuma longa* showed antifungal potential against *C. albicans* as well as other *Candida* species, by increasing reactive oxygen species (ROS) levels and inducing early apoptosis (Sharma et al., 2010). This compound is also known to act on **farnesol** (Fig. 4.59), a quorum-sensing molecule that inhibits hyphal development by targeting the TUP1 gene involved in the hyphal development (Kebaara et al., 2008).

Bisbibenzyl are newly isolated phenolic compounds that inhibit the growth of *C. albicans* by inhibiting the morphogenetic transformation and inhibiting biofilm formation by regulation of DPP3 gene (Zhang et al., 2011).

The biological activity of apigenin-7-*O*-glucoside connected to its antifungal effect on *Candida* spp. and cytotoxic potential on colon cancer cells was evaluated and compared with that of apigenin. The evaluation of antifungal activity took place against fourteen different isolates of *Candida* spp. using membrane

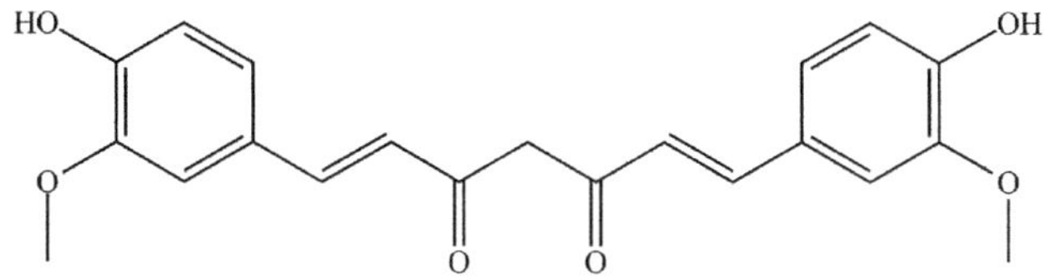

Curcurmin

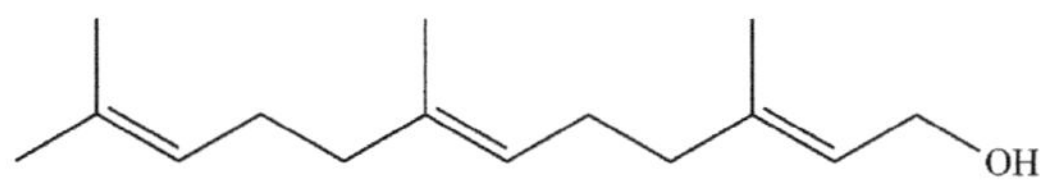

Farnesol

FIG. 4.59 Chemical structure of **farnesol**.

permeability assay, measuring inhibition of reactive oxidative species and inhibition of CYP51 of the strains. The cytotoxic effect of apigenin-7-*O*-glucoside was studied on colon cancer HCT116 cells by measuring cell viability, apoptosis rate, and apoptosis-gene expression that can be connected to colon cancer. From the results of the study, considerable antifungal activity of apigenin-7-*O*-glucoside was observed against all the tested *Candida* isolates. Breakdown of *C. albicans* plasma membrane took place after treatment with apigenin-7-*O*-glucoside in shorter period of time than with apigenin. Inhibition of intra- and extracellular reactive oxidative species was accomplished at minimum inhibitory concentrations for both tested compounds, supporting the hypothesis that reactive oxidative species inhibition could take part as an antifungal mechanism of action. Moreover, none of the evaluated compounds revealed binding affinity to *C. albicans* CYP51 protein. It should be mentioned that apigenin-7-*O*-glucoside was found to be more potent than apigenin regarding reduction of cell's viability and induction of cell death of HCT116 cells. When both compounds were applied simultaneously, it resulted in chromatin condensation, apoptotic bodies formation, and apoptotic genes expression in HCT116 cells, but the apigenin-7-*O*-glucoside required lower concentrations for exerting the same result. Apigenin-7-*O*-glucoside and apigenin revealed noteworthy antifungal activity and cytotoxic effect on HCT116 cells. Nevertheless, it was observed that apigenin-7-*O*-glucoside was in general more potent than apigenin in all performed assays (Smiljkovic et al., 2017).

C. albicans has the ability to generate significant amounts of ROS, which is related to its capability of invading host tissues by inducing oxidative damage in cells at MIC and sub-MIC concentrations of flavonoids, e.g., apigenin (Schröter et al., 2000). In particular, ROS react with polyunsaturated fatty acids of cellular membranes, sulfhydryl group in proteins and nucleotides, promoting that way tissue injury in fungal infections (Machlin and Bendich, 1987;

Nishikawa et al., 1997). Intracellular and extracellular ROS production by *C. albicans* was evaluated in the presence of AP7Glu, apigenin, and marketed drug ketoconazole. From the obtained results can be suggested that extracellular and intracellular ROS levels decreased in the presence of AP7Glu and apigenin. It would be interesting to point out that this effect was found to be dose dependent. Similar profile was observed for extracellular ROS, except that intracellular ROS was inhibited in a greater scale compared to extracellular ROS, at MIC values for both previously mentioned compounds. On the other hand, application of ketoconazole did not cause such changes concerning extracellular and intracellular ROS levels. AP7Glu and apigenin exhibited generally similar activity, with AP7Glu being a more potent ROS inhibitor, in lower concentrations (Smiljkovic et al., 2017).

It is widely known that CYP51 is an enzyme that is essential for *Candida albicans* as it is involved in ergosterol biosynthetic pathway and it is the main target of azole antifungal drugs. From the tested compounds none was found to bind to CYP51 of *C. albicans*. On the contrary, ketoconazole possesses a mechanism of action that involves CYP51 inhibition since it has the ability to bind at candidal enzyme, with Kd CaCYP51 $< 0.05\,\mu M$, demonstrating also affinity toward human enzyme, with Kd hCYP51 $< 0.05\,\mu M$. From these results it can be suggested that AP7Glu and apigenin possess a different mechanism of anticandidal activity compared to marketed antifungal drugs of the azoles group, taking also into account the previously mentioned nonselective binding affinities of ketoconazole (Smiljkovic et al., 2017).

Nonselective nature of ketoconazole CYP51 binding may be cause of some of the known ketoconazole side effects (Lee et al., 2014).

The biological activity of AP7Glu was studied and compared with that of apigenin. From the obtained results it is revealed that the sugar moiety in AP7Glu had a crucial effect on the biological potential of apigenin.

Study reports concerning anticandidal activity were in accordance with previously presented results (Mamadalieva et al., 2011) showing that AP7Glu is a more effective growth inhibitor of *Candida albicans* and *C. glabrata* compared to apigenin. On the other hand, both compounds demonstrated lower inhibitory activity than ketoconazole, but their fungicidal capacity can compete with ketoconazole due to drug's primarily fungistatic effect.

Regarding the effect of apigenin and AP7Glu on the cytoplasmic membrane of *C. albicans*, reported data are in agreement with recent literature where it is stated that flavonoids are localized either in the hydrophobic core of the lipid bilayer or at the membrane interface, leading to corresponding alterations in the membrane (Selvaraj et al., 2015). Commercial antifungal drug ketonazole was found to have no effect on the structure of fungal membrane when tested at same concentrations with previously mentioned studies. Moreover, the chemical structures of the flavonoids, alongside with their ability to alter the membrane, are two of the main factors that can influence the nature and extent of their antifungal activity (Selvaraj et al., 2015). Accordingly, it should be noted that AP7Glu caused more disturbances at the integrity of the membrane of *C. albicans* cells, which could be attributed to the difference in the chemical structures of the two mentioned compounds.

It is well established that ROS generation by *C. albicans* is connected to morphogenesis and their highest amounts can be found in hyphal form of fungal cells. Therefore, ROS generation play an important role in tissue invasion and infection (Sander et al., 2002).

From the obtained results it is suggested that the main mechanism of anticandidal action of AP7Glu and apigenin is closely connected to the disruption of fungal cell membrane. Reported data from in vitro assays regarding the inhibition of extra- and intracellular ROS in *C. albicans* cells are leading us to the assumption that both compounds might lower the host tissue invasion and consequently decrease the pathogenicity of the mentioned strain. In particular, AP7Glu exhibited in general stronger antifungal activity, being able to inhibit the growth of studied strains at lower concentration, cause membrane disruption in shorter period of time, and lead to lower cellular amounts of ROS regarding *C. albicans*, which can promote lower tissue invasion by this particular strain.

The cytotoxic effect of both compounds on HCT116 cells was also studied. Both AP7Glu and apigenin demonstrated in vitro cytotoxic effect on colon cancer cells, revealing the ability to diminish cell's viability and promote cell death. It has been reported (Lee et al. 2014) that apigenin effectively decreases viability of HCT116 cells and that treatment with this compound leads to chromosomal condensation and apoptotic body formation (Nelson et al., 2013). It was demonstrated that AP7Glu treatment of HCT116 cells was even more promising concerning the reduction of cell's viability and promotion of cell death. In particular, cell death induction was elucidated by flow cytometry. It was revealed that after treatment with AP7Glu, HCT116 cells experienced significant necrosis. As it has been already proposed, apigenin is able to induce apoptosis through p53-dependent pathway (Seo et al., 2012) and cells treated with apigenin demonstrate raised levels of p53 and Bax (Lu et al., 2010). Additionally, results revealed that AP7Glu is even more effective in inducing apoptosis-associated p53 and Bax genes expression in HCT116 cells.

Changes provoked by natural compounds toward *Aspergillus* spp.

Phenolic compounds are very potent in showing their antifungal activity against such pathogens due to their structural ability to be diffused through the microbial membrane and penetrate into the cell. Once they penetrated into the cells, they can interfere in the metabolic pathways by interfering with the synthesis of chitin, ergosterol, glucan, glucosamine, and proteins in fungi. *Spirulina* spp is a cyanobacterium used as a food supplement among humans as a good source of phenolic compounds (**caffeic**, **chlorogenic**, **salicylic**, **synaptic**, and **trans-cinnamic acids**, Figs. 4.42 and 4.60).

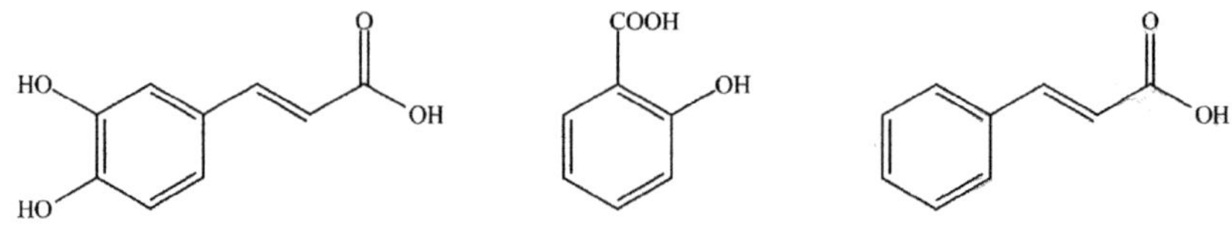

Caffeic acid **Salicylic acid** **Trans-cinnamic acid**

FIG. 4.60 Chemical structures of **caffeic**, **salicylic**, and **trans-cinnamic acids**.

Phenols deriving from *Spirulina* exhibited antifungal activity by inhibiting ergosterol, component of cell membrane, glucosamine, a growth indicator and other proteins (Sparringa and Owens, 1999). Some of the phenolic compounds (anthocyanins, anthraquinones, catechins, chalcones, phenolic acids, flavonoids, xanthones, stilbenes, tannins, and naphthoquinones) are able to inhibit the activity of transporters (Meschini et al., 2003; You et al., 1995).

Changes provoked by natural compounds toward *Cryptococcus neoformans*

Cryptococcus infections were very rare until 1880 (Lewitz et al., 1994) and mostly in immune-deficient patients. Recently, cryptococcal meningitis has become the second most common opportunistic infection associated with AIDS in Africa and Asia. Four phenols isolated from *Cudrania cochinchinensis*, cudraxanthone S and cudraflavanone B, toxyloxanthone C and wighteone demonstrated activity against *C. neoformans*, *C. albicans*, and *Aspergillus fumigatus* (Fukai et al., 2003). Benzophenones extracted from *C. cochinchinensis* exhibit antifungal activity against *C. neoformans* (Hou et al., 2001). Many other studies have demonstrated that the phenolic compounds can also show synergistic effect with commercial antifungal agents. For example, **2,4-dihydroxybenzaldehyde** (Fig. 4.61) has shown synergistic effect with itraconazole, and amphotericin B against *C. neoformans* (Ansari et al., 2013).

Natural products should be extensively evaluated and potentially used as alternatives for the treatment of human fungal infections. Additionally, formulations of natural products could be a solution to a variety of side effects of

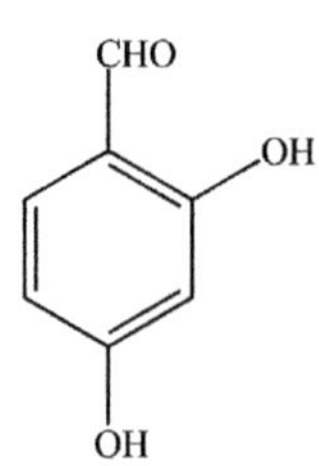

2,4-dihydroxybenzaldehyde

FIG. 4.61 Chemical structure of **2,4-dihydroxybenzaldehyde**.

commercial antifungal drugs, alongside with the emerging resistance of microorganisms to many synthetic fungicides. Taking into account all the previously reported data, it could be concluded that products based on secondary metabolites (i.e., essential oils extracted from aromatic plants) demonstrated antifungal efficacy and were found to be promising therapeutic agents while having limited side effects on humans and animals. Consequently, the future strategies regarding management of fungicides could be influenced by promoting research in the field of natural products and their derivatives. Technological progress concerning instrumentation has simplified the processes of isolation and identification of lead compounds that can form effective fungicides. Moreover, from the reviewed studies it is obvious that natural products of plant origin can be a great opportunity for pharmaceutical, nutritional, cosmetic, and agricultural purposes.

The ideal target of research in this field could be the discovery of prototypes of antifungal agents that have their basis on the essential oils. These innovative prototypes would represent new chemical classes, with diverse mechanisms of action compared to existing antifungals, aiming to overcome the resistance of pathogens to marketed drugs. The ultimate purpose remains the development of safer, more effective, and low-cost formulations and methods that are based on natural products, in order to address the serious health issue of fungal diseases in humans.

References

Adam, K., Sivropoulou, A., Kokkini, S., Lanares, T., Arsenakis, M., 1998. Antifungal activities of *Origanum vulgare* subsp. *hirtum, Mentha spicata, Lavandula angustifolia* and *Salvia fruticosa* essential oils against human pathogenic fungi. J. Agric. Food Chem. 46, 1739–1745.

Akkol, E.K., Arif, R., Ergun, F., Yesilada, E., 2009. Sesquiterpene lactones with antinociceptive and antipyretic activity from two *Centaurea* species. J. Ethnopharmacol. 122 (2), 210–215.

Aljancic, I., Vajs, V., Menkovic, N., Karadzic, I., Juranic, N., Milosavljevic, S., Mcura, S., 1999. Flavones and sesquiterpene lactones from *Achillea atrata* subsp. multifida: antimicrobial activity. J. Nat. Prod. 62, 909–911.

Al-Saghir, J., Al-Ashi, R., Salloum, R., Saliba, N., Talhouk, R.S., Homaidan, F.R., 2009. Anti-inflammatory properties of Salograviolide a purified from Lebanese plant *Centaurea ainetensis*. BMC Complement. Altern. Med. 9 (36), 1–10.

Ando, K., Suzuki, S., Saeki, T., Tamura, G., Arima, K., 1969. Funiculosin, a new antibiotic. Isolation, biological and chemical properties. J. Antibiot. 22, 189–194.

Ansari, M.A., Anurag, A., Fatima, Z., Hameed, S., 2013. Natural phenolic compounds: a potential antifungal agent. In: Méndez-Vilas, A. (Ed.), Microbial Pathogens and Strategies for Combating Them: Science, Technology and Education. Formatex, pp. 1189–1195.

Bach, S.M., Fortuna, M.A., Attarian, R., de Trimarco, J.T., Catalán, C.A., Av-Gay, Y., Bach, H., 2011. Antibacterial and cytotoxic activities of the sesquiterpene lactones cnicin andonopordopicrin. Nat. Prod. Commun. 6, 163–166.

Bajpai, V.K., Yoon, J.I., Kang, S.C., 2009. Antioxidant and antidermatophytic activities of essential oil and extracts of *Metasequoia glyptostroboides* Miki ex Hu. Food Chem. Toxicol. 47, 1355–1361.

Bakkali, F., Averbeck, S., Averbeck, D., Idaomar, M., 2008. Biological effects of essential oils. Rev. Food Chem. Toxicol. 46, 446–475.

Baratta, M.T., Dorman, H.J.D., Deans, S.G., Figueiredo, A.C., Barroso, J.G., Ruberto, G., 1998. Antimicrobial and antioxidant properties of some commercial essential oils. Flav. Fragr. J. 13 (4), 235–244.

Barrero, A.F., Oltra, J.E., Álvarez, M., Raslan, D.S., Saúde, D.A., Akssira, M., 2000. New sources and antifungal activity of sesquiterpene lactones. Fitoterapia 71 (1), 60–64.

Barros, L., Baptista, P., Correia, D.M., Sá Morais, J., Ferreira, I.C.F.R., 2007. Effects of conservation treatment and cooking on the chemical composition and antioxidant activity of Portuguese wild edible mushrooms. J. Agric. Food Chem. 55, 4781–4788.

Basak, S., Guha, P., 2017. Use of predictive model to describe sporicidal and cell viability efficacy of betel leaf (*Piper betle* L.) essential oil on *Aspergillus flavus* and *Penicillium expansum* and its antifungal activity in raw apple juice. LWT 80, 510–516.

Berdy, J., 2005. Bioactive microbial metabolites: a personal view. J. Antibiot. 58, 1–26.

Bhaskaran, N., Shukla, S., Srivastava, J.K., Gupta, S., 2010. Chamomile: an anti-inflammatory agent inhibits inducible nitric oxide synthase expression by blocking RelA/p65 activity. Int. J. Mol. Med. 226, 935–940.

Bishop, C.D., Thornton, I.B., 1997. Evaluation of the antifungal activity of the essential oils of *Monarda citriodora* var. *citriodora* and *Melaleuca alternifolia* on post-harvest pathogens. J. Essent. Oil Res. 9 (1), 77–82.

Bissett, N.G., 1994. Herbal Drugs and Phytopharmaceuticals. Stuttgart. Pharm. CRC Press, p. 566.

Božunović, J., Živković, S., Gašić, U., Glamočlija, J., Ćirić, A., Matekalo, D., Šiler, B., Soković, M., Tešić, Ž., Mišić, D., 2018. *In vitro* and *in vivo* transformations of *Centaurium erythraea* secoiridoid glucosides alternate their antioxidant and antimicrobial capacity. Ind. Crop. Prod. 111, 705–721.

Bremnes, L., 1994. Herbs. In: Eyewitness-Handbooks. DK Publishing, New York, USA.

Briskin, D.P., 2000. Medicinal plants and phytomedicines. Linking plant biochemistry and physiology to human health. Plant Physiol. 124, 507–514.

Brown, G.D., 2006. Dectin-1: a signalling non-TLR pattern-recognition receptor. Nat. Rev. Immunol. 6, 33–43.

Brul, S., Coote, P., 1999. Mode of action and microbial resistance mechanisms. Int. J. Food Microbiol. 50, 1–17.

Bruno, M., Bancheva, S., Rosselli, S., Maggio, A., 2013. Sesquiterpenoids in subtribe Centaureinae (Cass.) Dumort (tribe Cardueae, Asteraceae): distribution, 13C NMR spectral data and biological properties. Phytochemistry 95, 19–33.

Butler, M.S., 2005. Natural products to drugs: natural products derived compounds in clinical trials. Nat. Prod. Rep. 22, 162–195.

Cai, J., Zhao, X., Liu, A., Nian, H., Zhang, S., 2011. Apigenin inhibits hepatoma cell growth through alteration of gene expression patterns. Phytomedicine 18 (5), 366–373.

Carson, C.F., Hammer, K.A., Riley, T.V., 2006. *Melaleuca alternifolia* (tea tree) oil: a review of antimicrobial and other medicinal properties. Clin. Microbiol. Rev. 19, 50–62.

Cavaleiro, C., Pinto, E., Goncalves, M.J., Salgueiro, L., 2006. Antifungal activity of Juniperus essential oils against dermatophyte, *Aspergillus* and *Candida* strains. J. Appl. Microbiol. 100, 1333–1338.

Cavallito, C.J., Bailey, J.H., 1949. An antibacterial principle from *Centaurea maculosa*. J. Bacteriol. 57, 207–212.

Chadwick, M., Trewin, H., Gawthrop, F., Wagstaff, C., 2013. Sesquiterpenoids lactones: benefits to plants and people. Int. J. Mol. Sci. 14, 12780–12805.

Chaturvedi, D., 2011. 10. Sesquiterpene lactones: structural diversity and their biological activities. In: Opportunity, Challenge and Scope of Natural Products in Medicinal Chemistry., ISBN: 978-81-308-0448-4, pp. 313–334.

Chaturvedi, D., Dwivedi, P.K., 2016. Sesquiterpene lactones: a versatile class of structurally diverse biologically potent scaffolds in drug discovery research. Res. Rev. J. Chem. 5 (1), 5–9.

Chen, H.C., Chou, C.K., Lee, S.D., Wang, J.C., Yeh, S.F., 1995. Active compounds from Saussurea lappa Clarks that suppress hepatitis B virus surface antigen gene expression in human hepatoma cells. Antivir. Res. 27, 99–109.

Ciric, A., Karioti, A., Glamoclija, J., Sokovic, M., Skaltsa, H., 2011. Antimicrobial activity of secondary metabolites isolated from *Centaurea spruneri* Boiss. & Heldr. J. Serb. Chem. Soc. 76 (1), 27–34.

Ćirić, A., Karioti, A., Koukoulitsa, C., Soković, M., Skaltsa, H., 2012. Sesquiterpene lactones from *Centaurea zuccariniana and their a*ntimicrobial activity. Chem. Biodivers. 9 (12), 2843–2853.

Clinical and Laboratory Standards Institute (CLSI), 2009. Methods for Dilution Antimicrobial Susceptibility Tests for Bacteria that Grow Aerobically. Approved Standard. CLSI Publication M07-A8, Wayne, PA.

Conner, D.E., Beuchat, R.E., 1984. Effects of essential oils from plants on growth of food spoilage yeasts. J. Food Sci. 49 (2), 429–434.

Cowan, M.M., 1999. Plant products as antimicrobial agents. Clin. Microbiol. Rev. 12 (4), 564–582.

Daniel, T.C., Muharb, A., Arnbergerb, A., Aznarc, O., Boydd, J.W., Chane, K.M.A., Costanza, R., Elmqvistg, T., Flinth, C.G., Gobsteri, P.H., Grêt-Regame, A., Lavek, R., Muharl, S., Penkerm, M., Riben, R.G., Schauppenlehnerb, T., Sikoro, T., Soloviyp, I., Spierenburg, M., Taczanowska, K., Tame, J., von der Dunk, A., 2012. Contributions of cultural services to the ecosystemservices agenda. PNAS 109 (23), 8812–8819

Daouk, R., Dagher, S.M., Sattout, E.J., 1995. Antifungal activity of the essential oil of *Origanum syriacum* L. J. Food Prot. 58 (10), 1147–1149.

D'Auria, F.D., Tecca, M., Strippoli, V., Salvatore, G., Battinelli, L., Mazzanti, G., 2005. Antifungal activity of *Lavandula angustifolia* essential oil against *Candida albicans* yeast and mycelial form. Med. Mycol. 43 (5), 391–396.

Davidson, P.M., Parish, E.M., 1989. Methods for testing the efficacy of food antimicrobials. Food Technol. 43, 148–155.

de Kraker, J.W., Franssen, M.C.R., de Groot, A., Konig, W.A., Bouwmeester, H.J., 1998. (+)-Germacrene a biosynthesis – The committed step in the biosynthesis of bitter sesquiterpene lactones in chicory. Plant Physiol. 117, 1381–1392.

de Kraker, J.W., Franssen, M.C.R., Joerink, M., de Groot, A., Bouwmeester, H.J., 2002. Biosynthesis of costunolide, dihydrocostunolide, and leucodin. Demonstration of cytochrome P450-catalyzed formation of the lactone ring present in sesquiterpene lactones of chicory. Plant Physiol. 129, 257–268.

de Melo, J.O., Bitencourt, T.A., Fachin, A.L., Oliveira-Cruz, E.M., de Jesus-Ramos, H.C., Alves, P.B., Arrigoni-Blank, M.F., Franca, S.C., Beleboni, R.O., Fernandes, R.P.M., Blank, A.F., Scher, R., 2013. Antidermatophytic and antileishmanial activities of essential oils from *Lippia gracilis* Schauer genotypes. Acta Trop. 128, 110–115.

Deans, S.G., Svoboda, K.P., 1990. The antimicrobial properties of marjoram (*Origanum majorana* L.) volatile oil. Flav. Fragr. J. 5 (3), 187–190.

Dembitsky, V.M., Tolstikov, G.A., 2002. Chlorine containing sesquiterpenes of higher plants. Chem. Sust. Dev. 10, 363–370.

Dembitsky, V.M., Rezanka, T., Srebnik, M., 2003. Lipid compounds of freshwater sponges: family Spongillidae, class Demospongiae. Chem. Phys. Lipids 123, 117–155.

Dewick, P.M., 2002. Medicinal Natural Products. A Biosynthetic Approach, Second Edition. John Wiley & Sons Ltd., pp. 191–202.

Djeddi, S., Karioti, A., Soković, M., Stojković, D., Seridi, R., Skaltsa, H., 2007. Minor sesquiterpene lactones from *Centaurea pullata* and their antimicrobial activity. J. Nat. Prod. 70, 1796–1799.

Djeddi, S., Karioti, A., Soković, M., Koukoulitisa, C., Skaltsa, H., 2008. A novel sesquiterpene lactone from *Centaurea pullata*: structure elucidation, antimicrobial activity, and prediction of pharmacokinetic properties. Bioorg. Med. Chem. 16 (7), 3725–3731.

Du, Z., Zhu, N., Ze-Ren-Wang-Mu, N., Shen, Y., 2003. Two new antifungal saponins from the Tibetan herbalmedicine *Clematis tangutica*. Planta Med. 69 (6), 547–551.

Duke, S.O., Romagni, J.G., Dayan, F.E., 2000. Natural products as sources for new mechanisms of herbicidal action. Crop Prot. 19 (8–10), 583–589.

Dzamic, A., Sokovic, M., Ristic, M., Grujic-Jovanovic, S., Vukojevic, J., Marin, P.D., 2008. Chemical composition and antifungal activity of *Salvia sclarea* (Lamiaceae) essential oils. Arch. Biol. Sci. 60 (2), 233–237.

Dzamic, A., Sokovic, M., Ristic, M., Grijic-Jovanovic, S., Vukojevic, J., Marin, P.D., 2009. Chemical composition and antifungal activity of *Illicium verum* and *Eugenia caryophyllata* essential oils. Chem. Nat. Compd. 45 (2), 259–261.

Dzamic, A.M., Sokovic, M.D., Ristic, M.S., Novakovic, M., Grujic-Jovanovic, S., Tesević, V., Marin, P.D., 2010. Antifungal and antioxidant activity of *Mentha longifolia* (L.) Hudson (Lamiaceae) essential oil. Bot. Serb. 34 (1), 57–61.

Eisenman, H.C., Casadevall, A., 2012. Synthesis and assembly of fungal melanin. Appl. Microbiol. Biotechnol. 93 (3), 931–940.

Espinel-Ingroff, A., 2001. Comparison of the E-test with the NCCLS M38-P method for antifungal susceptibility testing of common and emerging pathogenic filamentous fungi. J. Clin. Microbiol. 39, 1360–1367.

Esteban, J.I.A., Pinela, J., Barros, L., Ćirić, A., Soković, M., Calhelha, R.C., Torija-Isasa, E., Cortes, M., Mata, S., Ferreira, I.C.F.R., 2019. Phenolic composition and antioxidant, antimicrobial and cytotoxic properties of hop (*Humulus lupulus* L.) seeds. Ind. Crop. Prod. 134, 154–159.

Faria, N.C.G., Kim, J.H., Goncalves, L.A.P., Martins, M.D.L., Chan, K.L., Campbell, B.C., 2012. Enhanced activity of antifungal drugs using natural phenolics against yeast strains of *Candida* and *Cryptococcus*. Lett. Appl. Microbiol. 52, 506–513.

Fukai, T., Yonekawa, M., Hou, A., Nomura, T., Sun, H., Uno, J., 2003. Antifungal agents from the roots of *Cudrania cochinchinensis* against *Candida, Cryptococcus*, and *Aspergillus* species. J. Nat. Prod. 66 (8), 1118–1120.

Garg, S.N., Siddiqui, M.S., Agarwal, S.K., 1992. New fatty acid esters and hydroxy ketones from fruits of *Laurus nobilis*. Nat. Prod. 55 (9), 1315–1319.

Giordani, R., Regli, P., Kaloustian, J., Mikaïl, C., Abou, L., Portugal, H., 2004. Potentiation of antifungal action of amphotericin B by essential oil from *Thymus vulgaris*. Phytother. Res. 18 (12), 990–995.

Giudici, A.M., Regente, M.C., la Canal, L., 2000. A potent antifungal protein from *Helianthus annuus* flowers is a trypsin inhibitor. Plant Physiol. Biochem. 38 (11), 881–888.

Glamočlija, J., 2006. Prevention and Treatment of *Mycogone perniciosa* Magn. In Mushroom Units of *Agaricus bisporus* (J. Lange) Imbach. Ph.D. thesis, Faculty of Biology, University of Belgrade, Belgrade, Serbia.

Glamočlija, J., Soković, M., Vukojević, J., Milenković, I., van Griensven, L.J.L.D., 2006. Chemical composition and antifungal activities of essential oils of *Satureja thymbra* L. and *Salvia pomifera* ssp. *calycina* (Sm.) Hayek. J. Essent. Oil Res. 18, 115–117.

Glamočlija, J., Soković, M., Grubišić, D., Vukojević, J., Milenković, I., Ristić, M., 2009. Antifungal activity of *Critmum maritimum* L. essential oil and its components against mushroom pathogen *Mycogone perniciosa*. Mang. Chem. Nat. Comp. 45, 96–97.

Glamočlija, J.M., Soković, M.D., Šiljegović, J.D., Ristić, M.S., Ćirić, A.D., Grubišić, D.V., 2011a. Chemical composition and antimicrobial activity of *Echinophora spinosa* L. (Apiaceae) essential oil. Rec. Nat. Prod. 5 (4), 319–323.

Glamočlija, J., Sokovic, M., Tesevic, V., Linde, G.A., Colauto, N.B., 2011b. Chemical characterization of *Lippia alba* essential oil: an alternative to control green molds. Braz. J. Microbiol. 42 (4), 1537–1546.

Glamočlija, J., Ćirić, A., Nikolić, M., Fernandes, Â., Barros, L., Calhelha, R.C., Ferreira, I.C., Soković, M., van Griensven, L.J., 2015a. Chemical characterization and biological activity of Chaga (*Inonotus obliquus*), a medicinal "mushroom". J. Ethnopharmacol. 162, 323–332.

Glamočlija, J., Stojković, D., Nikolić, M., Ćirić, A., Reis, F.S., Barros, L., Ferreira, I.C., Soković, M., 2015b. A comparative study on edible *Agaricus* mushrooms as functional foods. Food Funct. 6 (6), 1900–1910.

Goncalves, M.J., Cruz, M.T., Cavaleiro, C., Lopes, M.C., Salgueiro, L., 2010. Chemical, antifungal and cytotoxic evaluation of the essential oil of *Thymus zygis* subsp. *sylvestris*. Ind. Crop. Prod. 32, 70–75.

Govinden-Soulange, J., Magan, N., Gurib, A., Gauvin, A., Smadja, J., Kodja, H., 2004. Chemical composition and *in vitro* antimicrobial activities of the essential oils from endemic *Psiadia* species growing in Mauritius. Biol. Pharm. Bull. 27, 1814–1818.

Griffin, G.S., Wylie, G.S., Markham, L.J., Leach, N.D., 1999. The role of structure and molecular properties of terpenoids in determining their antimicrobial activity. Flavour Fragr. J. 14, 322–332.

Guha, P., Nandi, S., 2019. Essential oil of betel leaf (*Piper betle* L.): a novel addition to the world food sector. In: Malik, S. (Ed.), Essential Oil Research Trends in Biosynthesis, Analytics, Industrial Applications and Biotechnological Production. Springer Nature Switzerland AG.

Guo, N., Liu, J., Wu, X., Bi, X., Meng, R., Wang, X., Xiang, H., Deng, X., Yu, L., 2009. Antifungal activity of thymol against clinical isolates of fluconazole-sensitive and -resistant *Candida albicans*. J. Med. Microbiol. 58 (8), 1074–1079.

Harvey, A.L., 2008. Natural products in drug discovery. Drug Discov. Today 13, 894–901.

Heilmann, J., Wasescha, M.R., Schmidt, T.J., 2001. The influence of glutathione and cysteine levels on the cytotoxicity of helenanolide type sesquiterpene lactones against KB cells. Bioorg. Med. Chem. 9, 2189–2194.

Heleno, S.A., Ferreira, I.C.F.R., Esteves, A.P., Ciric, A., Glamoclija, J., Martins, A., Sokovic, M., Queiroz, M.-J.R.P., 2013. Antimicrobial and demelanizing activity of *Ganoderma lucidum* extract, p-hydroxybenzoic and cinnamic acids and their synthetic acetylated glucuronide methyl esters. Food Chem. Toxicol. 58, 95–100.

Hernandez, T., Canales, M., Garcia, A.M., Duran, A., Meraz, S., Davila, P., Gulermo Avila, J., 2008. Antifungal activity of the essential oils of two Verbenaceae: *Lantana achyranthifolia* and *Lippia graveolens* of Zapotitlan de las Salinas, Puebla (Mexico). Bol. Latino. Caribe. Plant. Medicin. Aromat. 7, 202–206.

Hou, A.J., Fukai, T., Shimazaki, M., Sakagami, H., Sun, H.D., Nomura, T., 2001. Benzophenones and xanthones with isoprenoid groups from *Cudrania cochinchinensis*. J. Nat. Prod. 64, 65–70.

Ilić, D., Nikolić, V., Ćirić, A., Soković, M., Stanojković, T., Kundaković, T., Stanković, M., Nikolić, L., 2012. Cytotoxicity and antimicrobial activity of allicin and its transformation products. J. Med. Plant Res. 6 (1), 59–65.

Inoue, S.B., Takewaki, N., Takasuka, T., Mio, T., Adachi, M., Fujii, Y., Miyamoto, C., Arisawa, M., Furuichi, Y., Watanabe, T., 1995. Characterization and gene cloning of 1,3-beta-D-glucan synthase from Saccharomyces cerevisiae. Eur. J. Biochem. 231 (3), 845–854.

Jacob, E.J., 2009. Natural products-based drug discovery: some bottlenecks and considerations. Curr. Sci. 96 (6), 753–754.

Jamalian, A., Shams-Ghahfarokhi, M., Jaimand, K., Pashootan, N., Amani, A., Razzaghi-Abyaneh, M., 2012. Chemical composition and antifungal activity of *Matricaria recutita* flower essential oil against medically important dermatophytes and soil-borne pathogens. J. Mycol. Med. 22, 308–315.

Janackovic, P., Tešević, V., Marin, P., Milosavljević, S., Petković, B., Soković, M., 2002. Polyacetylenes and a sesquiterpene lactone from *Ptilostemon strictus*. Biochem. Syst. Ecol. 30 (1), 69–71.

Janaćković, P., Soković, M., Vujisić, L., Vajs, V., Vucković, I., Krivošej, Z., Marin, P.D., 2011. Composition and antimicrobial activity of *Seseli globiferum* essential oil. Nat. Prod. Commun. 6 (8), 1163–1166.

Janssen, A.M., Scheffer, J.J.C., Baerheim Svendsen, A., Shelef, A.L., 1983. Antimicrobial effects of spice. J. Food Saf. 6, 29–44.

Jantan, I.B., Moharam, B.A.K., Santhanam, J., Jamal, J.A., 2008. Correlation between chemical composition and antifungal activity of the essential oils of eight *Cinnamomum* species. Pharm. Biol. 46, 406–412.

Jennings, D.H., 1977. In: Heywood, V.H., Harborne, J.B., Turner, B.L. (Eds.), The Biology and Chemistry of the Compositae. Academic Press, London, p. 1189.

Kabara, J.J., 1991. Phenols and chelators. In: Russell, N.J., Gould, G.W. (Eds.), Food Preservatives. Blackie, London, UK, pp. 200–214.

Kalemba, D., Kunicka, A., 2003. Antibacterial and antifungal properties of essential oils. Curr. Med. Chem. 10 (10), 813–829.

Kanazawa, K., Uehara, M., Yanagitani, H., Hashimoto, T., 2006. Bioavailable flavonoids to suppress the formation of 8-OHdG in HepG2 cells. Arch. Biochem. Biophys. 455 (2), 197–203.

Karabunarliev, S., Mekenyan, O.G., Karcher, W., Russom, C.L., Bradbury, S.P., 1996. Quantum-chemical descriptors for estimating the acute toxicity of electrophiles to the fathead minnow (*Pimephales promelas*). An analysis based on molecular mechanism. Quant. Struct. Activ. Relat. 15, 302–310.

Karioti, A., Skaltsa, H., Lazari, D., Soković, M., Garcia, B., Harvala, C., 2002. Secondary metabolites from *Centaurea deusta* with antifungal activity. Z. Naturforsch. 57 (1–2), 75–80.

Karioti, A., Bilia, A.R., Gabbiani, C., Messori, L., Skaltsa, H., 2009. Proanthocyanidin glycosides from the leaves of *Quercus ilex* L. (Fagaceae). Tetrahedron Lett. 50, 1771–1776.

Karioti, A., Sokovic, M., Ciric, A., Koukoulitsa, C., Rita Bilia, A., Skaltsa, H., 2011. Antimicrobial properties of *Quercus ilex* L. proanthocyanidin dimers and simple phenolics: evaluation of their synergistic activity with conventional antimicrobials and prediction of their pharmacokinetic profile. Agric. Food Chem. 59 (12), 6412–6422.

Kebaara, B.W., Langford, M.L., Navarathna, D.H., Dumitru, R., Nickerson, K.W., Atkin, A.L., 2008. *Candida albicans* Tup1 is involved in farnesol-mediated inhibition of filamentous-growth induction. Eukaryot. Cell 7, 980–987.

Kelsey, R.G., Locken, L.J., 1987. Phytotoxic properties of cnicin, a sesquiterpene lactone from *Centaurea muculosa* (spotted knapweed). J. Chem. Ecol. 13, 19–33.

Knobloch, K., Pauli, A., Iberl, B., Weis, N., Weigand, H., 1988. Mode of action of essential oil components on whole cells of bacteria and fungi in plate tests. In: Schreier, P. (Ed.), Bioflavor '87. Walterde Gruyter Verlag, Berlin, New York, USA, pp. 287–299.

Koehn, F.E., Carter, T.G., 2005. The evolving role of natural products in drug discovery. Nat. Rev. Drug Discov. 4, 206–220.

Kolli, E.H., León, F., Benayache, F., Estévez, S., Quintana, J., Estévez, F., Brouard, I., Bermejo, J., Benayache, S., 2012. Cytotoxic sesquiterpene lactones and other constituents of *Centaurea omphalotricha*. J. Braz. Chem. Soc. 23 (5), 977–983.

Konstantinopoulou, M., Karioti, A., Skaltsas, S., Skaltsa, H., 2003. Sesquiterpene lactones from *Anthemis altissima* and their anti-*Helicobacter pylori* activity. J. Nat. Prod. 66, 699–702.

Kostić, M., Smiljković, M., Petrović, J., Glamočilija, J., Barros, L., Ferreira, I.C.F.R., Ćirić, A., Soković, M., 2017. Chemical, nutritive composition and wide-broad bioactive properties of honey mushroom *Armillaria mellea* (Vahl: Fr.) Kummer. Food Funct. 38, 3239–3249.

Koukoulitsa, E., Skaltsa, H., Karioti, A., Demetzos, C., Dimas, K., 2002. Bioactive sesquiterpene lactones from *Centaurea* species and their cytotoxic/cytostatic activity against human cell lines *in vitro*. Planta Med. 68 (7), 649–652.

Koukoulitsa, C., Geromichalos, G.D., Skaltsa, H., 2005. VolSurf analysis of pharmacokinetic properties for several antifungalsesquiterpene lactones isolated from Greek *Centaurea* sp. J. Comput. Aided Mol. Des. 19 (8), 617–623.

Kowalski, J., Samojedny, A., Paul, M., Pietsz, G., Wilczok, T., 2005. Effect of apigenin, kaempferol and resveratrol on the expression of interleukin-1beta and tumor necrosis factor-alpha genes in J774.2 macrophages. Pharmacol. Rep. 57 (3), 390–394.

Kozarski, M., Klaus, A., Jakovljevic, D., Todorovic, N., Vunduk, J., Petrovic, P., Niksic, M., Vrvic, M.M., van Griensven, L., 2015. Antioxidants of edible mushrooms. Molecules 20, 19489–19525.

Krauze-Baranowska, M., Cisowski, W., Wiwart, M., Madziar, B., 1999. Antifungal biflavones from *Cupressocyparis leylandii*. Planta Med. 65 (6), 572–573.

Kukić, J., Popović, V., Petrović, S., Mucaji, P., Ćirić, A., Stojković, D., Soković, M., 2008. Antioxidant and antimicrobial activity of *Cynara cardunculus* extracts. Food Chem. 107 (2), 861–868.

Kumar, N., 2014. Inhibition of nail infecting fungi of peoples of north eastern U.P causing *Tinea unguium* through leaf essential oil of *Ageratum houstonianum* mill. IOSR J. Pharm. 4, 36–42.

Kumar, D., Bhat, Z.A., 2012. Anti-anxiety activity of methanolic extracts of different parts of *Angelica archangelica* Linn. J. Tradit. Complement. Med. 2, 235–241.

Kundaković, T.D., Ćirić, A.D., Soković, M.D., Milenković, M.T., Nikolić, V.D., Nikolić, G.S., 2011. Antimicrobial activity of lozenge with garlic bulb powder. Hem. Ind. 65 (5), 607–610

Kupchan, S.M., Fessler, D.C., Eakin, M.A., Giacobre, T.J., 1970. Reactions of alpha methylene lactone tumor inhibitors with model biological nucleophiles. Science 168, 376–378.

Kupchan, S.M., Eakin, M.A., Thomas, A.M., 1971. Tumor inhibitors. 69. Structure cytotoxicity relationship among the Sesquiterpene lactones. J. Med. Chem. 14, 1147–1152.

Kurobane, I., Hutchinson, C.R., Vining, L.C., 1981. The biosynthesis of fulvic acid, a fungal metabolite of heptaketide origin. Tetrahedron Lett. 22, 493–496.

Lahlou, M., 2013. The success of natural products in drug discovery. Pharmacol. Pharm. 4, 17–31.

Lam, K.S., 2007. New aspects of natural products in drug discovery. Trends Microbiol. 15 (6), 279–289

Lambert, R.J., Skandamis, P.N., Coote, P.J., Nychas, G.J., 2001. A study of the minimum inhibitory concentration and mode of action of oregano essential oil, thymol and carvacrol. J. Appl. Microbiol. 91, 453–462.

Lee, D.G., Park, Y., Kim, M.R., Jung, H.J., Seu, Y.B., Hahm, K.S., Woo, E.R., 2004. Anti-fungal effects of phenolic amides isolated from the root bark of *Lycium chinense*. Biotechnol. Lett. 14, 1125–1130.

Lee, Y., Sung, B., Kang, Y.J., Kim, D.H., Jang, J.Y., Hwang, S.Y., Kim, M., Lim, H.S., Yoon, J.H., Chung, H.Y., Kim, N.D., 2014. Apigenin-induced apoptosis is enhanced by inhibition of autophagy formation in HCT116 human colon cancer cells. Int. J. Oncol. 44 (5), 1599–1606.

Lewitz, S.M., Dupont, M.P., Smail, E.H., 1994. Direct activity of human T lymphocytes and natural killer cells against *Cryptococcus neoformans*. Infect. Immun. 62, 194–202.

Lis-Balchin, M., Deans, G.S., Eaglesham, E., 1998. Relationship between bioactivity and chemical composition of commercial essential oils. Flavour Frag. J. 13, 98–104.

Liu, Q., Majdi, M., Cankar, K., Goedbloed, M., Charnikhova, T., Verstappen, F.W., de Vos, R.C., Beekwilder, J., van der Krol, S., Bouwmeester, H.J., 2011. Reconstitution of the costunolide biosynthetic pathway in yeast and *Nicotiana benthamiana*. PLoS One 6 (8), e2325.

Lu, H.F., Chie, Y.J., Yang, M.S., Lee, C.S., Fu, J.J., Yang, J.S., Tan, T.W., Wu, S.H., Ma, Y.S., Ip, S.W., Chung, J.G., 2010. Apigenin induces caspase-dependent apoptosis in human lung cancer A549 cells through Bax- and Bcl-2-triggered mitochondrial pathway. Int. J. Oncol. 6, 1477–1484.

Lutz, M.K., 2003. Putting nature back into drug discovery. Nat. Biotechnol. 21 (3), 602–604.

Machlin, L.J., Bendich, A., 1987. Free radical tissue damage: protective role ofantioxidant nutrients. FASEB J. 1, 441–445.

Mahmoudvand, H., Sepahvand, A., Jahanbakhsh, S., Ezatpour, B., Ayatollahi Mousavi, A.A., 2012. Evaluation of antifungal activities of the essential oil and various extracts of *Nigella sativa* and its main component, thymoquinone against pathogenic dermatophyte strains. J. Mycol. Med. 24, 155–161.

Mamadalieva, N.Z., Herrmann, F., El-Readi, M.Z., Tahrani, A., Hamoud, R., Egamberdieva, D.R., Azimova, S.S., Wink, M., 2011. Flavonoids in *Scutellaria immaculata* and *S. ramosissima* (Lamiaceae) and their biological activity. J. Pharm. Pharmacol. 63 (10), 1346–1357.

Manach, C., Scalbert, A., Morand, C., Rémésy, C., Jiménez, L., 2004. Polyphenols: food sources and bioavailability. Am. J. Clin. Nutr. 79 (5), 727–747.

Marinković, B., Marin, P.D., Knežević-Vukčević, J., Soković, M.D., Brkić, D., 2002. Activity of essential oils of three *Micromeria* species (Lamiaceae) against micromycetes and bacteria. Phytother. Res. 16 (4), 336–339.

Marković, T., Chatzopoulou, P., Šiljegović, J., Nikolić, M., Glamočlija, J., Ćirić, A., Soković, M., 2011. Chemical analysis and antimicrobial activities of the essential oils of *Satureja thymbra* L. and *Thymbra spicata* L. and their main components. Arch. Biol. Sci. 63, 457–464.

Maruzzella, J.C., Balter, J., 1959. The action of essential oils on phytopathogenic fungi. Plant Dis. Rep. 43, 1143–1147.

Maskey, R.P., Grun-Wollny, I., Laatsch, H., 2003. Isolation, structure elucidation and biological activity of 8-O-methylaverufin and 1,8-O-dimethylaverantin as new antifungal agents from *Penicillium chrysogenum*. J. Antibiot. 56, 459–463.

Matsumoto, T., Tahara, S., 2001. Ampelopsin, a major antifungal constituent from *Salix sachalinensis*, and its methyl ethers. J. Agric. Chem. Soc. Japan 75 (6), 659–667.

Medjroubi, K., Benayache, F., Bermejo, J., 2005. Sesquiterpene lactones from *Centaurea musimomum*. Antiplasmodial and cytotoxic activities. Fitoterapia 76 (7–8), 744–746.

Meschini, S., Marra, M., Calcabrini, A., Federici, E., Galeffi, C., Arancia, G., 2003. Voacamine, a bisindolic alkaloid from *Peschiera fuchsiaefolia*, enhances the cytotoxic effect of doxorubicin on multidrug-resistant tumor cells. Int. J. Oncol. 23, 1505–1513.

Milosavljević, S., Tešević, V., Vučković, I., Jadranin, M., Vajs, V., Soković, M., Janaćković, P., Jovanović, A., 2007. Composition and antifungal activity of the essential oil of *Seseli annuum*wild-growing in Serbia. Fitoterapia 78 (4), 319–322.

Milošević-Ifantis, T., Solujić, S., Pavlović-Muratspahić, D., Skaltsa, H., 2013. Secondary metabolites from the aerial parts of *Centaurea pannonica* (Heuffel) Simonkai from Serbia. Chemotaxonomic importance. Phytochemistry 94, 159–170.

Mimica-Dukić, N., Kujundžić, S., Soković, M., Couladis, M., 2003. Essential oil composition and antifungal activity of *Foeniculum vulgare* mill. obtained by different distillation conditions. Phytother. Res. 17 (4), 368–371.

Mishra, A.K., Dubey, N.K., 1994. Evaluation of some essential oils for their toxicity against fungi causing deterioration of stored food commodities. Appl. Environ. Microbiol. 60 (4), 1101–1105.

Mukherjee, P.K., Saha, K., Saha, B.P., Pal, M., Das, J., 1996. Antifungal activities of the leaf extract of *Cassia tora* Linn. (Fam. Leguminosae). Phytother. Res. 10 (6), 521–522.

Muller-Riebau, F., Berger, B., Yegen, O., 1995. Chemical compositionand fungitoxic properties to phytopathogenic fungi of essential oils of selected aromatic plants growing wild in Turkey. J. Agric. Food Chem. 43, 2262–2266.

Nagarajkumar, M., Jayaraj, J., Muthukrishnan, S., Bhaskaran, R., Velazhahan, R., 2005. Detoxification of oxalic acid by Pseudomonas fluorescens strain PfMDU2: implications for the biological control of rice sheath blight caused by *Rhizoctonia solani*. Microbiol. Res. 160, 291–298.

Nakamura, C.V., Ishida, K., Faccin, L.C., Filho, B.P., Cortez, D.A., Rozental, S., De Souza, W., Ueda, T., 2004. *In vitro* activity of essential oil from *Ocimum gratissimum* L. against four *Candida* species. Res. Microbiol. 155, 579–586.

Nam, J.Y., Kim, H.K., Kwon, J.Y., Han, M.Y., Son, K.H., Lee, U.C., Choi, J.D., Kwon, B.M., 2000. 8-O-Methylsclerotiorinamine, antagonist of the Grb2 – SH2 domain, isolated from *Penicillium multicolour*. J. Nat. Prod. 63 (9), 1303–1305.

Nelson, C., Buttrick, B., Isoherranen, N., 2013. Therapeutic potential of the inhibition of the retinoic acid hydroxylases CYP26A1 and CYP26B1 by Xenobiotics. Curr. Top. Med. Chem. 13 (12), 1402–1428.

Nenadić, M., Soković, M., Glamočlija, J., Ćirić, A., Perić-Mataruga, V., Ilijin, V., Tešević, V., Vujisić, L., Todosijević, M., Vesović, N., Ćurčić, S., 2016a. Antimicrobial activity of the pygidial gland secretion of three ground beetle species (Insecta: Coleoptera: Carabidae). Sci. Nat. 103, 34.

Nenadić, M., Soković, M., Glamočlija, J., Ćirić, A., Perić-Mataruga, V., Tešević, V., Vujisić, L., Todosijević, M., Vesović, N., Ćurčić, S., 2016b. Antimicrobial activity of the pygidial gland secretion of the troglophilic ground beetle *Laemostenus* (Pristonychus) punctatus (Dejean, 1828) (Insecta: Coleoptera: Carabidae). Bull. Entomol. Res. 106 (4), 474–480.

Nenadić, M., Soković, M., Glamočlija, J., Ćirić, A., Perić, M.V., Ilijin, L., Tešević, V., Todosijević, M., Vujisić, L., Vesović, N., Ćurčić, S., 2017a. The pygidial gland secretion of the forest caterpillar hunter, *Calosoma* (Calosoma) sycophanta: the antimicrobial properties against human pathogens. Appl. Microbiol. Biotechnol. 101 (3), 977–985.

Nenadić, M., Ljaljevic-Grbic, M., Stupar, M., Vukojevic, J., Ciric, A., Tesevic, V., Vujisic, L., Todosijevic, M., Vesovic, N., Zivkovic, N., Curcic, S., 2017b. Antifungal activity of the pygidial gland secretion of *Laemostenus punctatus* (Coleoptera: Carabidae) against cave-dwelling micromycetes (article). Sci. Nat. 104, 5–6.

Nestorović, J., Mišić, D., Šiler, B., Soković, M., Glamočlija, J., Ćirić, A., Maksimović, V., Grubišić, D., 2010. Nepetalactone content in shoot cultures of three endemic *Nepeta* species and the evaluation of their antimicrobial activity. Fitoterapia 81 (6), 621–626.

Newman, D.J., Cragg, G.M., Snader, K.M., 2003. Natural products as sources of new drugs over the period 1981-2002. J. Nat. Prod. 66, 1022–1037.

Nielsen, K.F., Smedsgaard, J., 2003. Fungal metabolite screening: database of 474 mycotoxins and fungal metabolites for dereplication by standardised liquid chromatography-UV-mass spectrometry methodology. J. Chromatogr. 1002, 111–136.

Nikolić, M., Marković, T., Marković, D., Perić, T., Glamočlija, J., Stojković, D., Soković, M., 2012. Screening of antimicrobial and antioxidant activity of commercial *Melaleuca alternifolia* (tea tree) essential oils. J. Med. Plant Res. 6, 3852–3858.

Nikolić, M., Jovanović, K.K., Marković, T., Marković, D., Gligorijević, N., Radulović, S., Soković, M., 2014. Chemical composition, antimicrobial, and cytotoxic properties of five Lamiaceae essential oils. Ind. Crop. Prod. 61, 225–232.

Nikolić, M., Stojković, D., Glamočlija, J., Ćirić, A., Marković, T., Smiljković, M., Soković, M., 2015. Could essential oils of green and black pepper be used as food preservatives? J. Food Sci. Technol. 52 (10), 6565–6573.

Nishikawa, T., Tokunaga, S., Fuse, F., Takashima, M., Noda, T., Ohkawa, M., Nakamura, S., Namiki, M., 1997. Experimental study of ascending *Candida albicans* pyelonephritis focusing on the hyphal form and oxidant injury. Urol. Int. 58, 131–136.

Osbourn, A.E., 1996. Preformed antimicrobial compounds and plant defense against fungal attack. Plant Cell 8 (10), 1821–1831.

Özcan, M., Boyraz, N., 2000. Antifungal properties of some herb decoctions. Eur. Food Res. Technol. 212 (2), 86–88.

Padhan, D.K., Pattnaik, S., 2014. *In vivo* antifungal activity of *Acmella* essential oil on a dermatomycotic strain *Trichophyton mentagrophytes* (MTCC-7687). Der Pharm. Sin. 5, 40–44.

Panagouleas, C., Skaltsa, H., Lazari, D., Skaltsounis, A.L., Soković, M., 2003. Antifungal activity of secondary metabolites of *Centaurea raphanina* ssp. *mixta* growing wild in Greece. Pharm. Biol. 41 (4), 266–270.

Park, M.J., Gwak, K.S., Yang, I., Choi, W.S., Jo, H.J., Chang, J.W., 2007. Antifungal activities of the essential oils in *Syzygium aromaticum* (L.) Merr. Et Perry and *Leptospermum petersonii* Bailey and their constituents against various dermatophytes. J. Microbiol. 45, 460–465.

Patridge, E., Gareiss, P., Kinch, M.S., Hoyer, D., 2016. An analysis of FDA-approved drugs: natural products and their derivatives. Drug Discov. Today 21 (2), 204–207.

Pawar, V.C., Thaker, V.S., 2006. *In vitro* efficacy of 75 essential oils against *Aspergillus niger*. Mycoses 49, 316–323.

Pejin, B., Sabovljević, A., Soković, M., Glamočlija, J., Ćirić, A., Vujičić, M., Sabovljević, M., 2012. Antimicrobial activity of *Rhodobryum ontariense*. Hem. Ind. 66 (3), 381–384.

Pejin, B., Glamoclija, J., Ciric, A., Kien-Thai, Y., Sokovic, M., 2013. The moss *Rhodobryum ontariense* tea, a good source of natural antifungals against *Candida albicans*. Rev. Chim. (Bucharest) 64 (5), 552–554.

Pejin, B., Iodice, C., Tommonaro, G., Stanimirovic, B., Ciric, A., Glamoclija, J., Nikolic, M., De Rosa, S., Sokovic, M., 2014a. Further *in vitro* evaluation of antimicrobial activity of themarine sesquiterpene Hhydroquinone avarol. Curr. Pharm. Biotechnol. 15, 583–588.

Pejin, B., Talevski, A., Ciric, A., Glamoclija, J., Nikolic, M., Talevski, T., Sokovic, M., 2014b. *In vitro* evaluation of antimicrobial activity of thefreshwater sponge *Ochridaspongia rotunda* (Arndt, 1937). Nat. Prod. Res. 28 (18), 1489–1494.

Pereira de Oliveira, F., Wunderlay, P.A., Viana, A.C., de Lima, R.B., De Sousa, F.B., De Oliveira Lima, E., 2011. Growth inhibition and morphological alterations of *Trichophyton rubrum* induced by essential oil from *Cymbopogon winterianus* Jowitt ex Bor. Braz. J. Microbiol. 42, 233–242.

Petrović, J., Papandreou, M., Glamočlija, J., Ćirić, A., Baskakis, C., Proestos, C., Lamari, F., Zoumpoulakis, P., Soković, M., 2014. Different extraction methodologies and their influence on the bioactivity of the wild edible mushroom *Laetiporus sulphureus* (bull.) Murrill. Food Funct. 5, 2948–2960.

Picman, A.K., Schneider, F.E., 1993. Inhibition of fungal growth by selected sesquiterpene lactones. Biochem. Syst. Ecol. 21, 307–314.

Picman, A.K., Schneider, E.F., Picman, J., 1995. Effect of flavonoids on mycelial growth of *Verticillium albo-atrum*. Biochem. Syst. Ecol. 23 (7–8), 683–693.

Pinto, E., Pina-Vaz, C., Salgueiro, L., Gonçalves, M.J., Costa-de-Oliveira, S., Cavaleiro, C., Palmeira, A., Rodrigues, A., Martinez-de-Oliveira, J., 2006. Antifungal activity of the essential oil of *Thymus pulegioides* on *Candida, Aspergillus* and dermatophyte species. J. Med. Microbiol. 55, 1367–1373.

Prakash, B., Shukla, R., Singh, P., Kumar, A., Mishra, P.K., Dubey, N.K., 2010. Efficacy of chemically characterized *Piper betle* L. essential oil against fungal and aflatoxin contamination of some edible commodities and its antioxidant activity. Int. J. Food Microbiol. 142 (1–2), 114–119.

Prasad, C.S., Shukla, R., Kumar, A., Dubey, N.K., 2010. *In vitro* and *in vivo* antifungal activity of essential oils of *Cymbopogon martini* and *Chenopodium ambrosioides* and their synergism against dermatophytes. Mycoses 53, 123–129.

Pyun, M.S., Shin, S., 2006. Antifungal effects of the volatile oils from *Allium* plants against *Trichophyton* species and synergism of the oils with ketoconazole. Phytomedicine 13, 394–400.

Rančić, A., Soković, M., van Griensven, L.J.L.D., Vukojević, J., Brkić, B., Ristić, D.M.S., 2003. Antimicrobial activity of limonene. Mater. Med. 23, 83–88.

Rančić, A., Soković, M., Vukojević, J., Simić, A., Marin, P., Duletić-Lašević, S., Djoković, D., 2005. Chemical composition and antimicrobial activities of essential oils of *Myrrhis odorata* (L.) Scop, *Hypericum perforatum* L and *Helichrysum arenarium* (L.) Moench. J. Essent. Oil Res. 17 (3), 341–345.

Rancic, A., Sokovic, M., Karioti, A., Vukojevic, J., Skaltsa, H., 2006. Isolation and structural elucidation of two secondary metabolitesfrom the filamentous fungus *Penicillium ochrochloron* with anti-microbial activity. Environ. Toxicol. Pharmacol. 22, 80–84.

Rashed, K.N., Ciric, A., Glamoclija, J., Calhelha, R.C., Ferreira, I.C.F.R., Sokovic, M., 2013. Antimicrobial activity, growth inhibition of human tumour cell lines, and phytochemical characterization of the hydromethanolic extract obtained from *Sapindus saponaria* L. aerial parts. Biomed. Res. Int. 2013. https://doi.org/10.1155/2013/659183.

Raskin, I., Ribnicky, D.M., Komarnytsky, S., Ilic, N., Poulev, A., Borisjuk, N., Brinker, A., Moreno, D.A., Ripoll, C., Yakoby, N., O'Neal, J.M., Cornwell, T., Pastor, I., Fridlender, B., 2002. Plants and human health in the twenty-first century. Trends Biotechnol. 20 (12), 522–531.

Rates, S.M.K., 2001. Plants as source of drugs. Toxicon 39 (5), 603–613.

Ravichandran, S., Wahidullah, S., D'Souza, L., Anbuchezhian, R.M., 2011. Antimicrobial activity of marine sponge Clathria indica (Dendy, 1889). Bioorg. Khim. 37, 483–489.

Reichling, J., Schnitzler, P., Suschke, U., Saller, R., 2009. Essential oils of aromatic plants with antibacterial, antifungal, antiviral and cytotoxic properties – An overview. Forsch. Komplement. Med. 16, 79–90.

Reis, F.S., Barros, L., Martins, A., Ferreira, I.C., 2012. Chemical composition and nutritional value of the most widely appreciated cultivated mushrooms: an inter-species comparative study. Food Chem. Toxicol. 50 (2), 191–197.

Reis, F.S., Barros, L., Calhelha, R.C., Cirić, A., van Griensven, L.J., Soković, M., Ferreira, I.C., 2013. The methanolic extract of *Cordyceps militaris* (L.) link fruiting body shows antioxidant, antibacterial, antifungal and antihuman tumor cell lines properties. Food Chem. Toxicol. 62, 91–98.

Rios, J.L., Recio, M.C., 2005. Medicinal plants and antimicrobial activity. J. Ethnopharmacol. 100 (1–2), 80–84.

Ristić, M.D., Duletić-Lausević, S., Knezević-Vukcević, J., Marin, P.D., Simić, D., Vukojević, J., Janaćković, P., Vajs, V., 2000. Antimicrobial activity of essential oils and ethanol extract of *Phlomis fruticosa* L. (Lamiaceae). Phytother. Res. 14 (4), 267–271.

Robles, M., Aregullin, M., West, J., Rodriguez, E., 1995. Recent studies on the zoopharmacognosy, pharmacology and neurotoxicology of sesquiterpene lactones. Planta Med. 61 (3), 199–203.

Rosa, L.H., Vieira, L.M.A., Santiago, I.F., Rosa, C.A., 2010. Endophytic fungi community associated with the dicotyledonous plant *Colobanthus quitensis* (Kunth) Bartl. (Caryophyllaceae) in Antarctica. FEMS Microbiol. Ecol. 73, 178–189.

Rowntree, J.K., Cowan, R.S., Leggett, M., Ramsay, M.M., Fay, M.F., 2010. Which moss is which? Identification of the threatened moss *Orthodontium gracile* using molecular and morphological techniques. Conserv. Genet. 11 (3), 1033–1042.

Sabovljevic, A., Sabovljevic, M., 2008. Bryophytes: A Source of Bioactive and New Compounds. Studium Press, Houston, pp. 9–25.

Sabovljevic, A., Sokovic, M., Sabovljevic, M., Grubisic, D., 2006. Antimicrobial activity of *Bryum argenteum*. Fitoterapia 77 (2), 144–145.

Sabovljevic, A., Sokovic, M., Glamoclija, J., Ciric, A., Vujicic, M., 2011. Bio-activities of extracts from some axenically farmed and naturally grown bryophytes. J. Med. Plant Res. 5 (4), 565–571.

Saiki, H., Petersen, A., Scott, C.G., Bailey, K.R., Dunlay, S.M., Finley, R.R., Ruddy, K.J., Yan, E., Redfield, M.M., 2017. Risk of heart failure with preserved ejection fraction in older women after contemporary radiotherapy for breast cancer. Circulation 135, 1388–1396.

Salgueiro, L.R., Pinto, E., Gonçalves, M.J., Pina-Vaz, C., Cavaleiro, C., Rodrigues, A.G., Palmeira, A., Tavares, C., Costa-de-Oliveira, S., Martinez-de-Oliveira, J., 2004. Chemical composition and antifungal activity of the essential oil of *Thymbra capitata*. Planta Med. 70, 572–575.

Sander, C.S., Hipler, U.C., Wollina, U., Elsner, P., 2002. Inhibitory effect of terbinafine on reactive oxygen species (ROS) generation by *Candida albicans*. Mycoses 45 (5–6), 152–155.

Saroglou, V., Karioti, A., Demetzos, C., Dimas, K., Skaltsa, H., 2005. Bioactive sesquiterpene lactones and flavonoids from *Centaurea spinosa*. Antibacterial activity cytotoxic/cytostatic activity against human cell lines *in vitro*. J. Nat. Prod. 68 (9), 1404–1407.

Sato, J., Goto, K., Nanjo, F., Murata, S.K., 2000. Antifungal activity of plant extracts against *Arthrinium sacchari* and *Chaetomium funicola*. J. Biosci. Bioeng. 90 (4), 442–446.

Schmitz, S., Weidenbgner, M., Kunaz, B., 1993. Herbs and spices as selective inhibitors of mould growth. Chem. Mikrobiol. Technol. Lebensmit. 15, 175–177.

Schröter, C., Hipler, U.-C., Wilmer, A., Künkel, W., Wollina, U., 2000. Generation of reactive oxygen species by *Candida albicans* in relation to morphogenesis. Arch. Dermatol. Res. 292 (5), 260–264.

Selvaraj, S., Krishnaswamy, S., Devashya, V., Sethuraman, S., Maheswari Krishnan, U., 2015. Influence of membrane lipid composition on flavonoid–membrane interactions: implications on their biological activity. Prog. Lipid Res. 58, 1–13.

Senio, S., Pereira, C., Vaz, J., Soković, M., Barros, L., Ferreira, I.C.F.R., 2018. Dehydration process influences the phenolic profile, antioxidant and antimicrobial properties of *Galium aparine* L. Ind. Crop. Prod. 120, 97–103.

Seo, H.S., Choi, H.S., Kim, S.R., Choi, Y.K., Woo, S.M., Shin, I., Woo, J.K., Park, S.Y., Shin, Y.C., Ko, S.G., 2012. Apigenin induces apoptosis via extrinsic pathway, inducing p53 and inhibiting STAT3 and NFκB signaling in HER2-overexpressing breast cancer cells. Mol. Cell. Biochem. 336, 319–334.

Sharma, M., Mnoharlal, R., Puri, N., Prasad, R., 2010. Membrane biology laboratory, School of Life Sciences, Jawaharlal Nehru University, New Delhi-110067, India. Biosci. Rep. 30, 391–404.

Shimoni, M., Putievsky, E., Ravid, U., Reuveni, R., 1993. Antifungal activity of volatile fractions of essential oils from four aromatic wild plants in Israel. J. Chem. Ecol. 19 (6), 1129–1133.

Shin, S., Lim, S., 2004. Antifungal effects of herbal essential oils alone and in combination with ketoconazole against *Trichophyton* spp. J. Appl. Microbiol. 97, 1289–1296.

Šiler, B., Mišić, D., Nestorović, J., Banjanac, T., Glamočlija, J., Soković, M., Ćirić, A., 2010. Antibacterial and antifungal screening of *Centaurium pulchellum* crude extracts and main secoiridoid compounds. Nat. Prod. Commun. 5, 1525–1530.

Siljegovic, J.D., Stojkovic, D.S., Nikolic, M.M., Glamoclija, J.M., Sokovic, M.D., Ciric, A.M., 2011a. Antimicrobial activity of aqueous extract of *L. sulphureus* (Bull: Fr). Mur. Proc. Nat. Sci. 120, 297–303.

Siljegovic, J., Glamočlija, J., Soković, M., Vučković, I.M., Tešević, V., Milosavljević, S.M., Stesevic, D., 2011b. Composition and antimicrobial activity of *Seseli montanum* subsp tommasinii essential oil. Nat. Prod. Commun. 6 (2), 263–266.

Silva, S.M.A., Weidenbörner, M., Cavaleiro, A.S.J., 1998. Growth control of different Fusarium species by selected flavones and flavonid mixtures. Mycol. Res. 102 (5), 638–640.

Simic, A., Sokovic, M.D., Ristic, M., Grujic-Jovanovic, S., Vukojevic, J., Marin, P.D., 2004. Chemical composition of some Lauraceae essential oils and their antifungal activities. Phytother. Res. 18, 713–717.

Simic, A., Rančic, A., Sokovic, M.D., Ristic, M., Grujic-Jovanovic, S., Vukojevic, J., Marin, P.D., 2008. Essential oil composition of *Cymbopogon winterianus* and *Carum carvi* and their antimicrobial activities. Pharm. Biol. 46 (6), 437–441.

Sinanoglou, V.J., Zoumpoulakis, P., Heropoulos, F., Proestos, C., Ćirić, A., Petrovic, J., Glamoclija, J., Sokovic, M., 2015. Lipid and fatty acid profile of the edible fungus *Laetiporus sulphurous*. Antifungal and antibacterial properties. J. Food Sci. Technol. 52 (6), 3264–3272.

Singh, S.B., Barrett, J.F., 2006. Empirical antibacterial drug discovery-foundation in natural products. Biochem. Pharmacol. 71 (7), 1006–1015.

Sipkema, D., Franssen, M.C.R., Osinga, R., Tramper, J., Wijffels, R.H., 2005. Marine sponges as pharmacy. Mar. Biotechnol. (NY) 7, 142–162.

Skaltsa, H., Lazari, D., Georgiadou, E., Kakavas, S., Constantinidis, T., 1999. Sesquiterpene lactones from *Centaurea* species: *C. thessala* Hausskn. Subsp. *drakiensis* (Freyn & Sint.) Georg and *C. attica* Nym. subsp. *attica*. Planta Med. 65 (4), 393.

Skaltsa, H., Lazari, D., Garcia, B., Pedro, J.R., Soković, M., Constantinidis, T., 2000a. Sesquiterpene lactones from *Centaurea achaia*, a Greek endemic species: antifungal activity. Z. Naturforsch. 55 (7–8), 534–539.

Skaltsa, H., Lazari, D., Panagouleas, C., Georgiadou, E., Garcia, B., Soković, M., 2000b. Sesquiterpene lactones from *Centaurea* species: *C. thessala* and *C. attica*. Antifungal activity. Phytochemistry 55 (8), 903–908.

Skaltsa, H., Demetzos, C., Lazari, D., Sokovic, M., 2003. Essential oil analysis and antimicrobial activity of eight *Stachys* species from Greece. Phytochemistry 64 (3), 743–752.

Skandamis, P.N., Nychas, J.G., 2000. Development and evaluation of a model predicting the survival of *Escherichia coli* O157:H7 NCTC 12900 in homemade eggplant salad at various temperatures, pHs, and oregano essential oil concentrations. Appl. Environ. Microbiol. 66, 1646–1653.

Smiderle, F.R., Ruthes, A.C., van Arkel, J., Chanput, W., Iacomini, M., Wichers, L.J., Van Griensven, H.J., 2011. Polysaccharides from *Agaricus bisporus* and *Agaricus brasiliensis* show similarities in their structures and their immunomodulatory effects on human monocytic THP-1 cells. BMC Complement. Altern. Med. 11, 58.

Smiljkovic, M., Stanisavljevic, D., Stojkovic, D., Petrovic, I., Vicentic, J.M., Popovic, J., Golic Grdadolnik, S., Markovic, D., Sankovic-Babice, S., Glamoclija, J., Stevanovic, M., Sokovic, M., 2017. Apigenin-7-O-glucoside versus apigenin: insight into the modes of anticandidal and cytotoxic actions. EXCLI J. 16, 795–807.

Soković, D.M., 2001. Antifungal Activities of Essential Oil of Selected Aromatic and Medicinal Plants *in vitro* and *in vivo*. Ph.D. Thesis, Faculty of Biology, University of Belgrade, Yugoslavia.

Soković, M.D., van Griensven, L.J.L.D., 2006. Antimicrobial activity of essential oils and their components against the three major pathogens of the cultivated button mushroom, *Agaricus bisporus*. Eur. J. Plant Pathol. 116, 211–224.

Soković, M., Tzakou, O., Pitarokili, D., Couladis, M., 2002. Antifungal activities of selected aromatic plants growing wild in Greece. Food Nahrung 5, 317–320.

Sokovic, M., Grubisic, D., Ristic, M., 2005. Chemical composition and antifungal activities of essential oils from leaves, calyx and corolla of *Salvia brachyodon* Vandas. J. Essent. Oil Res. 17, 227–229.

Soković, M.D., Glamočlija, J., Marin, P.D., Brkić, D., Vukojević, J., Jovanović, D., Bulajić, N., Kataranovski, D., 2006. Antifungal activity of the essential oil of *Mentha x piperita*. Pharm. Biol. 44, 511–515.

Soković, M., Marin, P.D., Brkić, D., van Griensven, L.J.L.D., 2007. Chemical composition and antibacterial activity of essential oils of ten aromatic plants against human pathogenic bacteria. Food Glob. Sci. Books 1 (1), 1–10.

Soković, M., Ristić, M., Grubišić, D., 2008a. Chemical composition and antifungal activity of the essential oil from *Juniperus excelsa* berries. Pharm. Biol. 342 (4–5), 328–331.

Soković, M., Glamoclija, J., Ciric, A., Grubišić, D., Stojković, D., Ristić, M., 2008b. Antimicrobial activity of essential oils isolated from different parts of endemic plant *Portenschlagiella ramosissima* Tutin. J. Essent. Oil Res. 20 (4), 369–372.

Soković, M., Glamočlija, J., Ćirić, A., Kataranovski, D., Marin, P., Vukojević, J., Brkić, D., 2008c. Antifungal activity of the essential oil of *Thymus vulgaris* L. and thymol on experimentally induced dermatomycoses. Drug Dev. Ind. Pharm. 34, 1388–1393.

Soković, M., Vukojević, J., Marin, P.D., Brkić, D.D., Vajs, V., van Griensven, L.J.L.D., 2009a. Chemical composition of essential oils of *Thymus* and *Mentha* species and their antifungal activities. Molecules 14, 238–249.

Soković, M., Brkić, D., Džamić, A., Ristić, M., Marin, P., 2009b. Chemical composition and antifungal activity of *Salvia desoleana* Atzei & Picci essential oil and its major components. Flav. Fragr. J. 24 (2), 83–87.

Soković, M., Stojković, D., Glamočlija, J., Ćirić, A., Ristić, M., Grubišić, D., 2009c. Susceptibility of pathogenic bacteria and fungi to essential oils of wild *Daucus carota*. Pharm. Biol. 47 (1), 38–43.

Soković, M., Glamočlija, J., Marin, P.D., Brkić, D., van Griensven, L.J., 2010. Antibacterial effects of the essential oils of commonly consumed medicinal herbs using an *in vitro* model. Molecules 15, 7532–7546.

Soković, M., Glamoclija, J., Ciric, A., Kataranovski, D., Marin, P., Vukojevic, J., Brkić, D., 2012. Antifungal activity of the essential oils and components *in vitro* and *in vivo* on experimentally induced dermatomycoses at rats. Dig. J. Nanomater. Biostruct. 7 (3), 959–966.

Soković, M., Glamočlija, J., Ćirić, A., 2013. Natural products from plants and fungi as fungicides. In: Nita, M. (Ed.), Showcases of Integrated Plant Disease Management from around the World, pp. 185–232.

Sokovic, M., Ciric, A., Glamoclija, J., Skaltsa, H., 2017. Biological activities of sesquiterpene lactones isolated from the genus *Centaurea* L. (Asteraceae). Cur. Pharm. Des. 23 (19), 2767–2786.

Sonboli, A., Salehi, P., Yousefzadi, M., 2004. Antimicrobial activity and chemical composition of the essential oil of *Nepeta crispa* Willd. From Iran. Z. Naturforsch. 59, 653–656.

Sparringa, R.A., Owens, J.D., 1999. Glucosamine content of tempemould, *Rhizopus oligosporus*. Int. J. Food Microbiol. 47, 153–157.

Srivastava, J.K., Gupta, S., 2009. Extraction, characterization, stability and biological activity of flavonoids isolated from chamomile flowers. Mol. Cell. Pharmacol. 1 (3), 138–147.

Stojković, D., Glamočlija, J., Soković, M., Grubišić, D., Petrović, S., Kukić, J., Ristić, M., 2008a. Chemical composition, antimicrobial and antiradical properties of the essential oils of *Seseli globiferum*. Fruits. Nat. Prod. Commun. 3 (11), 1935–1938.

Stojković, D., Soković, M., Glamočlija, J., Džamić, A., Ristić, M., Fahal, S., Khalid, S., Djuić, S., Petrović, S., 2008b. Susceptibility of three clinical isolates of *Actinomodura maduraeto* α-pinene, the bioactive agent of *Pinus pinaster* turpentine oil. Arch. Biol. Sci. 60 (4), 697–701.

Stojkovic, S., Petrovic, S., Kukic, J., Dzamic, A., Ristic, M., Milenkovic, M., Glamoclija, J., Sokovic, M., Stojkovic, D., 2009. Chemical composition and antimicrobial and antioxidant activity of *Seseli rigidum* flower essential oil. Chem. Nat. Comp. 45 (2), 253–256.

Stojković, D., Soković, M., Glamočlija, J., Džamić, A., Ciric, A., Ristić, M., Grubišić, D., 2011. Chemical composition and antimicrobial activity of *Vitex agnus-castus* L. fruits and leaves essential oils. Food Chem. 128 (4), 1017–1022.

Stojković, D.S., Barros, L., Calhelha, R.C., Glamoclija, J., Ciric, A., van Griensven, L.J.L.D., Sokovic, M., Ferreira, I.C.F.R., 2013b. A detailed comparative study between chemical and bioactive properties of *Ganoderma lucidum* from different origins. Int. J. Food Sci. Nutr. 65 (1), 42–47.

Stojkovic, D., Reis, F.S., Glamoclija, J., Ćirić, A., Barros, L., van Griensven, L.J.L.D., Ferreira, C.F.R.I., Soković, M., 2014. Cultivated strains of *Agaricus bisporus* and *A. brasiliensis*: chemical characterization and evaluation of antioxidant and antimicrobial properties for the final healthy product–natural preservatives in yoghurt. Food Funct. 5, 1602–1612.

Suabjakyong, P., Saiki, R., van Griensven, L.J., Higashi, K., Nishimura, K., Igarashi, K., Toida, T., 2015. Polyphenol extract from *Phellinus igniarius* protects against acrolein toxicity *in vitro* and provides protection in a mouse stroke model. PLoS One 10, e0122733.

Taira, J., Nanbu, H., Ueda, K., 2009. Nitric oxide-scavengingcompounds in *Agrimonia pilosa* Ledeb on LPS-induced RAW264. 7 macrophages. Food Chem. 115, 1221–1227.

Tangarife-Castaño, V., Roa-Linares, V., Betançur-Galvis, L.A., Durán García, D.C., Stashenko, E., Mesa-Arango, A.C., 2012. Antifungal activity of Verbenaceae and Labiatae families essential oils. Pharmacology 1, 133–145.

Tavares, A.C., Gonçalves, M.J., Cruz, M.T., Cavaleiro, C., Lopes, M.C., Canhoto, J., Salgueiro, L.R., 2010. Essential oils from *Distichoselinum tenuifolium*: chemical composition, cytotoxicity, antifungal and anti-inflammatory properties. J. Ethnopharmacol. 130, 593–598.

Tešević, V., Aljančić, I., Milosavljević, S., Vajs, V., Djordjević, I., Jadranin, M., Menković, N., Matevski, V., 2014. Secondary metabolites of three endemic *Centaurea* L. species. J. Serb. Chem. Soc. 79 (11), 1355–1362.

Thompson, D.P., 1989. Fungitoxic activity of essential oil components on food storage fungi. Mycologia 81 (1), 151–153.

Thongson, C., Davidson, P.M., Mahakarnchanakul, W., Weiss, J., 2004. Antimicrobial activity of ultrasound assisted solventextracted spices. Lett. Appl. Microbiol. 39, 401–406.

Tiwari, T.N., Chansouria, J.P.N., Dubey, N.K., 2003. Antimycotic potency of some essential oils in the treatment of induced dermatomycosis of an experimental animal. Pharm. Biol. 41, 351–356.

Toledo Jr., J.C., Augusto, O., 2012. Connecting the chemical and biological properties of nitric oxide. Chem. Res. Toxicol. 25, 975–989.

Treutter, D., 2005. Significance of flavonoids in plant resistance and enhancement of their biosynthesis. Plant Biol. 7 (6), 581–591.

Trillini, B., Velasquez, E.R., Pellegrino, R., 1996. Chemical composition and antimicrobial activity of essential oil of *Piper angustifolium*. Planta Med. 62, 372–373.

Tshikalange, T.E., Meyer, J.J.M., Hussein, A.A., 2005. Antimicrobial activity, toxicity and the isolation of a bioactive compound from plants used to treat sexually transmitted diseases. J. Ethnopharmacol. 96, 515–519.

Tsukatani, T., Suenaga, H., Shiga, M., Noguchi, K., Ishiyama, M., Ezoe, T., Matsumoto, K., 2012. Comparison of the WST-8 colorimetric method and the CLSI broth microdilution method for susceptibility testing against drug-bacteria. J. Microbiol. Methods 90, 160–166.

Turkoglu, M.E., Durub, N., Mercan, I., Kivrak, I., Gezera, K., 2007. Antioxidant and antimicrobial activities of Laetiporus sulphureus (bull.) Murrill. Food Chem. 101 (1), 267–273.

Usjak, L., Petrović, S., Drobac, M., Soković, M., Stanojković, T., Ćirić, A., Niketić, M., 2016a. Chemical composition and bioactivity of the essential oils of *Heracleum pyrenaicum* subsp. *pollinianum* and *Heracleum orphanidis*. Nat. Prod. Commun. 11 (4), 529–534.

Usjak, L.J., Petrovic, S.D., Drobac, M.M., Sokovic, M.D., Stanojkovic, T.P., Ciric, A.D., Grozdanić, N.Đ., Niketić, M.S., 2016b. Chemical composition, antimicrobial and cytotoxic activity of *Heracleum verticillatum* Pančić and *H. ternatum* Velen. (Apiaceae) essential oils. Chem. Biodivers. 13 (4), 466–476.

Uysal, S., Aumeeruddy-Elalfi, Z., Zengin, G., Aktumsek, A., Mocan, A., Custodio, L., Neng, N.R., Nogueira, J.M.F., Ćirić, A., Glamočlija, J., Soković, M., Mahomoodally, M.F., 2018. Insight into the biological properties and phytochemical composition of *Ballota macrodonta* Boiss. Et Balansa,—An endemic medicinal plant from Turkey. Ind. Crop. Prod. 113, 422–428.

Vajs, V., Todorović, N., Ristić, M., Tesević, V., Todorović, B., Janaćković, P., Marin, P., Milosavljević, S., 1999. Guaianolides from *Centaurea nicolai*: antifungal activity. Phytochemistry 52 (3), 383–386.

Vajs, V., Trifunović, S., Janaćković, P., Soković, M., Milosavljević, S., Tešević, V., 2004. Antifungal activity of davanone-type sesquiterpenes from *Artemisia lobelii* var. Canescens. J. Serb. Chem. Soc. 69, 969–972.

Veljic, M., Tarbuk, M., Marin, P.D., Ciric, A., Sokovic, M., Marin, M., 2008. Antimicrobial activity of methanol extracts of mosses from Serbia. Pharm. Biol. 46 (12), 871–875.

Veljic, M., Djuric, A., Sokovic, M., Ciric, A., Glamoclija, J., Marin, P.D., 2009. Antimicrobial activity of methanol extracts of *Fontinalis antipyretica*, *Hypnum cupressiforme* and *Ctenidium molluscum*. Archiv. Biol. Sci. Belgrade 61 (2), 225–229.

Veljic, M., Ciric, A., Sokolovic, M., Janackovic, P., Marin, P.D., 2010. Antibacterial and antifungal activity of the liverwort (*Ptilidium pulcherrimum*) methanol extract. Archiv. Biol. Sci. Belgrade 62 (2), 381–395.

Ververidis, F., Trantas, E., Douglas, C., Vollmer, G., Kretzschmar, G., Panopoulos, N., 2007. Biotechnology of flavonoids and other phenylpropanoid-derived natural products. Part I: Chemical diversity, impacts on plant biology and human health. Biotechnol. J. 2 (10), 1214–1234.

Wedge, D.E., Galindo, J.C., Macías, F.A., 2000. Fungicidal activity of natural and synthetic sesquiterpene lactone analogs. Phytochemistry 53 (7), 747–757.

Wei, S., Helsper, J.P.F.G., van Griensven, L.J., 2008. Phenolic compounds present in medicinal mushroom extracts generate reactive oxygen species in human cells *in vitro*. Int. J. Med. Mushrooms 10 (1), 1–13.

Weidenbörner, M., Jha, H.C., 1997. Antifungal spectrum of flavone and flavanone tested against 34 different fungi. Mycol. Res. 101 (6), 733–736.

Weidenbörner, M., Hindorf, H., Jha, H.C., Tsotsonos, P., Egge, H., 1989. Antifungal activity of isoflavonoids against storage fungi of the genus *Aspergillus*. Phytochemistry 28, 3317–3319.

Williamson, G., Barron, D., Shimoi, K., Terao, J., 2005. *In vitro* biological properties of flavonoid conjugates found *in vivo*. Free Radic. Res. 39 (5), 457–469.

Willis, S.R., 2017. World Health Organization NLM classification summer edition. NLM Tech. Bull. 418, e1 (ISBN 978 92 4 159444 8 (NLM classification: QV 766).

Xiao, J., Kai, G., Yamamoto, K., Chen, X., 2013a. Advance in dietary polyphenols as α-glucosidases inhibitors: a review on structure-activity relationship aspect. Crit. Rev. Food Sci. Nutr. 53 (8), 818–836.

Xiao, J., Ni, X., Kai, G., Chen, X., 2013b. A review on structure–activity relationship of dietary polyphenols inhibiting α-amylase. Crit. Rev. Food Sci. Nutr. 53 (5), 497–506.

Yamada, Y., Azuma, K., 1997. Evaluation of the *in vitro* antifungal activity of allicin. Antimicrob. Agents Chemother. 11, 743–749.

You, M., Wickramaratne, D.B., Silva, G.L., Chai, H., Chagwedera, T.E., Farnsworth, N.R., Cordell, G.A., Kinghorn, A.D., Pezzuto, J.M., 1995. Roemerine, anaporphine alkaloid from *Annonas enegalens* is that reverse the multi drug resistance phenotype with cultured cells. J. Nat. Prod. 58, 598–604.

Zengin, G., Aktumsek, A., Ceylan, R., Uysal, S., Mocan, A., Guler, G.O., Mahomoodally, M.F., Glamočlija, J., Ćirić, A., Soković, M., 2017. Shedding light on the biological and chemical fingerprints of three *Achillea* species (*A. biebersteinii*, *A. millefolium* and *A. teretifolia*). Food Funct. 8 (3), 1152–1165.

Zengin, G., Mahomoodally, M.F., Aktumsek, A., Ceylan, R., Uysal, S., Mocan, A., Abdulla, M., Carene, Y., Nancy, M., Allainb, P., Ćirić, A., Glamočlija, J., Soković, M., 2018. Functional constituents of six wild edible *Silene* species: a focus on their phytochemical profiles and bioactive properties. Food Biosci. 23, 75–82.

Zhang, L., Chang, W., Sun, B., Groh, M., Speicher, A., Lou, H., 2011. Bisbibenzyls, a new type of antifungal agent, inhibit morphogenesis switch and biofilm formation through upregulation of DPP3 in *Candida albicans*. PLoS One 6 (12), e28953.

Zhu, X., Zhang, H., Lo, R., Lu, Y., 2005. Antimicrobial activities of *Cynara scolymus* L. leaf, head, and stem extracts. J. Food Sci. 70, M149–M152.

Zjawiony, J.K., 2004. Biologically active compounds from Aphyllophorales (polypore) fungi. J. Nat. Prod. 67 (2), 300–310.

Chapter 5

Synthetic antifungal compounds

Konstantinos Liaras[a] and Marina Soković[b]

[a]*Department of Pharmaceutical Chemistry, School of Pharmacy, Aristotle University of Thessaloniki, Thessaloniki, Greece,* [b]*Mycological Laboratory, Institute for Biological Research "Siniša Stanković", University of Belgrade, National Institute of Republic of Serbia, Belgrade, Serbia*

Chapter outline

The class of **synthetic antifungal compounds** consists of chemical derivatives, which are intended to be used for the treatment of different kinds of fungal infections in humans. A variety of synthetic antifungals have been already approved and are being administered in clinical practice against several types of mycoses, while continuous research efforts around the globe are taking place for the discovery of new lead compounds that will be more potent and with less side effects than the existing marketed drugs. Many of the compounds that belong in this group are derivatives of molecules that can be found in the nature and the rest of them were synthesized after rational drug design processes, including conventional and/or computational methods.

5.1 Allylamines

Allylamines are antifungal agents that exhibit their activity by inhibiting selectively fungal squalene epoxidase. The increased levels of squalene in the cell membrane are harmful for the fungal cell, leading to pH imbalance and disturbance of normal function of the membrane. Allylamines are mainly effective against dermatomycetes and are usually administered for the treatment of skin and nail infections. **Tolnaftate** and **terbinafine** (Fig. 5.1) are the most prominent compounds of this class (Bennet, 2006).

Antifungal Compounds Discovery. https://doi.org/10.1016/B978-0-12-815824-1.00005-6

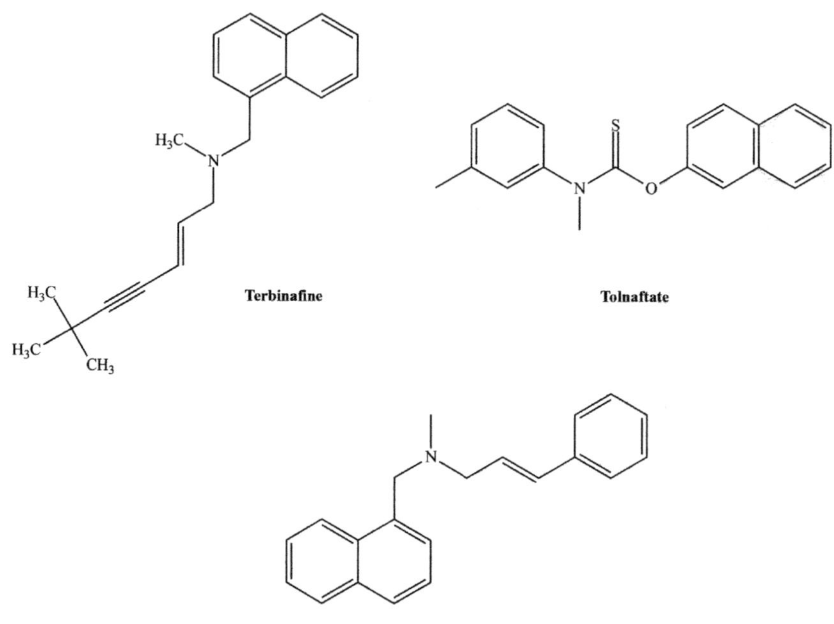

Naftifine

FIG. 5.1 Chemical structures of tolnaftate, terbinafine, and naftifine.

Terbinafine can be used topically or per os. It can be hepatotoxic and also can cause headache, dizziness, diarrhea, heartburn, stomach pain, loss of appetite, vomiting, fever, breathing problems, etc. Terbinafine is much more potent than its precursor, **naftifine** (Fig. 5.1), in both in vitro and in vivo studies, exhibiting much higher selectivity for fungal squalene epoxidase over the mammalian enzyme (Nussbaumer et al., 1991a; Petranyi et al., 1987; Ryder, 1987).

Tonaftate is employed mostly for topical applications to treat tinea skin infections. It is considered to be a safe drug since its adverse effects are rare.

Other allylamine derivatives

Taking into account the results about the activity of terbinafine, Nussbaumer et al. (1991b) synthesized novel terbinafine derivatives by replacing naphthalene with benzo[*b*]thiophene and its derivatives and screened them for their antifungal potential in vitro against *Microsporum canis*, *Sporothrix schenckii*, *Aspergillus fumigatus*, *Candida albicans*, *Candida parapsilosis*, and *Trichophyton mentagrophytes*. After several structural modifications on the most active analogues, the authors came to the conclusion that, concerning the antifungal effect, naphthalene could be replaced with benzo[*b*]thiophene that has a side chain at position 3, 4, or 7. It should be pointed out that from all the synthesized derivatives,

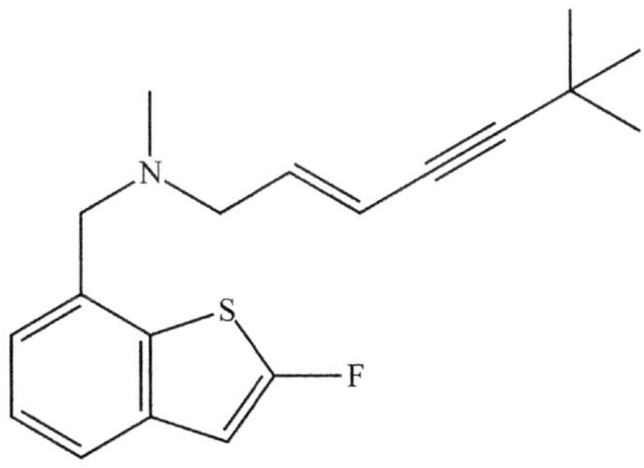

S1

FIG. 5.2 Chemical structure of compound **S1**.

compound **S1** (Fig. 5.2) exhibited the best antifungal potential (MICs range of 0.0015–0.80 μg/L) being about 30 times more potent than terbinafine (MICs range of 0.003–25.0 μg/L).

Inspired by their previous findings, Nussbaumer et al. (1991a) presented the synthesis of a large amount of compounds containing carbon analogues of terbinafine, terbinafine metabolites, and benzo[*b*]thienyl allylamines. Metabolite **S2** (Fig. 5.3) revealed a very strong in vitro antifungal potential compared to terbinafine, leading to the design and synthesis of a new series of derivatives based on the scaffold of this compound, having as most active against *C. albicans* the derivatives **S3**, **S4**, and **S5** (MICs = 0.8 mg/L). Furthermore, compounds **S6** and **S7** with a phenyl ring at the side chain were also prepared, with the first of them lacking activity while the second being very potent as antifungal with MICs ranging at 0.006–25.0 mg/L.

Taking into consideration the positive results regarding compound **S7** (Fig. 5.3), Nussbaumer et al. (1993) designed and synthesized a series of analogues of this terbinafine derivative containing para-substituted benzylamine side chains. The in vitro evaluation of the antifungal activity of the synthesized compounds took place with the methods that were used in the previously mentioned papers of these authors. From the obtained results was observed that compounds **S8** and **S9** (Fig. 5.4) with a carbon atom linking their phenyl rings exhibited improved antifungal potential, while introducing a quaternary carbon atom in the same position led to derivative **S10** that exhibited strong activity against all the selected fungi. It should be pointed out that compound **S10** demonstrated stronger activity than terbinafine and derivative **S7** against *A. fumigatus* and all *Candida* strains with MIC = 0.1 mg/L and MICs = 0.1–1.56 mg/L, respectively. Moreover, replacement of naphthalene moiety benzo[*b*]thiophene gave derivative **S11** with activity similar to that of **S9**. The authors came to the conclusion that introduction of an additional phenyl ring at the benzylamine derivatives of this class not only favors antifungal potential, but also could lead to the discovery of a compound that would potentially be candidate for systemic administration.

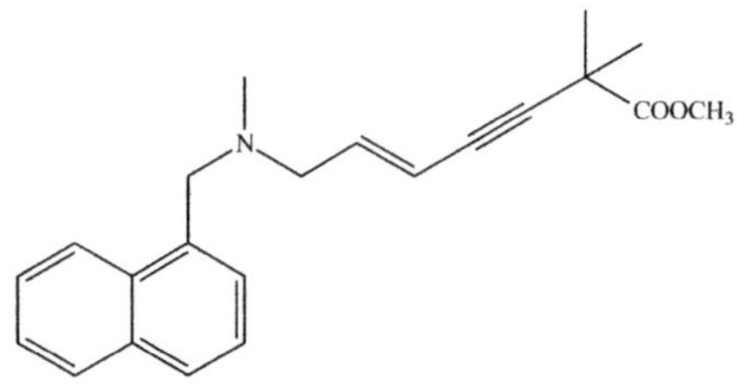

S2

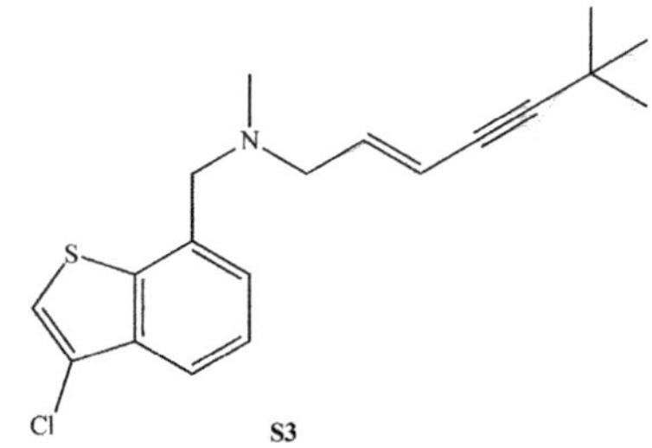

S3

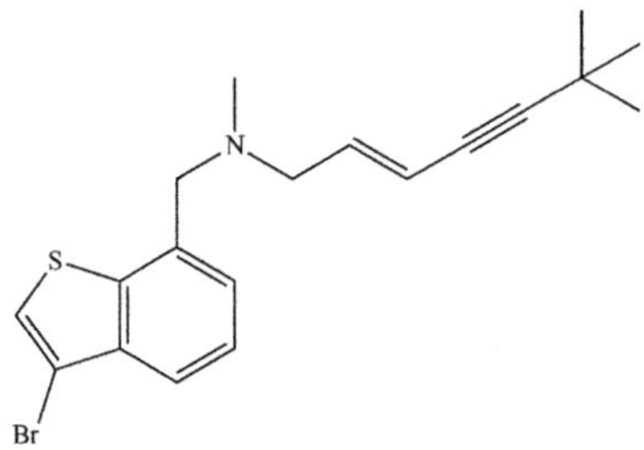

S4

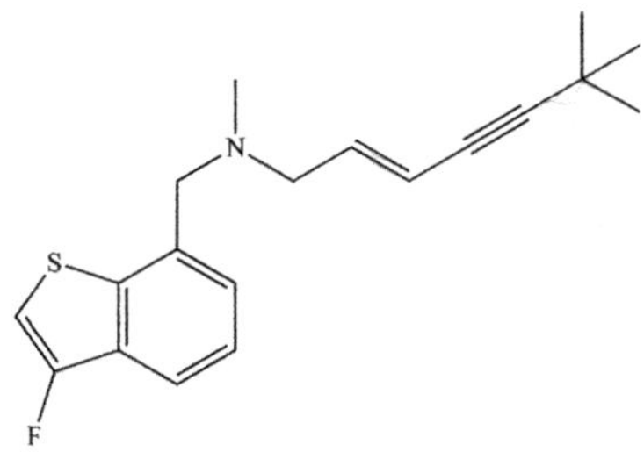

S5

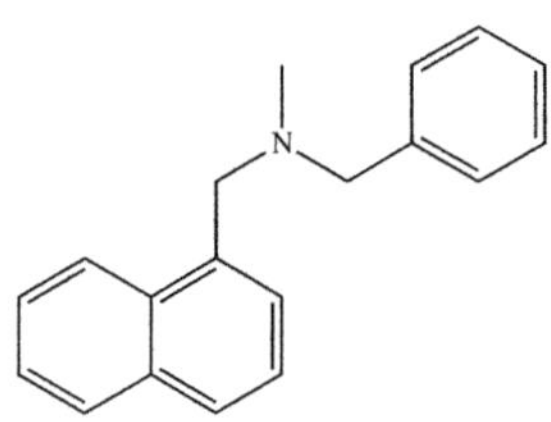

S6

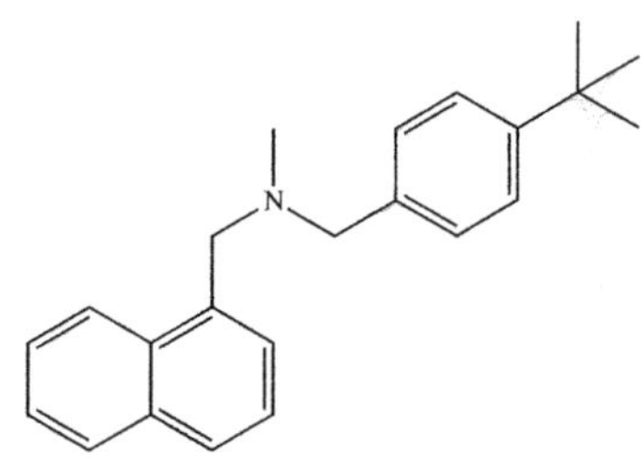

S7

FIG. 5.3 Chemical structures of compounds **S2–S7**.

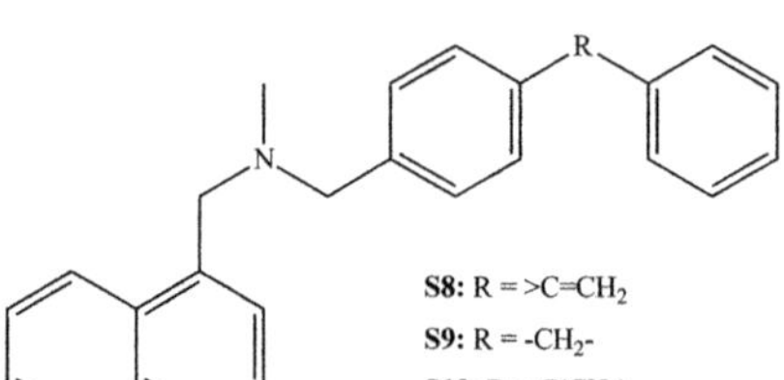

S8: R = >C=CH$_2$
S9: R = -CH$_2$-
S10: R = C(CH$_3$)$_2$

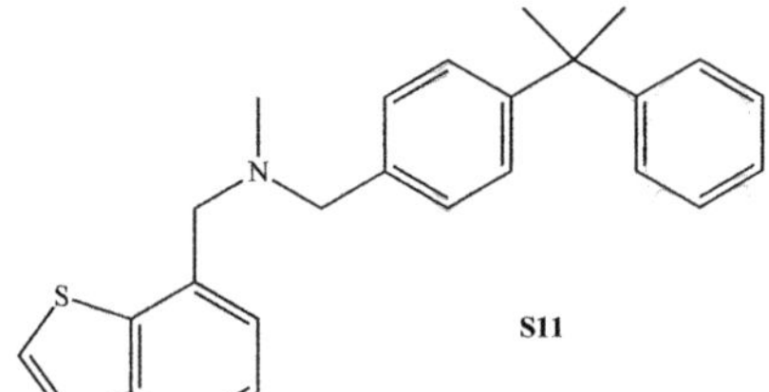

S11

FIG. 5.4 Chemical structures of compounds **S8–S11**.

5.2 Azoles

Imidazoles and triazoles

Azoles, since 1944, have been one of the most important classes of antifungal agents for the treatment of fungal infections. Compounds that belong to the class of azoles can be categorized chemically, according to the number of nitrogen atoms in their azole ring in imidazoles (two nitrogens) and triazoles (three nitrogens). **Chlormidazole** (Fig. 5.5) was the first azole that was developed to be used in clinical practice topically for the treatment of mycoses such as skin and nail infections (Herrling et al., 1959; Seeliger, 1958).

Ketoconazole, **bifonazole**, **miconazole**, and **clotrimazole** (Fig. 5.6) are some of the most prominent imidazole antifungals, while **itraconazole**, **fluconazole**, and **voriconazole** (Fig. 5.7) are widely known triazoles.

In general, azoles exhibit fungicidal activity at higher concentrations (μM range) and fungistatic at lowers ones (nM range) (Ballington and Laughlin, 2008; Barar, 2006).

Azoles demonstrate fungistatic activity in vitro, blocking the cytochrome P450-dependent enzyme lanosterol 14α-demethylase (CYP51A1) that is essential for the synthesis of ergosterol, an important constituent of fungal plasma membrane. Azoles show higher affinity for the fungal enzyme over the one that can be found in mammals and takes part in the synthesis of cholesterol, making them effective for treating fungal infections in humans. More specifically, they act by binding of the heterocyclic nitrogen atom (*N*-3 of imidazole and *N*-4 of 1,2,4-triazole) to heme iron atom of the porphyrin in the binding site of 14α-demethylase (Hitchcock et al., 1990). The presence of the azole aromatic ring is important for the antifungal activity and, in particular, replacement of imidazole by a bioisosteric triazole ring retains antifungal potential while exhibiting increased selectivity on fungi over mammals.

Beside their beneficial effects in treating fungal infections, azoles are known to possess a series of side effects such as hepatic toxicity, gastrointenstinal intolerance, dizziness, rashes, psychosis, adrenal and testicular suppression (e.g., ketoconazole), visual disturbances (e.g., voriconazole), etc. (Wu et al., 2004).

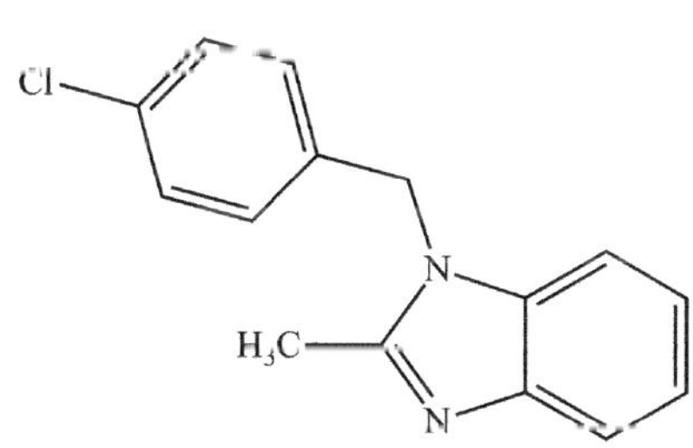

Chlormidazole

FIG. 5.5 Chemical structure of **chlormidazole**.

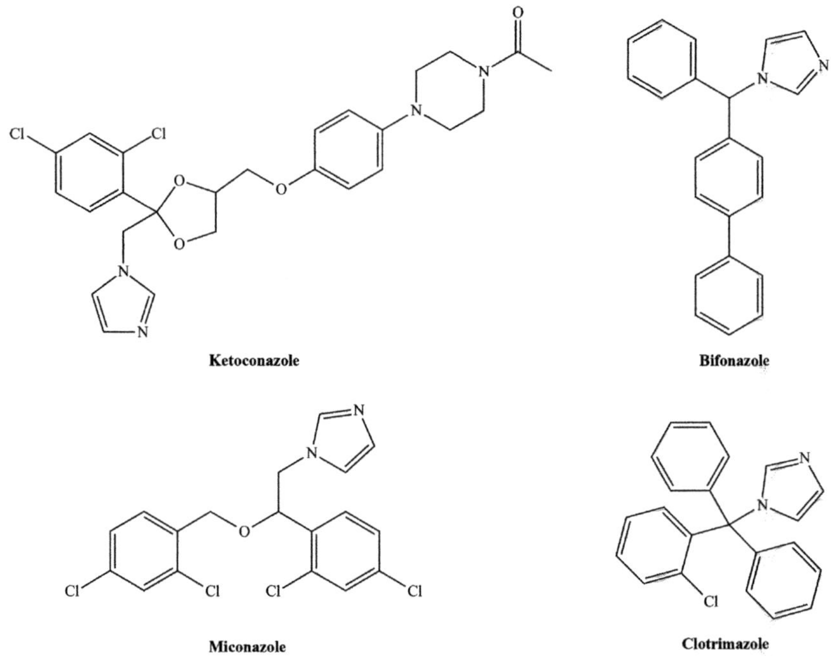

FIG. 5.6 Chemical structures of **ketoconazole**, **bifonazole**, **miconazole**, and **clotrimazole**.

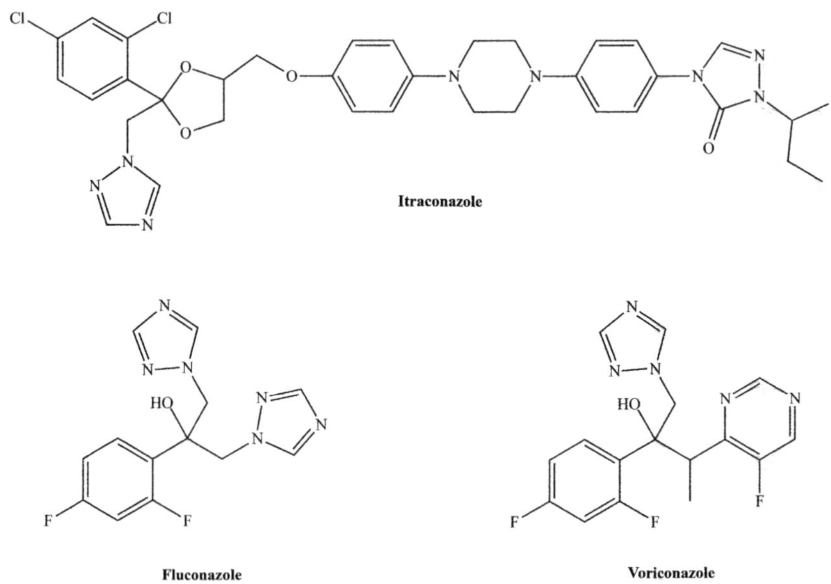

FIG. 5.7 Chemical structures of **itraconazole**, **fluconazole**, and **voriconazole**.

In particular, ketoconazole and miconazole are considered significantly toxic when they are administered systemically, while fluconazole, itraconazole, posaconazole, voriconazole, and isavuconazole demonstrate a safer profile (Como and Dismukes, 1994).

In general, miconazole and clotrimazole are used mostly for the treatment of superficial mycoses and not for systemic ones (Como and Dismukes, 1994). Bifonazole exhibits activity in vitro against dermatomycetes, yeasts, molds, dimorphic fungi and is widely used in formulations for topical application having as possible side effects redness, burning, skin irritation, itching, etc. (Lackner and Clissold, 1989; Watanabe et al., 2006).

Miconazole shares a general chemical structure with **econazol**e, **oxiconazole**, **sulconazole**, and **tioconazole** (Fig. 5.8). All these compounds are administered only topically as antifungals for the treatment of several kinds of fungal infections such as tinea corporis, tinea pedis, tinea cruris, superficial candidiasis, etc.

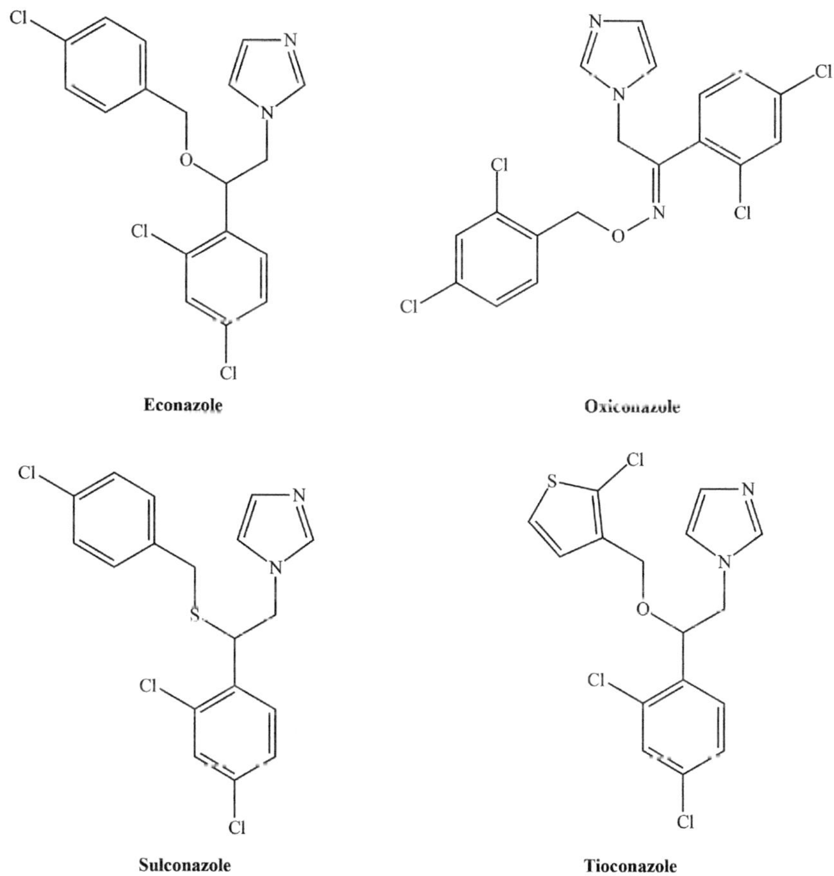

FIG. 5.8 Chemical structures of **econazole**, **oxiconazole**, **sulconazole**, and **tioconazole**.

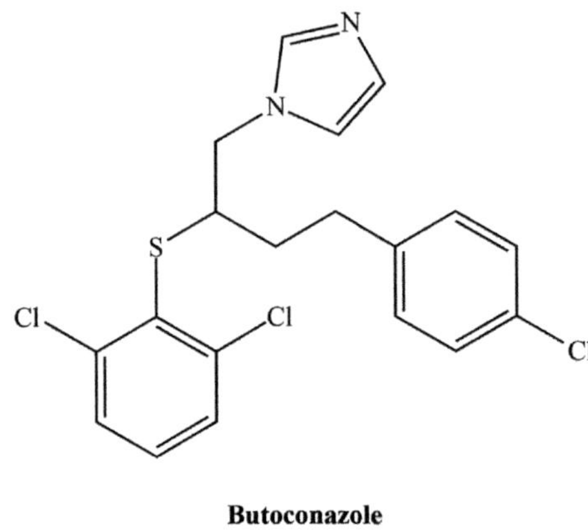

FIG. 5.9 Chemical structure of **butoconazole**.

Butoconazole (Fig. 5.9) is also used topically for the treatment of vulvovaginitis but can cause vaginal irritation, foul smell, stomach pain, and other unwanted effects.

The other marketed azoles that are being used for the treatment of systemic infections (itraconazole, ketoconazole, voriconazole) exhibit low solubility in water, variable absorption per os, more complex hepatic metabolism, and numerous drug interactions. It should be noted that fluconazole and itraconazole are generally better tolerated than ketoconazole (Como and Dismukes, 1994).

Triazoles are used in clinical practice as first-line drugs for the treatment of mucosal and blood-borne candidiasis, sporotrichosis, penicilliosis, coccidioidomycosis, paracoccidioidomycosis, blastomycosis, various melanized hyphomycetes, and mild-to-moderate histoplasmosis (Wu et al., 2004).

Fluconazole is one of the most prescribed azoles, exhibiting good oral absorption. Having an expensive intravenous formulation, is usually administered per os, unless patients are able to accept only parenteral drugs. It demonstrates a relatively narrow activity spectrum focusing mainly on yeasts and some endemic mycoses, while having only a few interactions with other drugs (Como and Dismukes, 1994). Fluconazole is active against several *Candida* spp. excluding *C. krusei* and exhibits also excellent activity against *Cryptococcus neoformans* (Table 5.1) (Nett and Andes, 2016). It is widely used for the treatment of oropharyngeal and esophageal candidiasis and is the first choice against cryptococcal meningitis and meningitis that is caused by *Coccidioides* spp. (Galgiani et al., 1993). Fluconazole is also effective against endemic mycoses but less potent than itraconazole in treating histoplasmosis, blastomycosis, sporotrichosis, paracoccidioidomycosis, and *Penicillium marneffei* caused penicilliosis (Jacobs et al., 1996). Like the majority of azoles, it can be hepatotoxic and also cause fever, chills, rashes, diarrhea, headache, heartburns, dizziness, etc. (Uchida et al., 2001).

Voriconazole is a derivative of fluconazole, having a triazole replaced by a fluoropyrimidine moiety and also a methyl group introduced at the propyl

TABLE 5.1 Activity spectrum of some systemic antifungals.

	AMF	FLC	FLU	ITR	VOR	POS	ISA	CAS	MIF	ANF
C. albicans	✓✓	✓✓	✓✓	✓✓	✓✓	✓✓	✓✓	✓✓	✓✓	✓✓
C. glabrata	✓✓	✓✓	✓	✓	✓✓	✓✓	✓✓	✓	✓	✓
C. parapsilosis	✓✓	✓✓	✓✓	✓✓	✓✓	✓✓	✓✓	✓✓	✓✓	✓✓
C. tropicalis	✓✓	✓✓	✓✓	✓✓	✓✓	✓✓	✓✓	✓✓	✓✓	✓✓
C. krusei	✓✓	✓	–	✓	✓✓	✓✓	✓✓	✓✓	✓✓	✓✓
C. lusitaniae	–	✓✓	✓✓	✓✓	✓✓	✓✓	✓✓	✓✓	✓✓	✓✓
A. fumigatus	✓✓	–	–	✓	✓✓	✓✓	✓✓	✓	✓	✓
C. neoformans	✓✓	✓✓	✓✓	✓✓	✓✓	✓✓	✓✓	–	–	–
Mucorales	✓✓	–	–	–	–	✓✓	✓✓	–	–	–
Fusarium spp	✓	–	–	✓	✓✓	✓✓	✓✓	–	–	–
Scedosporium spp.	✓	–	–	✓	✓	✓	✓	–	–	–
B. dermatitidis	✓✓	–	✓	✓✓	✓✓	✓✓	✓✓	–	–	–
C. immitis	✓✓	–	✓✓	✓✓	✓✓	✓✓	✓✓	–	–	–
H. capsulatum	✓✓	–	✓	✓✓	✓✓	✓✓	✓✓	–	–	–

AMF, amphotericin B; *FLC*, flucytosine; *FLU*, fluconazole; *ITR*, itraconazole; *VOR*, voriconazole; *POS*, posaconazole; *ISA*, isavuconazole; *CAS*, caspofungin; *MIF*, micafungin; *ANF*, anidulafungin.
C. albicans, *Candida albicans*; *C. glabrata*, *Candida glabrata*; *C. parapsilosis*, *Candida parapsilosis*; *C. tropicalis*, *Candida tropicalis*; *C. krusei*, *Candida krusei*; *C. lusitaniae*, *Candida lusitaniae*; *A. fumigatus*, *Aspergillus fumigatus*; *C. neoformans*, *Cryptococcus neoformans*; *B. dermatitidis*, *Blastomyces dermatitidis*; *C. immitis*, *Coccidioides immitis*; *H. capsulatum*, *Histoplasma capsulatum*.

backbone of molecule. These structural modifications led to an increase in the potency compared to fluconazole that can be attributed to higher affinity of voriconazole for lanosterol 14α-demethylase. Voriconazole can be found in oral as well as intravenous formulations and is being used in clinical practice for the treatment of esophageal candidiasis and fungal infections caused by *Fusarium* spp. and *Scedosporium apiospermum*. In general, voriconazole exhibits potential similar to fluconazole and itraconazole being additionally more effective against *C. glabrata* strains that are resistant to fluconazole. The activity of voriconazole against several fungal species is presented in Table 5.1. It should be noted that the potency and safety of voriconazole in immunosuppressed patients has been proven in clinical trials. Despite its increased potency compared to fluconazole, voriconazole can still be responsible for hepatotoxicity and other adverse effects that are common among azole antifungals (Denning et al., 2002; Nett and Andes, 2016; Pfaller et al., 2007; Sabo and Abdel-Rahman, 2000).

Ravuconazole (Fig. 5.10) is also a fluconazole analogue, having one of the triazoles of the molecule replaced by a thiazole. It exhibits activity against many pathogenic fungi such as *Aspergillus* spp., *Trichosporon* spp., *Candida neoformans,* and *Candida* spp. The in vitro anticandidal potential of ravuconazole is considered to be broader than that of fluconazole and itraconazole, respectively. It also demonstrates in vitro activity against dermatomycetes like *Trichophyton mentagrophytes*, *T. rubrum*, *Epidermophyton floccossum*, *Microsporum gypseum*, and *Microsporum canis*.

Ketoconazole is the most popular antifungal agent in third world, due to its low cost. For best absorption requires lipid-containing foods and gastric acidity. Ketoconazole is not so often the choice of preference for the treatment of systemic fungal infections and this can be explained by the efficacy of the other azole antifungals. It is prescribed for the treatment of tinea pedis, tinea cruris, tinea corporis among other kinds of fungal infections. It shares some of the common adverse effects of azoles such as itching, rashes, dry skin or scalp, etc. (Wu et al., 2004).

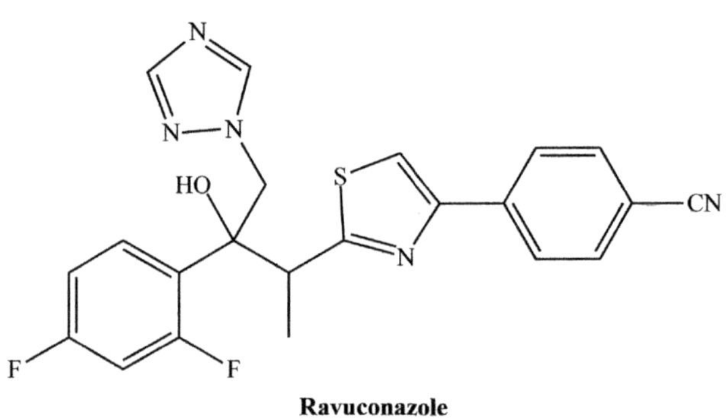

FIG. 5.10 Chemical structure of **ravuconazole**.

Itraconazole shares a similar absorption profile with ketoconazole, needing gastric acidity and lipid-containing foods for maximum result. The activity spectrum of itraconazole is shown in Table 5.1. It exhibits antifungal potential against most *Candida* spp. and it can be used for the treatment of paracoccidiomycosis, histoplasmosis, aspergillosis, and phaeohyphomycosis. Fluconazole shows activity also against several *Aspergillus* spp. such as *A. fumigatus*, *A. flavus*, *A. nidulans*, and *A. terreus*. Formulation of itraconazole as a solution in beta-hydrodextrin has as a result better absorption per os, overcoming the need for gastric acidity and foods rich in lipids. Itraconazole can also be administered parenterally, being licensed for intravenous therapies. Like many azoles, itraconazole is considered hepatotoxic and can cause diarrhea, constipation, heartburn nervousness, muscle pains, and many other undesired effects (Nett and Andes, 2016; Prentice et al., 1995; Sharkey et al., 1990).

Posaconazole (Fig. 5.11) is an antifungal drug that chemically is a derivative of itraconazole, developed by introducing a 1,3-dioxolone backbone. It possesses antifungal activity against a series of dermatomycetes, *Aspergillus* spp., *Candida* spp., *C. neoformans*, *Trichosporon* spp., and Zygomycetes. It should be noted that posaconazole is effective also against strains of *Candida* that have developed resistance to fluconazole (Table 5.1) while demonstrating a similar safety profile (Barchiesi et al., 2001; Paphitou et al., 2002; Pfaller et al., 2002; Sheehan et al., 1999; Sun et al., 2002; Wu et al., 2004). It would be interesting to point out that innovative capsule formulations of posaconazole, which have as target the controlled release of this agent, lead ultimately to higher bioavailability than that of conventional azoles (Merck and Co. Inc, 2014).

Isavuconazole (Fig. 5.12) is a broad-spectrum antifungal (Table 5.1), which was released in the recent years and shows in general a favorable pharmacological profile, better bioavailability, and not so many drug interactions compared to previous azole generations. It is active against most *Candida* spp., including *C. glabrata* and *C. krusei*, as well as against *A. fumigatus*, *A. flavus*, *A. terreus*, and *Cryptococcus neoformans* (Nett and Andes, 2016).

Terconazole (Fig. 5.13) is another triazole that can be applied topically for the treatment of vaginal yeast infections. Some of its adverse effects

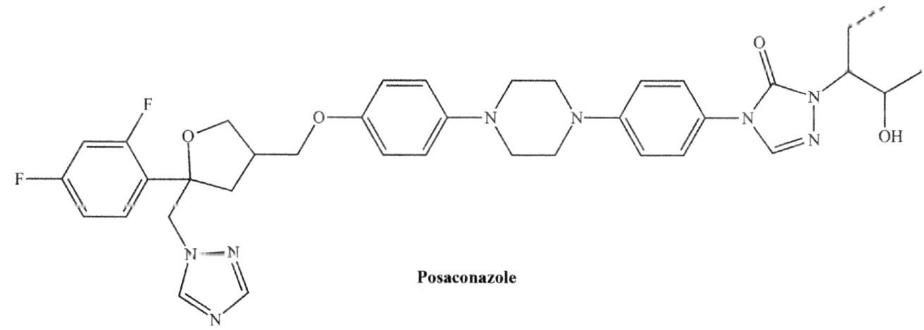

FIG. 5.11 Chemical structure of **posaconazole**.

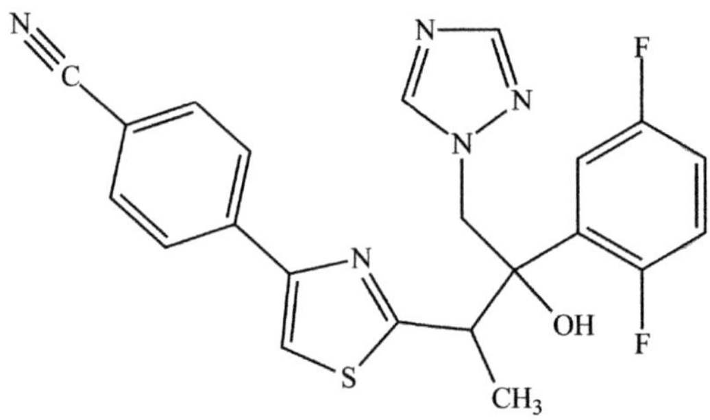

FIG. 5.12 Chemical structure of **isavuconazole**.

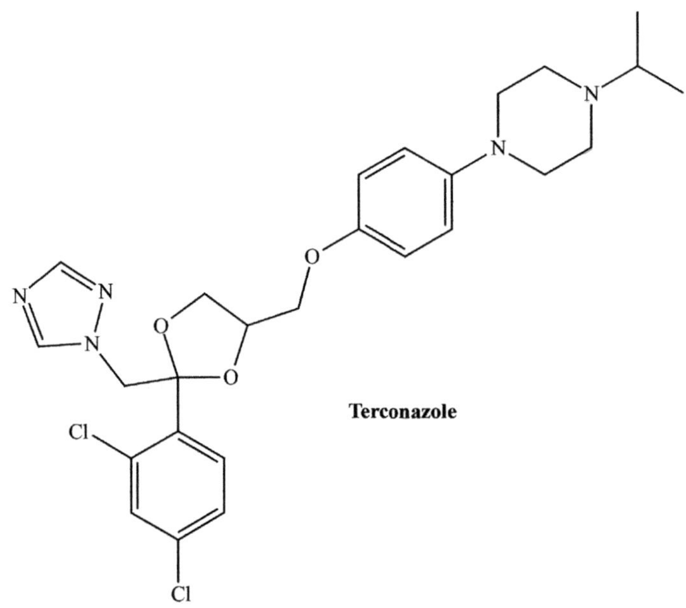

FIG. 5.13 Chemical structure of **terconazole**.

include headache, topical irritation, stomach pain, foul smell, disturbance of menstrual cycle, etc. (Godefroi et al., 1969; Kumar et al., 2014). It exhibits in vitro antifungal activity (Godefroi et al., 1967) against a series of species such as *Microsporum canis*, *Trichophyton mentagrophytes*, *Dychophyton rubrum*, *Cryptococcus neoformans*, *Phialophora verrucosa*, *Candida albicans*, *C. tropicalis*, *Aspergillus fumigatus*, *Sporothrix schenckii*, *Mucor* sp., and *Saprolegnia* sp.

Other imidazole derivatives

Bartlett et al. (1992, 1994) presented the results of their studies regarding the antifungal activity of **albendazole** (Fig. 5.14), a benzimidazole derivative known for its use to treat helminth infections, against *Pneumocystis carinii*. After confirming the in vitro inhibitory effect of albendazole against short-term cultures from rats, the authors evaluated the ability of this compound to treat and prevent the development of *P. carinii* pneumonia in inoculated immunosuppressed mice. From the obtained results was found that the tested compound was effective for both treatment (dose of 300 and 600 mg/kg/day) and prophylaxis (dose of 150 mg/kg/day), turning albendazole into a candidate for further studies.

Botta et al. (2001) reported the synthesis and evaluation of antifungal potential of both enantiomers of the imidazole derivatives **S12** and **S13** (Fig. 5.15). The compounds were tested in vitro against *Candida albicans* strains in comparison with the corresponding racemates by using the microbroth dilution method, having as reference drugs fluconazole and amphotericin B. It would be interesting to point out that although potent (MICs 1–2 μg/mL and 8–32 μg/mL for the sensitive and fluconazole-resistant strains respectively), tested derivatives showed no enantioselectivity.

As a continuation of their ongoing project regarding the synthesis of novel azole derivatives, Fringuelli et al. (2002) designed and synthesized two 1,4-benzoxazine derivatives as potential antifungals. Compound **S14** (Fig. 5.16)

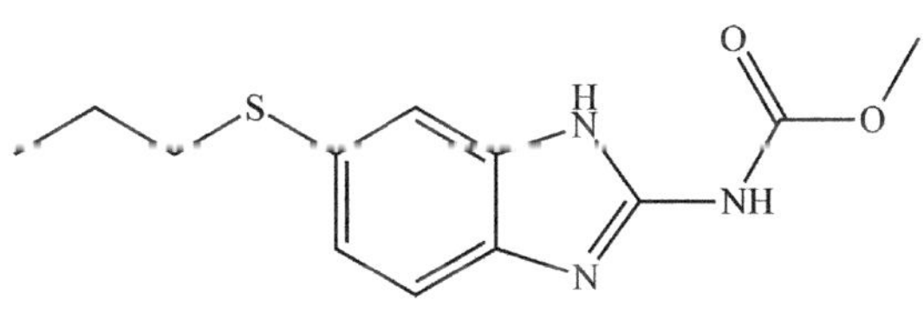

Albendazole

FIG. 5.14 Chemical structure of **albendazole**.

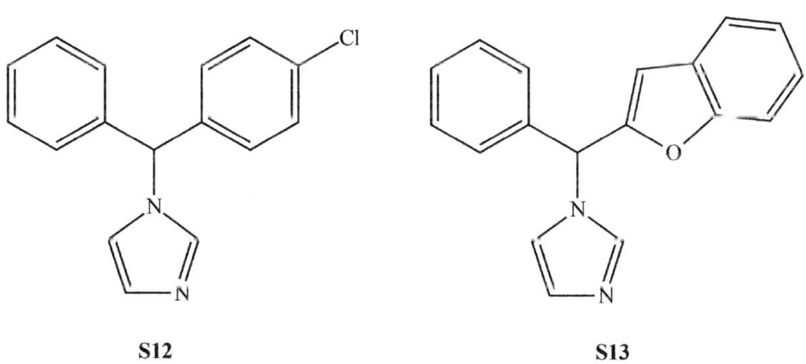

S12 **S13**

FIG. 5.15 Chemical structure of compounds **S12** and **S13**.

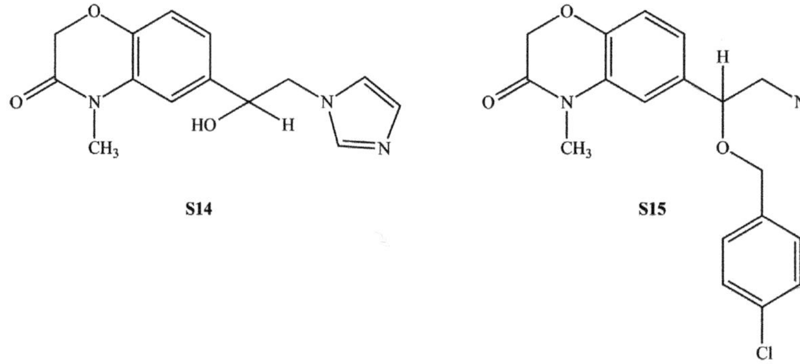

FIG. 5.16 Chemical structure of compounds **S14** and **S15**.

revealed a satisfactory in vitro antifungal profile against *C. albicans*, *C. krusei* and *Saccharomyces cerevisiae* with MIC range of 3.9–15.6 μg/mL, while derivative **S15** (Fig. 5.16) exhibited appreciable effect against *C. krusei* (MIC = 31.2 μg/mL) but no activity against *C. albicans* or *S. cerevisiae*. In vivo experiments confirmed the poor activity of compound **S15** against *C. albicans*, while imidazazole **S14** demonstrated alongside with the direct effect on *C. albicans* also an indirect effect on the immune system by promoting or enhancing its protective response against the previously mentioned strain.

Plachta et al. (2007) presented the design and synthesis of an analogue of ketoconazole, which contains a 1,3-dioxolane ring connected with a 4-(adamant-1-yl)phenoxymethyl group and a 4-bromophenyl group in position 2′ of the previously mentioned ring (**S16**, Fig. 5.17). Antifungal activity was tested by the disc-diffusion assay against *C. albicans*, *C. parapsilosis*, *C. tropicalis*, *C. krusei*, *C. guillermondii*, and *S. cerevisiae*. From the obtained results can be confirmed that ketoconazole analogue **S16** did not reveal a more favorable antifungal profile than parent drug.

Two groups of imidazole and benzimidazole derivatives related to clotrimazole were synthesized by Khabnadideh et al. (2008) as antifungal candidates. All the derivatives were docked into the active site of 14α-demethylase achieving scores comparable to these of clotrimazole which was used as reference

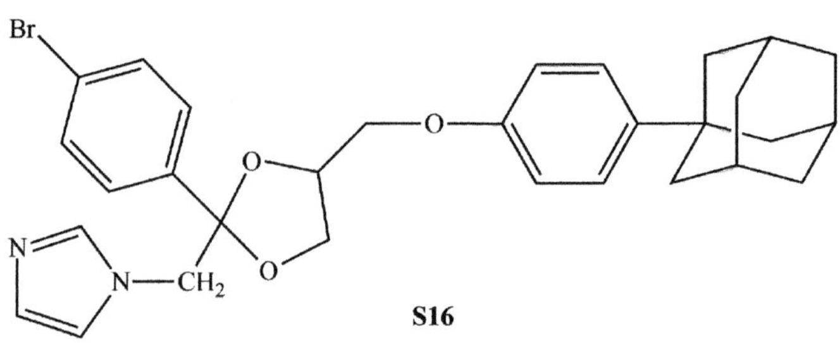

FIG. 5.17 Chemical structure of compound **S16**.

drug. The antifungal potential of the compounds was investigated in vitro against *Trichophyton mentagrophytes*, *Trichophyton rubrum*, *Microsporum canis*, *Microsporum gypseum*, *Candida albicans*, and *Epidermophyton floccosum* by using agar dilution and macrodilution methods. The strongest activity among the synthesized azoles exhibited compound **S17**, which was equipotent with clotrimazole (100% inhibition), followed by derivative **S18** (Fig. 5.18), which exhibited the same antifungal effect against all tested strains except *M. gypseum* (90% inhibition).

Sharma et al. (2009) presented the synthesis and evaluation of antifungal activity of a series of twelve 2-(substituted phenyl)-1*H*-imidazoles against *Aspergillus niger* and *Candida albicans*. For the biological activity assay the tube dilution method was employed. The authors found that derivatives **S19**, **S20**, **S21**, and **S22** (Fig. 5.19) were potent against *A. niger* with MIC $=4\times10^{-3}$ μM/mL, while the strongest inhibitory effect against *C. albicans* was exerted by compounds **S23**, **S24**, and **S22** (Fig. 5.19) with MIC $=16\times10^{-3}$ μM/mL. In all cases the novel imidazoles were less potent than fluconazole (MIC $=1\times10^{-3}$ μM/mL against both fungal strains). Furthermore, by comparing the MICs with the MFC values of the compounds, the authors proposed that the synthesized azoles could act not only as bacteriostatic but also as fungistatic agents.

The synthesis of eighteen benzimidazole derivatives was presented by Chandrika et al. (2016) with the aim to test the novel compounds for their antifungal effects against a panel of *Candida albicans* and *Aspergillus* strains. Subsequently, in vitro cytotoxicity tests were performed alongside with studies to determine the ability of the compounds to induce reactive oxygen species (ROS) in fungi. Concerning the results of antifungal assay, it was found that mono- and trisbenzimidazole derivatives were not effective against the used strains and in general the activity of alkylated analogues follows the order

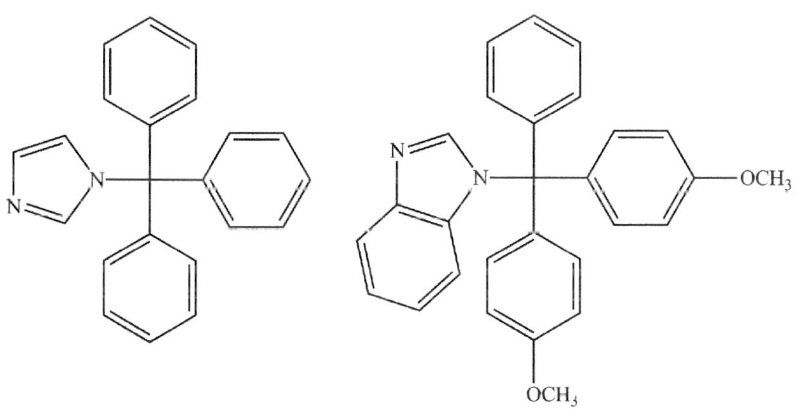

S17 **S18**

FIG. 5.18 Chemical structures of compounds **S17** and **S18**.

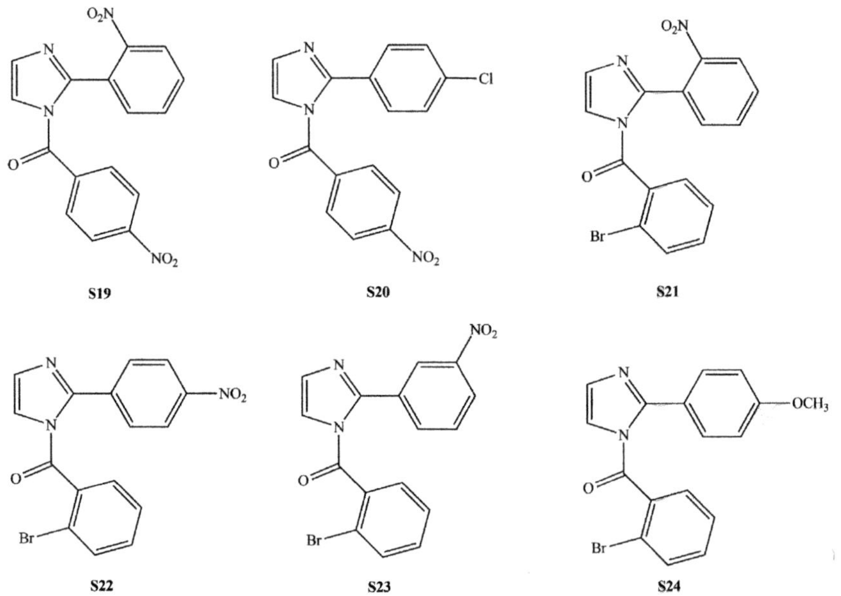

FIG. 5.19 Chemical structures of compounds **S19–S24**.

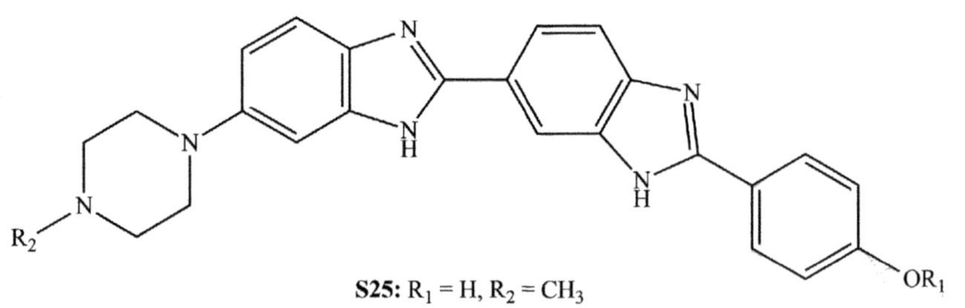

S25: $R_1 = H, R_2 = CH_3$

S26: $R_1 = C_2H_5, R_2 = CH_3$

S27: $R_1 = CH_2C{\equiv}CH, R_2 = CH_3$

FIG. 5.20 Chemical structures of compounds **S25–S27**.

$C_1 > C_2 > C_3 > C_5 > C_{12}$. Derivatives **S25** (MICs of 0.975–31.2 μg/mL), **S26** (MICs of 3.9–31.2 μg/mL) and **S27** (MICs of 1.95–31.2 μg/mL) (Fig. 5.20) demonstrated broad-spectrum activity against the used species being in most of the cases more potent or with comparable potential to reference commercial drugs fluconazole, itraconazole, posaconazole, voriconazole, and amphotericin B. Furthermore, it was mentioned by the authors that for these compounds, the accumulation of ROS that they induce in cells could be the cause of their inhibitory effect against fungi. Taking into account the results of the in vitro cytotoxicity assay against mammalian cell lines, the authors proposed that derivative **S25** with IC_{50} values ≥ 31.2 μg/mL (better than amphotericin B) and with low MICs, could be a candidate for further investigation in vivo as promising antifungal agent. It should also be taken into consideration that compound **S27** due to its significant antifungal potential and the presence of alkyne group in its scaffold, could turn into an ideal candidate for further structural modifications.

A collection of aromatic heterocyclic compounds were designed and synthesized by Zhao et al. (2017) as potential antifungal agents. The novel heterocyclics were tested for their inhibitory effect against a panel of five fungal strains and MICs were determined by employing the broth microdilution assay, with standards fluconazole (MICs of 0.25–>16 μg/mL) and itraconazole (MICs of 0.03125–1.0 μg/mL). It was found that derivatives **S28** (MICs of 0.03125–4.0 μg/mL) and **S29** (MICs of 0.03125–4.0 μg/mL) were the most promising (Fig. 5.21), especially against *Candida* spp., *C. neoformans*, *A. fumigatus,* and fluconazole-resistant *C. albicans* strains, exhibiting activity greater or comparable to that of reference antifungals. It should also be mentioned that these two derivatives revealed low inhibition for several CYP450 isoforms and excellent blood plasma stability, making them promising candidates for further studies.

Other triazole derivatives and combinations

Heeres et al. (1984) presented the synthesis and evaluation of antifungal effects of a series of novel triazol-3-ones against a panel of species including *Microsporum canis*, *Trichophyton mentagrophytes*, *T. rubrum*, *pHialophora verrucosa*, *Cryptococcus neoformans*, *Candida tropicalis*, *C. albicans*, *Mucor* sp., *Aspergillus fumigatus*, *Sporothrix schenckii*, and *Saprolegnia* sp., by employing the assay described by Godefroi et al. (1967). From the obtained results was observed that the novel triazol-3-ones were less potent than ketoconazole and the reason for that was probably the low solubility of synthesized compounds in the test medium. Derivative **S30** (Fig. 5.22) was active against *Candida* spp., *Cr. neoformans*, *A. fumigatus*, *Sp. schenckii*, *Ph. verrucosa* at 0.1 μg/mL and toward dermatophytes at 0.1–1.0 μg/mL. Additionally, compounds **S31**, **S32**, **S33**, and **S34** (Fig. 5.22) were found to be more potent against *T. rubrum*, T. *mentagrophytes*, *Cr. Neoformans,* and *C. tropicalis* (MIC values of 1–10 μg/mL). Regarding *Mucor* sp. would be interesting to point out that none of the tested derivates exerted noteworthy antifungal effect, even at concentration of 100 μg/mL.

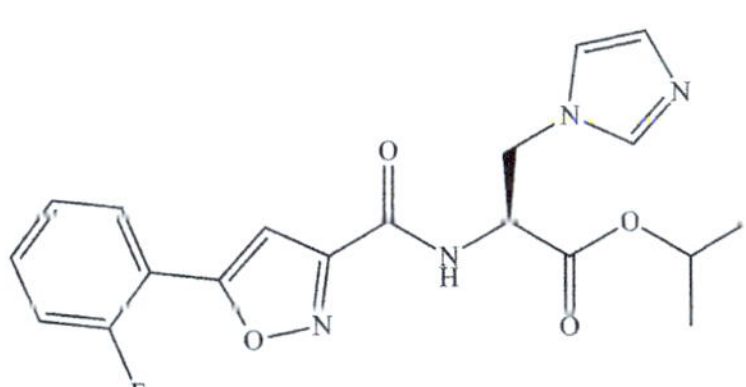

S28

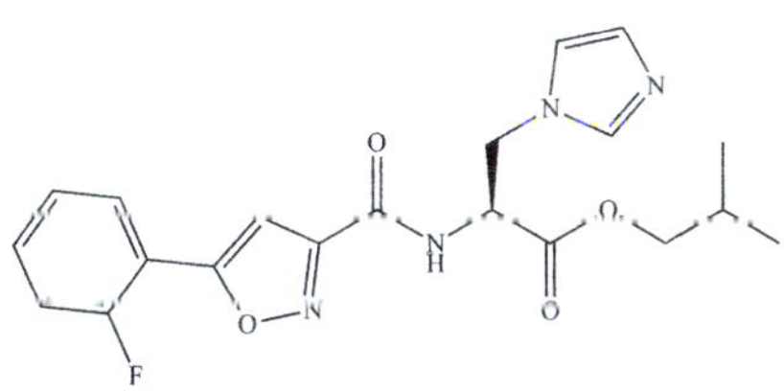

S29

FIG. 5.21 Chemical structures of compounds **S28–S29**.

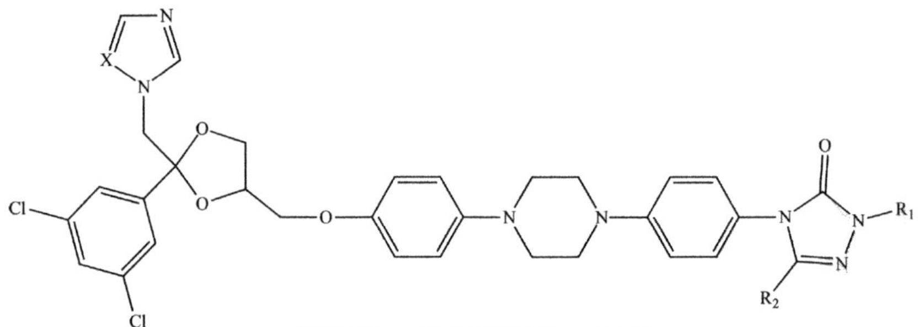

S30: X = N, R_1 = -CHCH=CH_2, R_2 = H
S31: X = CH, R_1 = CH_3, R_2 = H
S32: X = CH, R_1 = s-C_4H_9, R_2 = H
S33: X = N, R_1 = C_2H_5, R_2 = H
S34: X = N, R_1 = $(CH_3)_2CHCH_2$-, R_2 = H
S35: X = CH, R_1 = i-C_3H_7, R_2 = CH_3
S36: X = N, R_1 = i-C_3H_7, R_2 = H
S37: X = N, R_1 = $(CH_3)_2CHCH_2$-, R_2 = CH_3

FIG. 5.22 Chemical structures of compounds **S30–S37**.

The novel triazol-3-ones were also evaluated in vivo for their potential against microsporosis and vaginal candidiasis in rats. It should be noted that compounds **S31**, **S35**, **S36**, **S37**, **S34**, and **S30** exhibited good activity even in low doses of 2.5 mg/kg against experimental microsporosis, while they were found to be inactive in vitro against *M. canis* at 100 μg/mL. From the SAR studies on synthesized compounds, the authors observed that the nonalkylated triazol-3-ones showed no activity in established vaginal candidiasis and microsporosis at the indicated doses. On the contrary, the alkylated derivatives exhibited far better in vivo antifungal profile. Moreover, replacement of trazole moiety with imidazole led to enhanced in vivo potential, while 5′-methyl-substitution for the triazol-one derivatives led to diminished antifungal effect (especially in microsporosis). Furthermore, it was noticed that while elongation of N-R_1 alkyl chain had only a minor effect in the activity of the compounds, branching of -R_1 had as result a decrease in their potential against vaginal candidiasis and microsporosis.

Taking into consideration the well-known antifungal effects of triazoles, Ram Shankar et al. (2004a) designed and synthesized a series of triazole derivatives that contained 1-aryl-4-(2-[*1H*-tetrazol-5-yl]-ethyl)-piperazine moiety. The compounds were evaluated in vitro for their antifungal potential against several species (*Candida albicans*, *C. albicans* V-01-191A-261 (resistant strain), *C. tropicalis*, *C. krusei* ATCC 6528, *C. krusei*, *C. parapsilosis* ATCC 22019, *C. glabrata*, *Cryptococcus neoformans* LA314, *Aspergillus fumigatus*, and *A. niger* by using agar diffusion and broth dilution methods. In a preliminary screening with agar diffusion method the authors observed that at concentrations of 500 μg/mL derivatives **S38–S45** (Fig. 5.23) were potent against

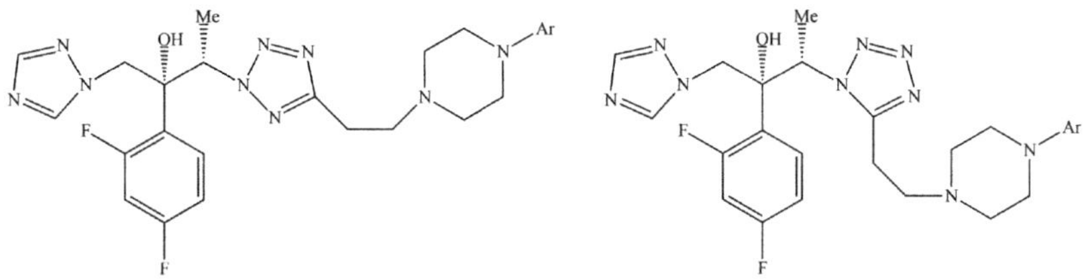

S38: Ar = 3-$CF_3C_6H_4$
S40: Ar = 2-n-$BuOC_6H_4$
S42: Ar = 4-ClC_6H_4
S44: Ar = $CHPh_2$

S39: Ar = 3-$CF_3C_6H_4$
S41: Ar = 2-n-$BuOC_6H_4$
S43: Ar = 4-ClC_6H_4
S45: Ar = $CHPh_2$

FIG. 5.23 Chemical structures of compounds **S38–S45**.

resistant *Candida* species and *C. neoformans*, with compounds **S38** and **S39** being effective also against *Aspergillus* species. Subsequently, the synthesized triazoles were assayed against *Candida*, *Cryptococcus*, and *Aspergillus* strains by using broth microdilution method in order to determine their MICs. It should be pointed out that compounds **S39** and **S38** were very potent against *Candida*-resistant strain with MIC values of 0.12 µg/mL and 0.25 µg/mL, respectively, exhibiting stronger activity than reference drugs (fluconazole and itraconazole).

As part of their research project regarding the development of antifungal agents, Ram Shankar et al. (2004b) reported the synthesis and structural modifications of a series of tetrazole-based triazoles, aiming to develop compounds that will combine potency against pathogenic fungi and appropriate physicochemical properties. The novel triazoles were evaluated in vitro with agar diffusion and broth microdilution methods against a series of fungal strains, namely, *C. albicans*, *C. albicans* V-01-191A-261 (resistant strain), *C. tropicalis*, *C. krusei* ATCC 6528, *C. krusei*, *C. parapsilosis* ATCC 22019, *C. glabrata*, *Cr. neoformans* LA314, *A. fumigatus*, and *A. niger*. Commercial drug fluconazole was used as reference drug. The majority of the compounds were found to be, in general, effective as antifungals with **S46** (MICs of 0.5–2.0 µg/mL) and **S47** (MICs of 1.0–2.0 µg/mL) (Fig. 5.24) exhibiting the best potential among the compounds of their groups.

A series of triazole-based compounds of the general structure presented in Fig. 5.25 were synthesized as potential antifungals in dermatology by Meerpoel et al. (2005). The novel triazoles were assayed in vitro against a panel of 128 dermatophytes and 24 *Candida* species, *Aspergillus* species and several other fungal strains by using microplate broth dilution method (Odds et al., 2004). Regarding the dermatomycetes, it was observed that imidazolidinones were more potent than the pyrimidinetrione derivatives, which were followed by the triazinetriones. It should be noted that optical isomerism had a crucial effect on the antifungal activity of the synthesized compounds, having as example the much greater potential of 2S-*cis* derivatives **S51a**, **S51b**, and **S51c** than the 2R-*cis* ones **S50a**, **S50b**, and **S50c**. Derivatives **S48c** and **S51c**with MICs ranging at 0.01–0.1 µg/mL exhibited activity similar to the ones of itraconazole

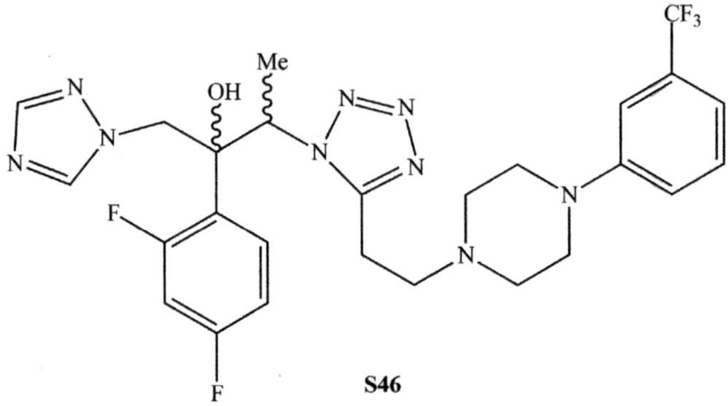

S46

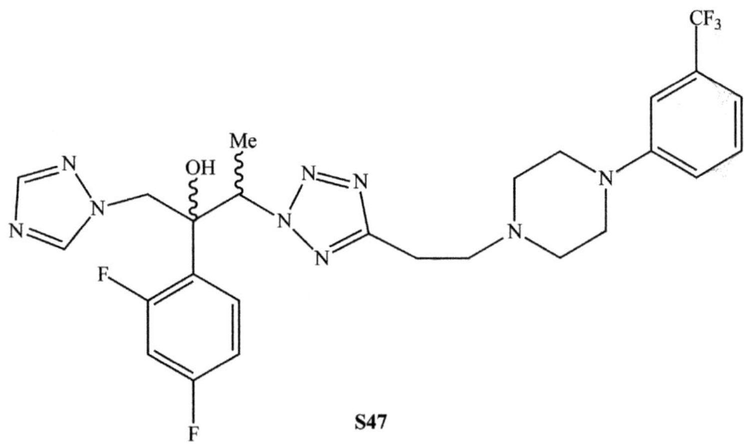

S47

FIG. 5.24 Chemical structures of compounds **S46** and **S47**.

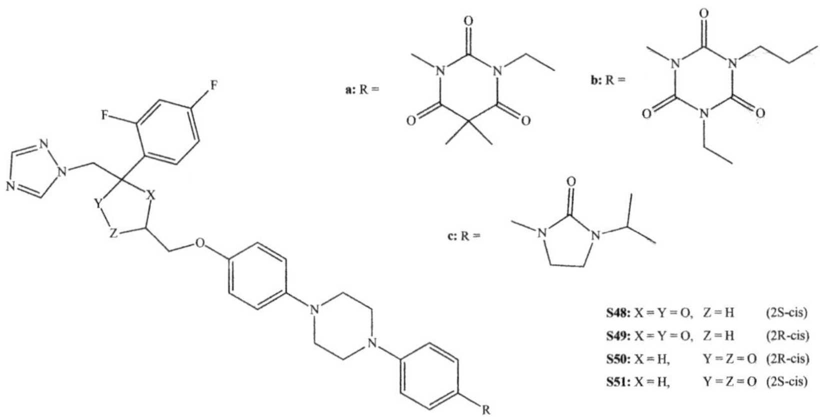

FIG. 5.25 Chemical structures of compounds **S48–S51**.

(MICs of 0.01–0.32 μg/mL) and terbinafine (MICs of 0.01–0.1 μg/mL) toward *Epidermophyton* species, but they were less potent against *Trichophyton* strains. Moreover, the two previously mentioned compounds alongside with **S49c** were found to be in general effective against the 24 *Candida* strains, less potent against *A. fumigatus* and showed no activity toward *Cryptococcus*, *Sporothrix*, and *Fusarium* species. Furthermore, in vivo tests were performed for the synthesized azoles by employing a guinea pig infected with *Microsporum canis* model and also a mouse infected with *Trichophyton mentagrophytes* model. From the obtained results was revealed that the mentioned azoles **S51c** and **S48c** (Fig. 5.25) were the most potent, with activities even stronger than reference drug, itraconazole (ED_{50} values of 0.84 mg/kg and 2.60 mg/kg for the two models, respectively).

In an effort to combine sulfonamide nucleus and *s*-triazole moiety in one frame, Ezabadi et al. (2008) synthesized 10 novel 5-[2-(substituted sulfamoyl)-4,5-dimethoxy-benzyl]-4aryl-s-triazole-3-thiones. The synthesized compounds were tested in vitro in order to elucidate their antifungal potential against several fungal strains, having as reference drug the commercial drug bifonazole. All derivatives exhibited significant antifungal effect against the selected fungi with *Trichoderma viride* being the most sensitive strain. It was proposed that the presence of N-$(CH_3)_2$ group in the structure of derivatives **S52** (MICs 50.0–100 μg/mL and MFCs 100.0–150.0 μg/mL) and **S53** (MICs 50.0–100 μg/mL and MFCs 100.0–150.0 μg/mL) (Fig. 5.26) is responsible for their enhanced antifungal effect compared to the rest of the tested compounds.

Emami et al. (2008) described the synthesis of some novel 2-hydroxyphenacyl azoles and related azolium derivatives as potential antifungals. All the compounds were subjected to in vitro evaluation of their antifungal effects against *Candida albicans*, *Saccharomyces cerevisiae*, *Microsporum gypseum*, and *Aspergillus niger* strains. For the determination of their MIC values the agar dilution assay was performed. According to the authors, the synthesized derivatives were, in general, effective against the tested fungi with potential comparable to fluconazole, which was used as reference agent. It was found that derivatives **S54** (MIC = 2 μg/mL) and **S55** (MIC = 4 μg/mL) (Fig. 5.27) were the most potent against *C. albicans* with activity stronger than that of fluconazole

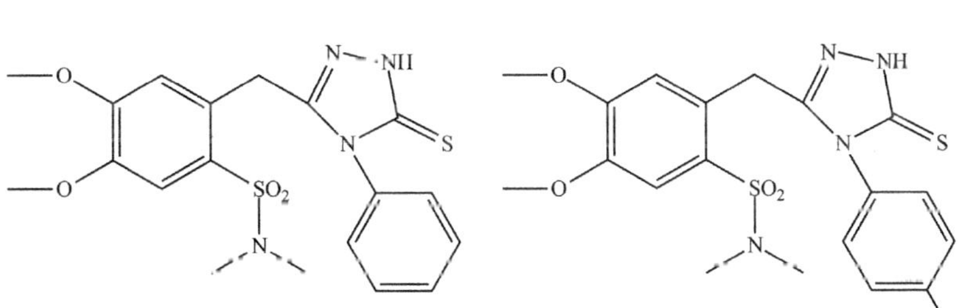

S52

S53

FIG. 5.26 Chemical structures of compounds **S52** and **S53**.

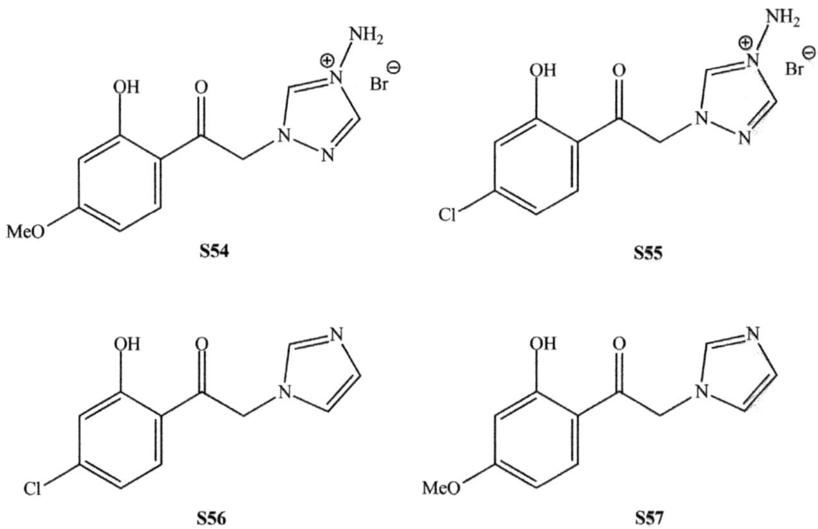

FIG. 5.27 Chemical structures of compounds **S54–S57**.

(MIC = 8 μg/mL). It should be pointed out that compounds **S56**, **S57**, and **S54** (MIC = 0.25 μg/mL) were very potent against *S. cerevisiae*, with activity 128-fold stronger than that of fluconazole and 4-fold stronger than that of miconazole. Furthermore, derivatives **S57** and **S54** exhibited very good activity toward *M. gypseum* (MIC = 8 μg/mL) being equipotent to miconazole and 4-fold more effective than fluconazole. After determining the cytotoxicity of selected compounds in normal mouse fibroblast cells, the authors concluded that the novel synthesized derivatives are, in general, effective as antifungals with no significant cytotoxicity.

As part of their project regarding the synthesis of novel antimicrobial compounds, Rezaei et al. (2009) presented the design, synthesis, and evaluation of antifungal activity of some triazole and benzotriazole derivatives against *Microsporum canis*, *Trichophyton mentagrophytes*, *T. rubrum*, *Epidermophyton floccosum*, and *Candida albicans* having as reference drugs fluconazole and clotrimazole. Furthermore, docking studies were performed and from the docking score was observed that all the compounds (as well as fluconazole) were successfully docked to the active site of cytochrome P450 14α-demethylase. From the determination of MICs of the synthesized compounds is clear that they exhibited, in general, moderate-to-good antifungal effect (MIC range at 1–64 μg/mL) and in some cases similar to that of fluconazole (MIC range at 0.25–32 μg/mL) against the used species. It should be noted that the novel triazoles and benzotriazoles were less potent that clotrimazole (MIC range at 0.05–4 μg/mL). By studying docking scores and biological activity results, the authors confirmed that for compounds **S58** and **S59** (Fig. 5.28) there is correlation between docking energies and antifungal inhibitory effect.

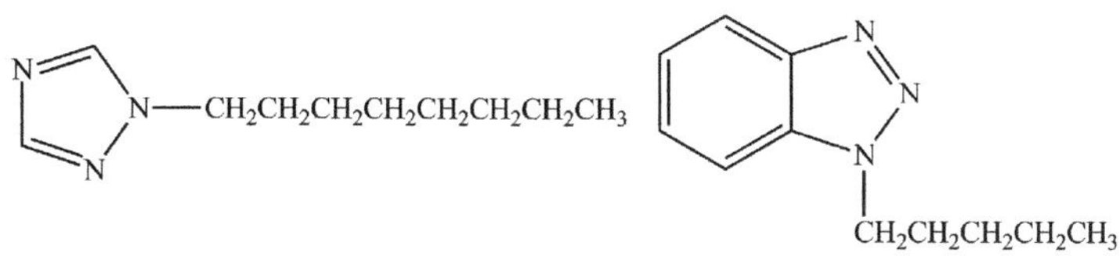

S58 S59

FIG. 5.28 Chemical structures of compounds **S58** and **S59**.

Rezaei et al. (2011) synthesized a series of novel 1,2,4-triazoles, imidazoles, benzimidazoles, and benzotriazoles, which were selected after docking a large collection of compounds into the active site of MT-CYP51 (*Mycobacterium tuberculosis* CYP51), which is similar to fungal lanosterol 14α-demethylase. Subsequently, all the synthesized derivatives were screened for their antifungal effect against a panel of *Candida* species, *Aspergillus fumigatus*, *A. flavus*, *Microsporum canis*, *M. gypseum*, *Trichophyton mentagrophytes*, and *Epidermophyton floccosum* having as reference drugs fluconazole, itraconazole, and/or griseofulvin. From the novel derivatives, imidazoles **S60** and **S61** (Fig. 5.29) were found to be potent against all the used fungi (MIC values of 0.5–64.0 μg/mL) and it should be noted that they were effective also against fluconazole- and griseofulvin-resistant *C. tropicalis* and *C. albicans* strains. It would also be interesting to mention that the authors found no correlation between FDE (Final Docking Energy), logp, and antifungal activity.

Inspired from the results of their previous efforts regarding the synthesis of sulfonamide-1,2,4-triazoles and 1,3,4-thiazoles as antifungal agents, Zoumpoulakis et al. (2012) reported the synthesis of a series of novel sulfonamide-1,2,4-triazoles, 1,3,4-thiadiazoles, and 1,3,4-oxadiazoles and evaluated their antifungal effects against several fungal strains. Moreover,

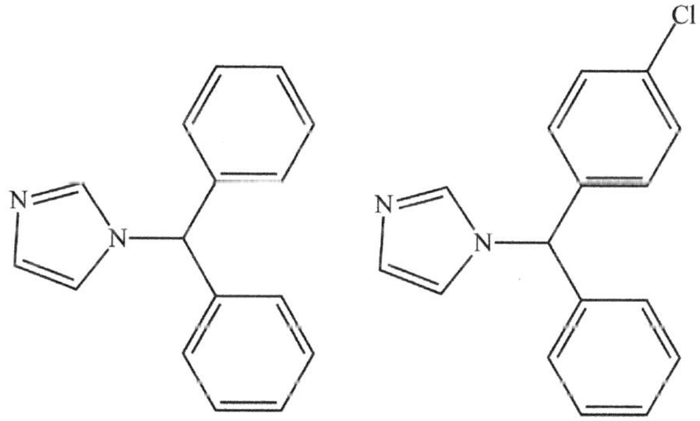

S60 S61

FIG. 5.29 Chemical structures of compounds **S60** and **S61**.

conformational studies have been performed using molecular modeling and NMR spectroscopy methods. It should be noted that the authors designed the compounds by using different aliphatic chains in order to study the effect of increased hydrophobic interactions on the pharmacological activity. All synthesized triazole derivatives exhibited very good antifungal potential (MICs of 0.01–0.27 μmoL/mL and MFCs of 0.06–0.50 μmoL/mL), better than that of bifonazole (MICs of 0.32–0.64 μmoL/mL and MFCs of 0.64–0.80 μmoL/mL) and ketoconazole (MICs of 0.38–4.75 μmoL/mL and MFCs of 0.95–5.70 μmoL/mL). Compound **S62** (Fig. 5.30) was found to be the most potent triazole derivative wit MIC values ranging between 0.01 and 0.25 μmoL/mL and MFC values between 0.03 and 0.38 μmoL/mL. The most sensitive fungus was *Trichoderma viride* and *Aspergillus versicolor* was found to be the most resistant strain. As far as the thiadiazole derivatives were concerned, they were all found to be very good antifungal agents (MICs of 0.06–0.50 μmoL/mL and MFCs of 0.07–0.24 μmoL/mL), more potent than ketoconazole and almost all of them more active than bifonazole. Compound **S63** was the most potent thiadiazole derivative with MICs of 0.06–0.12 μmoL/mL and MFCs of 0.12–0.24 μmoL/mL. It was observed that *Fulvia fulvum* was the most sensitive strain, while *Aspergillus versicolor* was the most resistant one. The inhibitory activity of oxadiazole derivatives against the selected fungi ranged between 0.03 and 0.28 μmoL/mL, while their fungicidal potential between 0.03 and 0.75 μmoL/mL. Almost all of the synthesized oxadiazoles were more potent than bifonazole, while all of them exhibited better activity than ketoconazole. Derivative **S64** (Fig. 5.30) was the most potent compound of this group with MICs range 0.03–0.27 μmoL/mL and MFCs of 0.03–0.40 μmoL/mL. *Trichoderma viride* was the most sensitive fungus concerning oxadiazole derivatives, while *Penicillium* species were the most resistant ones.

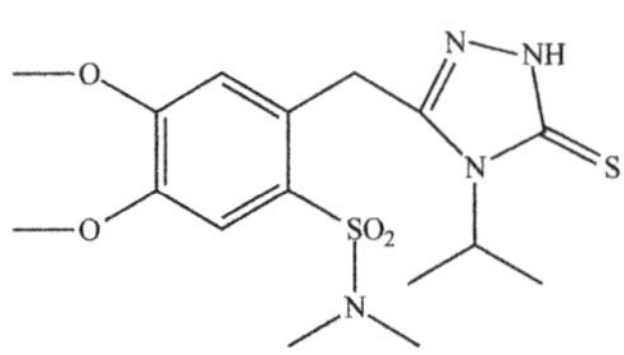

S62

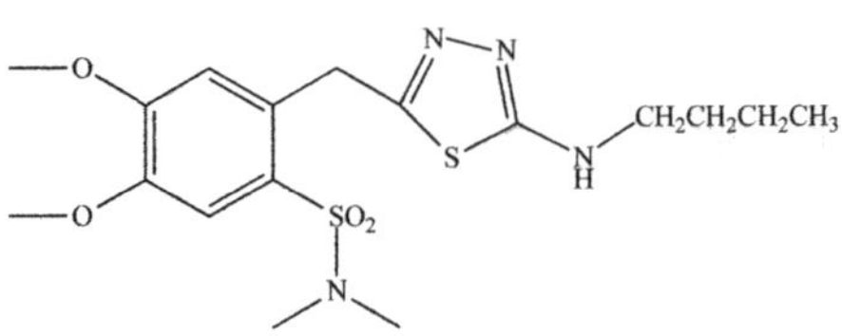

S63

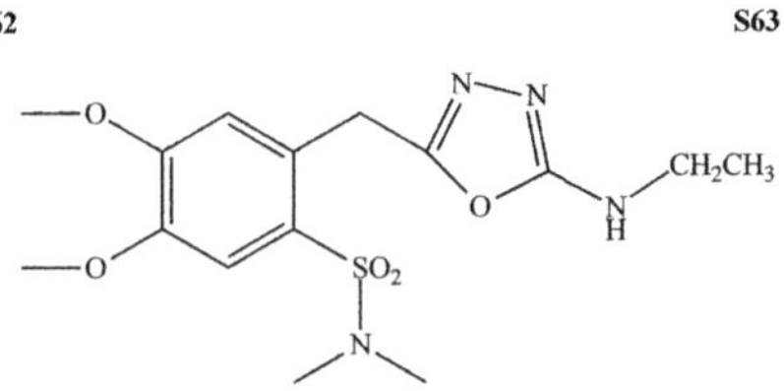

S64

FIG. 5.30 Chemical structures of compounds **S62–S64**.

Regarding SAR, it is noteworthy that among all tested groups, triazole-3-thiones were in general the most potent derivatives against the selected fungi, while large alkyl substituents (butyl, *tert*-butyl, isopropyl) were found to be favorable, in most of the cases, for the antifungal activity of the synthesized molecules. From the prediction of molecular properties of the synthesized compounds was revealed that the derivatives that were found to be potent against all the tested fungi are characterized by high dipole moments, increased number of hydrogen bond acceptors, increased polar contributions to the total solvent accessible surface area, high dip^2/V values that lead to high energy of salvation, increased electron affinity, high water/gas partition coefficients, and low ionization potential values.

Lima-Neto et al. (2012) described the synthesis and evaluation of antifungal activity of a series of compounds containing triazol-4-yl moiety. The novel derivatives were assayed by using the microdilution broth method against twenty *C. albicans*, five *C. krusei*, eleven *C. parapsilosis,* and five *C. tropicalis* species. From the group of derivatives that were synthesized initially, compounds **S65**, **S66**, **S67**, **S68**, and **S69** (Fig. 5.31) were more potent than reference drug fluconazole against the following *Candida* strains: 4990, 4987, 4986, 4820, 4819, 4817, 4609, 4606, 4388, 4124, 3719. Their MICs were also determined and were found to be in the range of 1.024–2.048 μg/mL, while fluconazole MIC values were at 4–64 μg/mL. Moreover, derivatives **S65**, **S66**, **S67**, and **S68** exerted better antifungal effect (MICs of 1.024–2048 μg/mL) than fluconazole (MICs of 16–64 μg/mL) toward the following *C. krusei* species: 4802, 4263, 934, 109. Similar activity exhibited compounds **S66**, **S67**, **S68**, and **S69** against eight *C. parapsilosis* strains, namely: 4984, 4987, 4889, 4818, 4804, 4608, 4607, 4261, with MICs of 1.024–2.048 μg/mL while the fluconazole MICs ranged at 2–16 μg/mL. Almost the same antifungal effect was exerted toward the *C. tropicalis* species with triazole derivatives **S66**, **S67**, **S68**, and **S69** being more potent than fluconazole. It should be noted that compound **S65** was found to possess no activity against *C. parapsilosis* and *C. tropicalis* strains.

A series of novel benzimidazole tertiary amine type of fluconazole analogues were designed and synthesized by Kankate et al. (2015). The compounds were docked into the active site of *Candida albicans* homology modeled lanosterol 14α-demethylase by using the ligand fit method. By studying the docking results, the authors observed that all the synthesized derivatives were

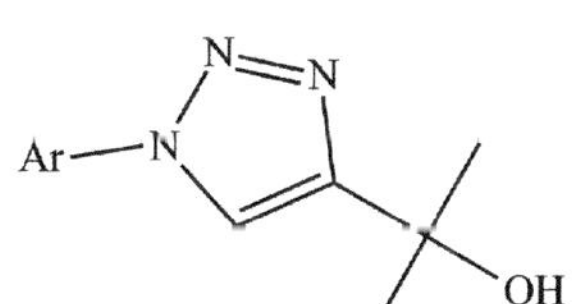

S65: Ar = phenyl

S66: Ar = p-tolyl

S67: Ar = p-Br-phenyl

S68: Ar = p-OCH_3-phenyl

S69: Ar = m-NO_2-phenyl

FIG. 5.31 Chemical structures of compounds **S65–S69**.

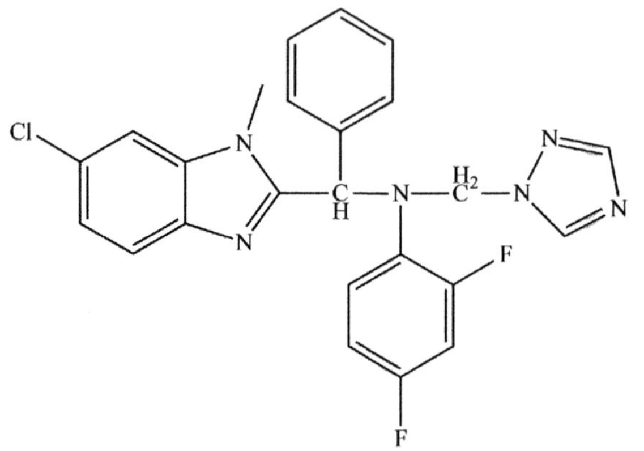

S70

FIG. 5.32 Chemical structure of compound **S70**.

found to be able to bind in the active site of the mentioned enzyme. The fluconazole analogues were screened for their antifungal potential against *Candida albicans* spores in vitro and in vivo using the tube dilution method and kidney burden test in rats, respectively. Regarding the in vitro assay, it should be pointed out that compound **S70** (Fig. 5.32) demonstrated the best activity with MIC = 0.0075 μmoL/mL being equipotent with reference drug, fluconazole. The in vitro assay findings were confirmed by the in vivo test, with derivative **S70** showing the best potential compared to the rest of the synthesized compounds.

Thiazoles

Abafungin

Abafungin (Fig. 5.33) is a synthetic thiazole antifungal, which acts in both resting and growing states of fungi. There were found two targets of abafungin in the fungal cell membrane: (1) inhibition of transmethylation at the C-24 position of the sterol side chain, catalyzed by the enzyme sterol-C-24-methyltransferase; (2) direct effect on the fungal cell membrane (Borelli et al., 2008). It exhibits fungistatic as well as fungicidal activities against several pathogenic strains and is currently being used as a topical cream formulation for the treatment of dermatomycoses.

Other thiazole derivatives

Taking into consideration the antimicrobial potential of several thiazole and isatine derivatives, Pandeya et al. (1999) reported the synthesis of a series of compounds in order to evaluate their antifungal activity in vitro by using the agar dilution method against a group of pathogenic fungal strains, having as reference drug clotrimazole. From the obtained results was revealed that compounds **S72–S78** and **S80–S82** are more potent than clotrimazole against *Microsporum*

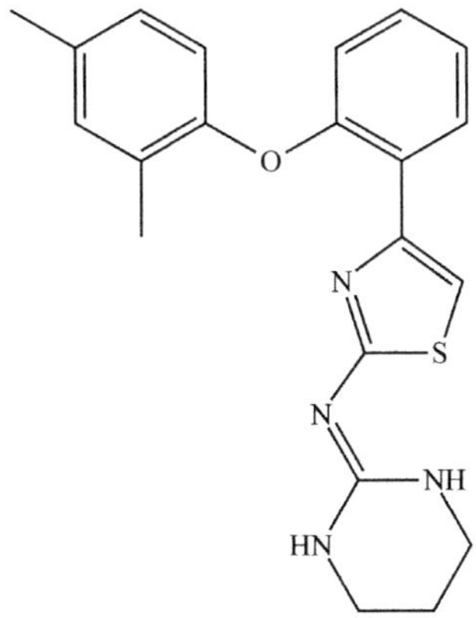

Abafungin

FIG. 5.33 Chemical structure of **abafungin**.

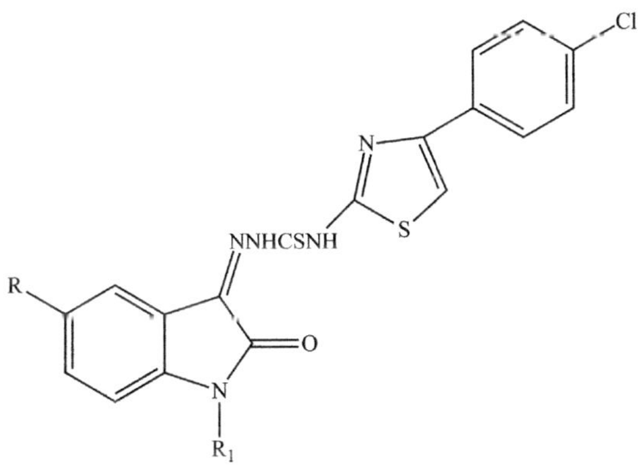

S71: R = R_1 = H
S72: R = Cl, R_1 = H
S73: R = Br, R_1 = H
S74: R = H, R_1 = $CH_2N(CH_3)_2$
S75: R = H, R_1 = CH_2-piperidine
S76: R = H, R_1 = CH_2-morpholine
S77: R = Cl, R_1 = $CH_2N(CH_3)_2$
S78: R = Cl, R_1 = CH_2-piperidine
S79: R = Cl, R_1 = CH_2-morpholine
S80: R = Br, R_1 = $CH_2N(CH_3)_2$
S81: R = Br, R_1 = CH_2-piperidine
S82: R = Br, R_1 = CH_2-morpholine

FIG. 5.34 Chemical structures of compounds **S71–S82**.

gypseum with MICs of 1.2 and 2.4 μg/mL, respectively. Moreover, compounds **S72**, **S73**, **S76**, **S77**, **S79**, and **S81** exhibited stronger antifungal effect than clotrimazole (MIC = 4.9 μg/mL) toward *Microsporum audouinii*. On the other hand, derivatives **S71**, **S72**, **S74**, **S77**, **S81**, and **S82** exhibited stronger antifungal effect (MIC = 1.2 μg/mL) than clotrimazole (MIC = 2.4 μg/mL) against *Epidermophyton floccosum*. Additionally, it should be pointed out that the synthesized compounds were in general effective against all tested fungi except *Candida albicans*, *Aspergillus niger*, and *Histoplasma capsulatum* (Fig. 5.34).

A series of novel thiazole derivatives, incorporated with benzimidazole or benzoxazole moieties, were designed and synthesized by Kaplancikli et al. (2004). For the evaluation of the antifungal potential of the compounds were used *Candida albicans* and *C. glabrata*, having as reference drug ketoconazole.

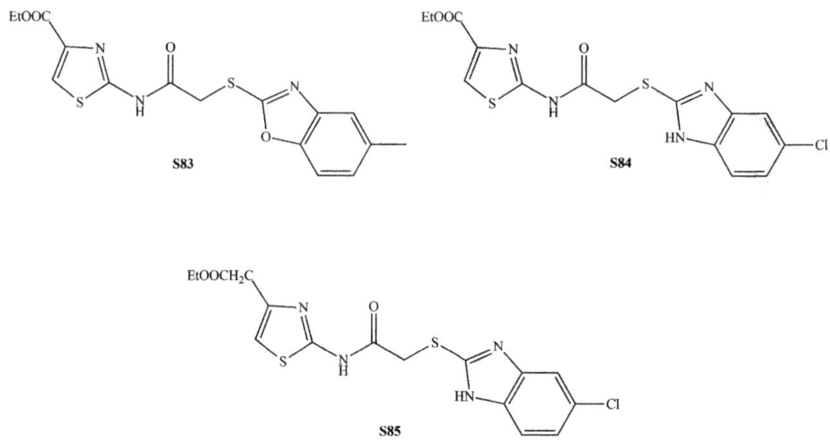

S83 S84 S85

FIG. 5.35 Chemical structures of compounds **S83–S85**.

From the screening results was observed that three of the synthesized thiazole derivatives, namely, **S83**, **S84**, and **S85** (Fig. 5.35) were effective against the two selected fungal strains with MICs of 62 μg/mL for *C. glabrata* and 125 μg/mL for *C. albicans,* respectively.

Ozdemir et al. (2007) presented the synthesis and evaluation of antimicrobial activity of a series of 1-(4-aryl-2-thiazolyl)-3-(2-thienyl)-5-aryl-2-pyrazolines. The chemical structures of compounds **S86–S100** are shown in Fig. 5.36. The antifungal potential of synthesized derivatives was assayed by microdilution broth susceptibility assay against *Candida albicans* and *C. glabrata*, having fluconazole as reference drug. All compounds exhibited good antifungal potential, exerting stronger or equal antifungal effect with fluconazole (MICs 62.5–250 μg/mL).

Mallikarjuna et al. (2009) described the synthesis and evaluation of antifungal activity of a series of 4-isopopylthiazole hydrazide derivatives against *Saccharomyces cerevisiae*, *Candida tropicalis*, and *Aspergillus niger* by employing the broth dilution assay. All synthesized derivatives were found to possess moderate activity, being less potent than fluconazole (MICs ≤ 1 μg/mL), which was used as reference antifungal. It should be noted though that derivative **S101** (Fig. 5.37) was the most potent among the tested compounds with MICs of 8–16 μg/mL.

A number of benzothiazole derivatives were prepared by Bondock et al. (2009) with the aim to screen them for their antifungal potential against *Botrytis fabae* and *Fusarium oxysporum* by using agar diffusion method. As reference drug was used fluconazole. From the obtained results can be observed that benzothiazole derivatives **S102**, **S103**, and **S104** (Fig. 5.38) exhibited similar activity with reference drug against *B. fabae* with MIC = 3.125 μg/mL. On the other hand, the same compounds were less strong as antifungals in comparison with fluconazole toward *F. oxysporum* with MIC = 6.25 μg/mL.

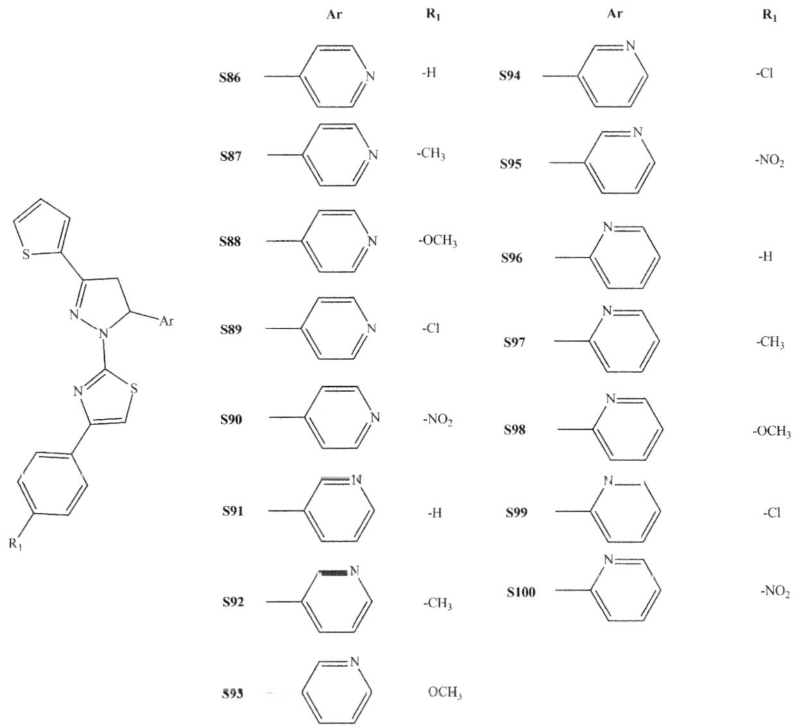

FIG. 5.36 Chemical structures of compounds **S86–S100**.

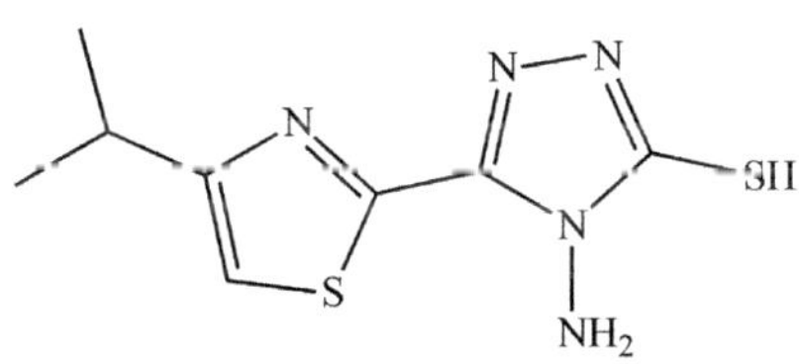

S101

FIG. 5.37 Chemical structure of compound **S101**.

As a continuation of their ongoing project regarding the synthesis of benzothiazole derivatives as antifungal agents, Bondock et al. (2010) prepared a group of novel compounds and examined their activity in vitro against *F. oxysporum* and *A. fumigatus* having as reference compound cycloheximide. However, only derivatives **S105** and **S106** (Fig. 5.39) revealed antifungal potential (MICS = 6.25 μg/mL) against the two selected fungi but still lower than that of reference compound (MIC = 3.125 μg/mL).

Liaras et al. (2011a) designed and synthesized ten novel compounds containing both thiazole and chalcone moieties in order to evaluate their antifungal

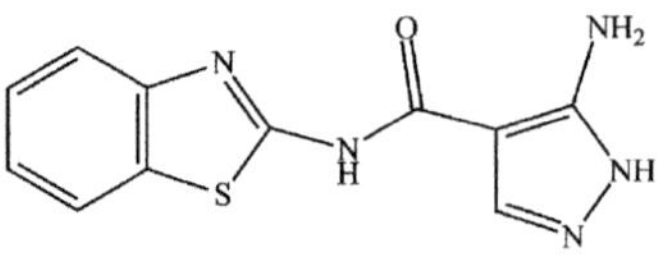

S102

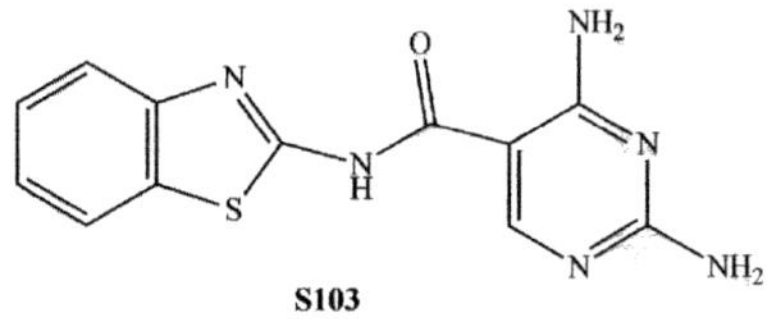

S103

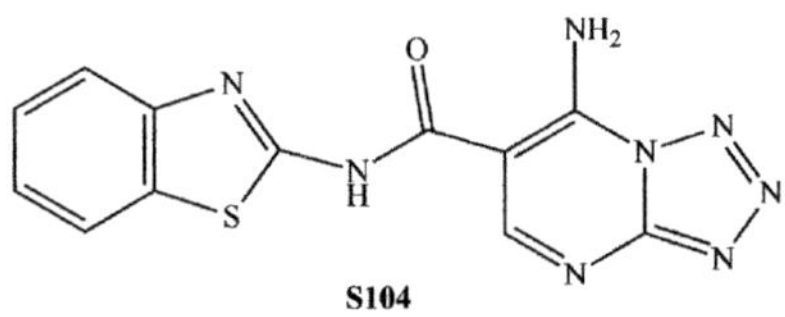

S104

FIG. 5.38 Chemical structures of compounds **S102–S104**.

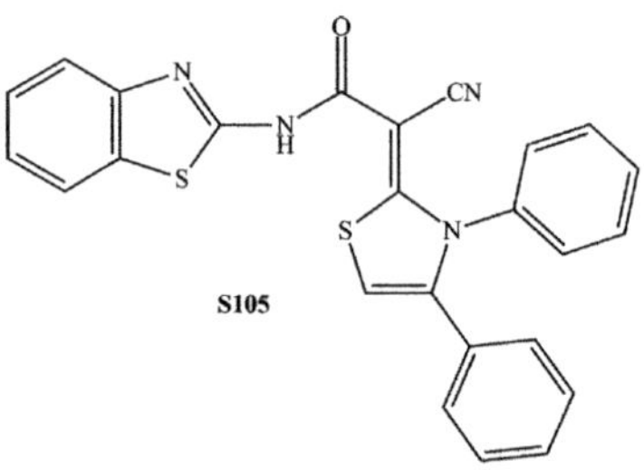

S105

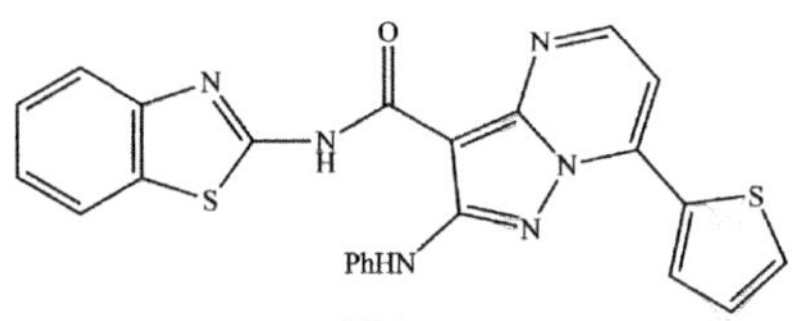

S106

FIG. 5.39 Chemical structures of compounds **S105** and **S106**.

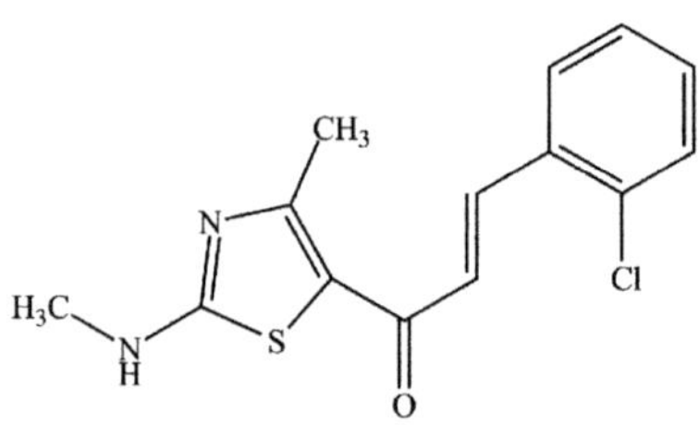

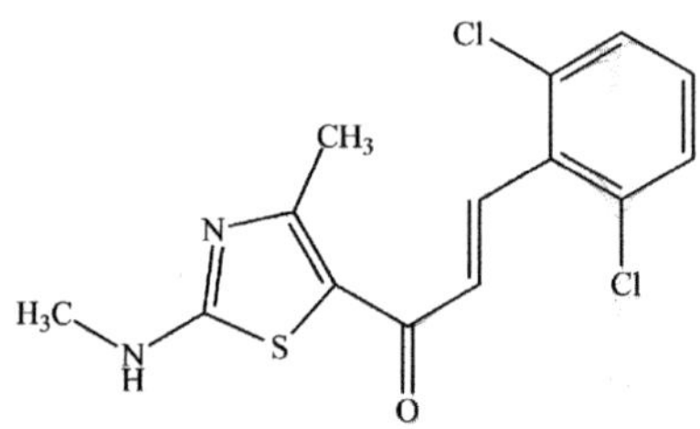

S107 S108

FIG. 5.40 Chemical structures of compounds **S107** and **S108**.

potential in vitro against a series of fungi. MICs and MFCs for the synthesized chalcones were determined by using the microdilution method. All tested compounds were found to possess moderate antifungal activity, being less potent than bifonazole and ketoconazole, which were used as reference antimycotics. It is noteworthy that derivative **S107** exhibited the strongest inhibitory effect against *Aspergillus versicolor*, *Trichoderma viride,* and *Fusarium sporotrichoides* with MIC = 15.29 μmol × 10^{-2}/ml, alongside with compound **S108** (Fig. 5.40), which expressed the same activity against *Trichoderma viride*, *Penicillium ochrochloron*, and *Fusarium sporotrichoides*.

Taking into consideration their previous findings (Liaras et al., 2011a), Liaras et al. (2011b) reported the design and synthesis of seventeen structurally modified thiazole-based chalcones aiming to achieve improved in vitro antifungal capacity against a panel of fungal strains. The MICs and MFCs were determined by employing microdilution method and as reference drugs were used bifonazole (MICs of 32.20–64.0 μmol × 10^{-2}/mL an MFCs of 48.0–80.0 μmol × 10^{-2}/mL) and ketoconazole (MICs of 38.0–475.0 μmol × 10^{-2}/mL and MFCs of 95.0–570.0 μmol × 10^{-2}/mL). All the synthesized chalcones were found to possess remarkable antifungal activity being more potent than reference antimycotics with MIC values of 2.62–7.35 μmol × 10^{-2}/mL and MFC values of 3.50–9.19 μmol × 10^{-2}/mL. The most potent compounds were **S109** of the methylamino group alongside with **S110** and **S111** (Fig. 5.41) of the ethylamino group. From the obtained results the authors observed that the most sensitive fungus was *Penicillium funiculosum*, while *Candida albicans* was the most resistant strain.

SAR studies were also performed by Liaras et al. (2011a,b) in order to elucidate the impact of structural modifications on the antifungal activity of the synthesized thiazole-based chalcones. The authors found that regarding the methylamino derivatives, introduction of fluoro- and bromo-substituents at the phenyl ring had as a result stronger antifungal potential than the unsubstituted chalcone. Moreover for the case of ethylamino derivatives was observed that while introduction of different groups at the phenyl ring enhanced the antifungal activity compared to parent chalcone, the biological result depended more on the type of substituents than their

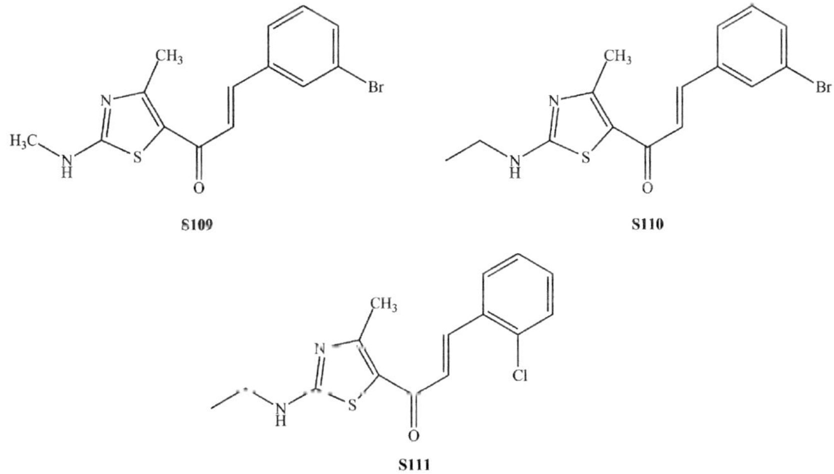

S109 S110 S111

FIG. 5.41 Chemical structures of compounds **S109–S111**.

position. It should be noted that the ethylamino chalcones exhibited stronger antifungal effect than the methylamino derivatives with the same phenyl ring substitution. This observation concerning elongation of alkylamino chain was also confirmed by comparing the activities of unsubstituted alkylamino derivatives, with their antifungal potential following the order methylamino < ethylamino < propylamino.

During an effort to combine thiazole and pyridine moieties in the same scaffold, Bondock et al. (2013) designed and synthesized a series of compounds containing both previously mentioned heterocycles. All the novel derivatives were assayed to elucidate their antifungal potential against four fungi, having as reference drug amphotericin B. Although the vast majority of the novel compounds were found to possess moderate-to-low antifungal activity against the used strains, it should be noted that derivative **S112** (Fig. 5.42) exhibited in general satisfactory inhibitory effect. This compound was equipotent with reference drug (amphotericin B) with MIC = 0.48 mg/mL against *Geotrichum candidum* and was found to be effective against *Aspergillus fumigatus* with MIC = 1.95 mg/mL, *Aspergillus clavatus* with MIC = 3.9 mg/mL and *Penicillium marneffei* with MIC = 3.9 mg/mL.

As part of their ongoing research project on thiazole derivatives as antimicrobial agents, Liaras et al. (2014) reported the synthesis and in vitro evaluation of antifungal activity of eight thiazole-based *N*-phenylpyrazolines and two aminopyrimidines. The MICs and MFCs of the compounds were determined by microdilution method against a panel of nine fungal species and were in the range of $1.45–47.80 \times 10^{-2}$ μmoL/mL and $3.37–95.5 \times 10^{-2}$ μmoL/mL, respectively. As reference drugs were used ketoconazole (MICs of $37.6–475.0 \times 10^{-2}$ μmoL/mL and MFCs of $94.0–570.0 \times 10^{-2}$ μmoL/mL) and bifonazole (MICs of $32.2–64.0 \times 10^{-2}$ μmoL/mL and MFCs of $48.3–80.0 \times 10^{-2}$ μmoL/mL). It should be noted that aminopyrimidine **S113** exhibited the best antifungal activity, followed by the *N*-phenylpyrazoline derivative **S114** (Fig. 5.43). Regarding the fungi which were used in the assay, *Aspergillus ochraceus* and *A. versicolor* were the most sensitive species, while *A. niger* and *Candida albicans* were found to be the most resistant ones against the tested compounds.

Pereira de Sá et al. (2015) evaluated the antifungal effects of thirteen thiazole derivatives in vitro against *C. neoformans* and *C. gattii* and determined

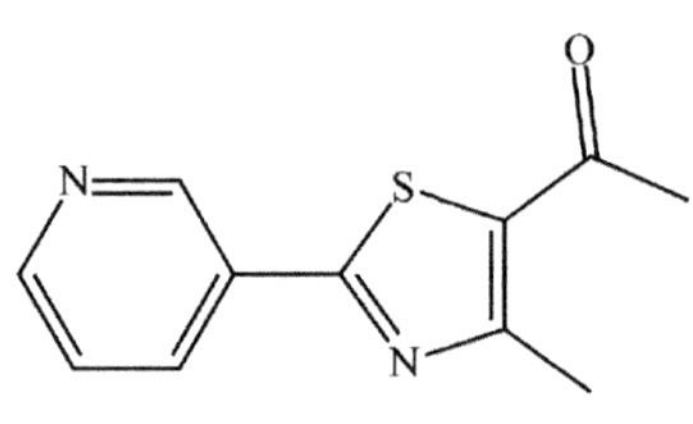

S112

FIG. 5.42 Chemical structure of compound **S112**.

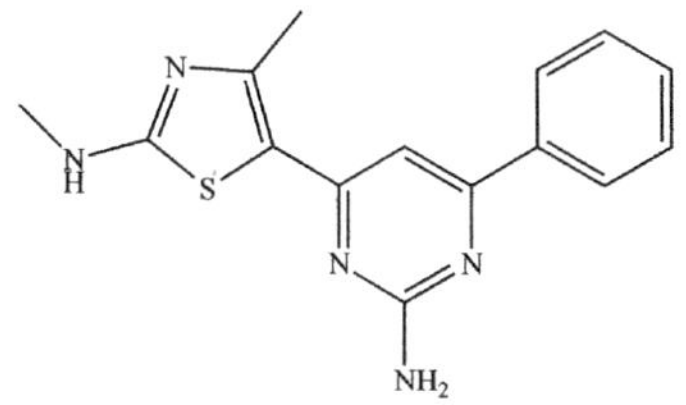

S113

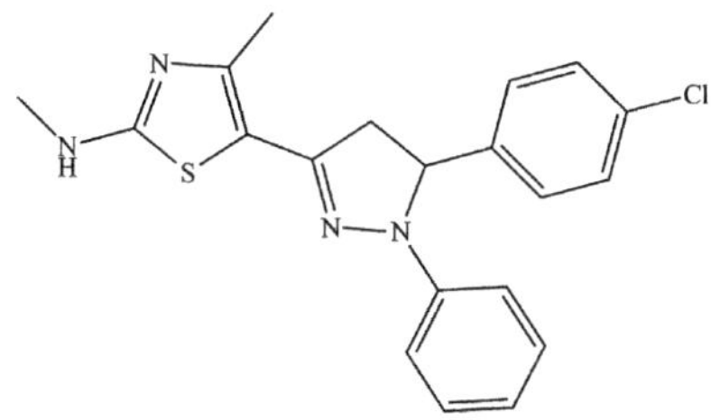

S114

FIG. 5.43 Chemical structures of compounds **S113** and **S114**.

their MICs by using microbroth dilution method, having as reference drugs fluconazole and amphotericin B. It was found that derivatives **S115** (MICs of 1.0–3.4 μM and 1.2–3.5 μM for *C. gattii* and *C. neoformans*, respectively), **S116** (MICs of 0.9–1.7 μM and 0.5–1.2 μM for *C. gattii* and *C. neoformans*, respectively), **S117** (MICs of 0.7–2.2 μM and 1.2–2.4 μM for *C. gattii* and *C. neoformans*, respectively), and **S118** (MICs of 1.4–4.5 μM and 1.9–3.5 μM for *C. gattii* and *C. neoformans*, respectively) (Fig. 5.44) were very potent against both fungal strains, exhibiting also stronger activity than fluconazole (MICs of 7.8–41.7 μM) against *C. gattii*. Furthermore, the authors observed that the death curves of the derivatives were comparable to the ones of reference, showing also no toxic effects in mammalian cells.

In an effort to develop novel antifungal agents, Lino et al. (2018) prepared a series of hydrazine-thiazole derivatives and evaluated their in vitro antifungal potential against a selection of strains, including *Candida* sp., *Cryptococcus* sp., and *Paracoccidioides brasiliensis*. The activity of the compounds was determined by using the microbroth dilution assay with three well-known antifungal drugs (fluconazole, amphotericin B, and itraconazole) as positive controls. It was observed that, despite the inactivity against *P. brasiliensis*, several compounds

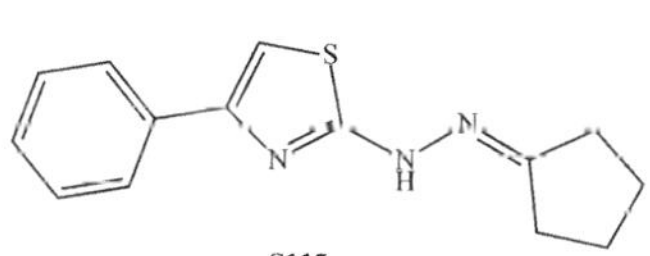

S115

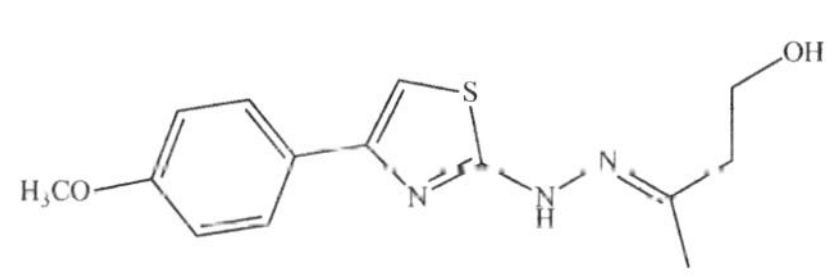

S116

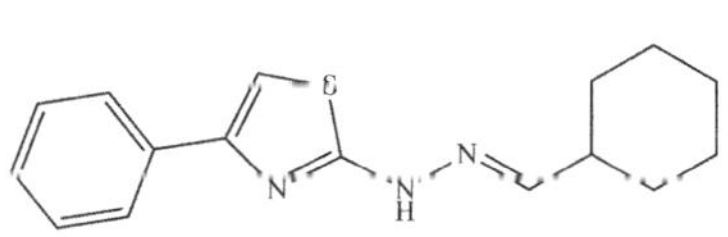

S117

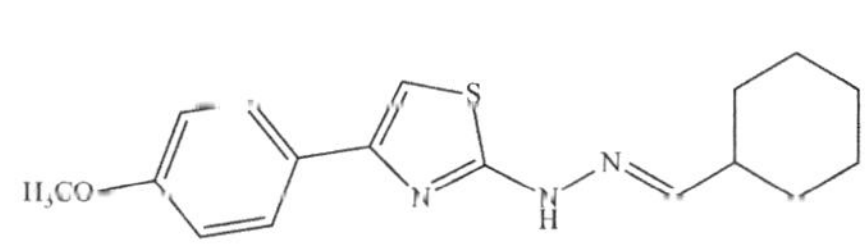

S118

FIG. 5.44 Chemical structures of compounds **S115–S118**.

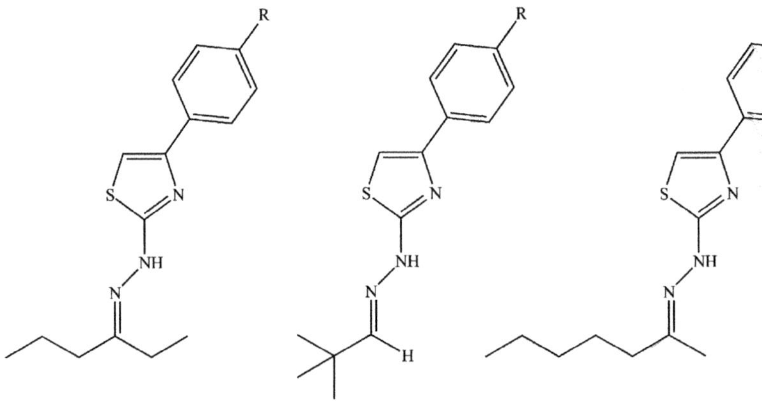

S119: R = H
S120: R = OCH_3
S121: R = Cl

S122: R = H
S123: R = OCH_3
S124: R = Cl

S125: R = OCH_3
S126: R = Cl
S127: R = CN

FIG. 5.45 Chemical structures of compounds **S119–S127**.

(**S119–S127**, Fig. 5.45) were potent against the vast majority of the fungi with MICs of 0.45–31.2 μM, with some of them being equipotent or even more potent than fluconazole and amphotericin B. Regarding the SAR, the authors found that the presence of thiazole ring was crucial for the activity, alongside with the presence of a hydrophobic aliphatic chain that is possibly also of high importance. It should also be mentioned that –Cl group favors, in general, the antifungal potential of this class of compounds. While the mode of action of this series of compounds is not clear yet, the authors related the activity of this class of thiazoles to their ability to induce oxidative stress in fungal cells. Furthermore, the novel heterocycles revealed a low-cytotoxicity profile against human cells with IC_{50} values higher than 50 μM turning them into candidates for further investigation.

Taking into account previous findings regarding the antifungal potential of thiazole-based chalcones, Tratrat et al. (2019) synthesized a series of novel compounds of this class and screened them in vitro for their activity against a panel of fungal strains. For the determination of the MIC and MFC values the microdilution method was employed and as reference drugs were chosen bifonazole and ketoconazole. The authors observed that all the chalcones were very potent against the tested fungi with MICs of 0.17–53.9 $\mu mol \times 10^{-2}$/mL and MFCs of 0.34–107.9 $\mu mol \times 10^{-2}$/mL. While all the compounds were more potent than reference drugs, the best activity was exhibited by chalcone **S131**, followed by **S130**, **S128**, and **S129** (Fig. 5.46). It should be noted that the most sensitive strain was *T. viride*, while the most resistant ones were found to be *A. fumigatus* and *C. albicans*. Furthermore, the authors found no correlation of lipophilicity with the activity observed, while docking studies showed that lanosterol 14α-demethylase and dihydrofolate reductase inhibitions are not straightly involved in the mode of action of this class of compounds.

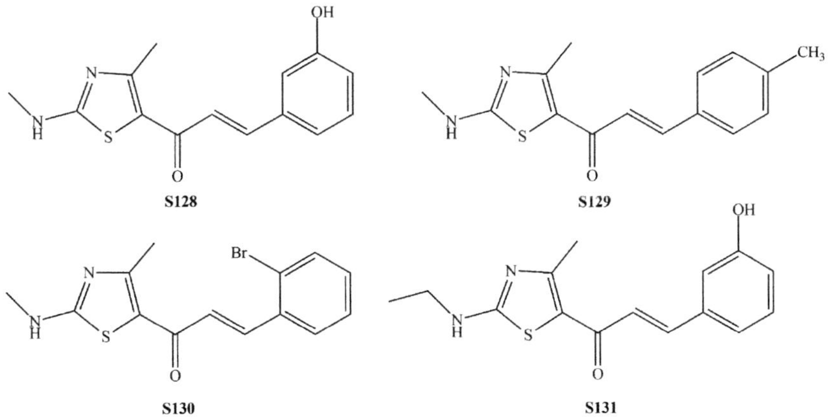

FIG. 5.46 Chemical structures of compounds **S128–S131**.

Althagafi et al. (2019) synthesized a series of derivatives, which contained di-, tri-, and tetrathiazole moieties in order to screen them for their antimicrobial activity. The antifungal activity of the novel compounds was investigated against *Aspergillus niger* and *Geotricum candidum* by measuring the inhibition zone in mm having as reference drug amphotericin B. Furthermore, the MICs of the compounds that exhibited significant antifungal potentials were also determined. From the obtained results was observed that derivatives **S132**, **S133**, **S134**, **S135** (Fig. 5.47) and were in general potent against *A. niger* exhibiting the same effect with amphotericin B (MIC=0.98 μg/mL), while compound **S133** was the only one that was found to be equipotent with reference drug against *G. candidum* (MIC=0.49 μg/mL).

Miscellaneous azole derivatives

A series of novel azoles containing pyrazoline moiety were synthesized and screened for their antifungal potential by Ozdemir et al. (2010) against a panel of *Candida* strains. The activity of the compounds was evaluated in vitro by the broth dilution method having ketoconazole as reference antifungal agent. The novel derivatives demonstrated, in general, medium-to very-good potential (MICs of 0.003–2.0 mg/mL) against the tested strains in comparison with ketoconazole (MICs of 0.003–0.5 mg/mL). It should be mentioned though that all the derivatives were inactive against *Candida tropicalis*, *C. utilis*, and *C. albicans*. It would be interesting to point out that pyrazolines **S136**, **S137**, **S138**, **S139**, and **S140** (Fig. 5.48) were the most potent among all tested compounds, including ketoconazole. Regarding SAR, the authors concluded that 3,4-methylenedioxyl substitution at the phenyl ring favors the antifungal potential of the synthesized compounds, while substitutional modifications on the benzimidazole/benzoxazole rings have no effect in the activity.

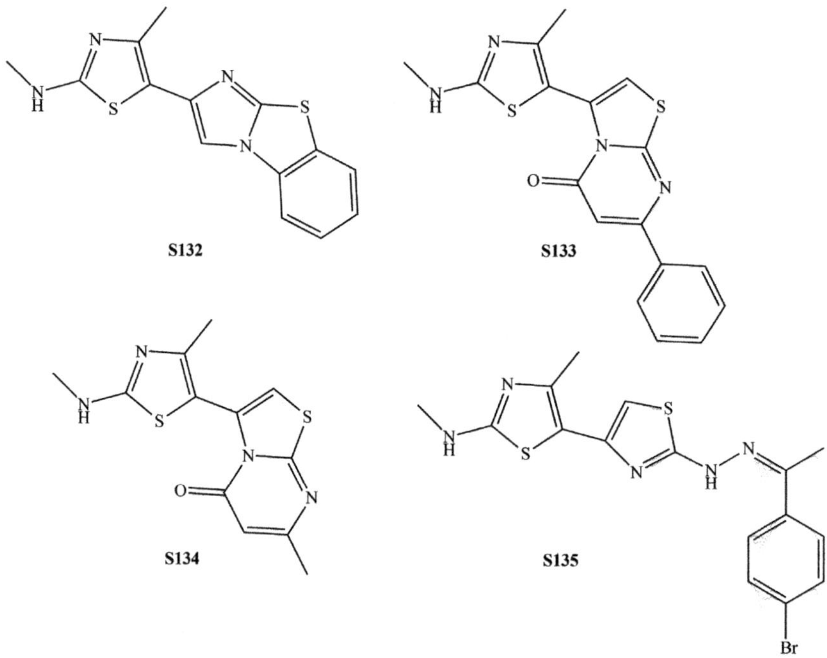

FIG. 5.47 Chemical structures of compounds **S132–S135**.

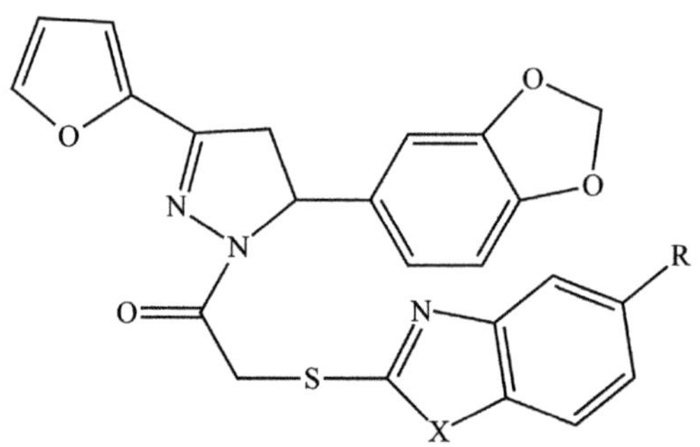

S136: R = H, X = NH

S137: R = Cl, X = NH

S138: R = H, X = O

S139: R = CH_3, X = O

S140: R = H, X = S

FIG. 5.48 Chemical structures of compounds **S136–S140**.

Sreedevi et al. (2013) synthesized six novel compounds, which incorporated both imidazole and thiazole moieties and screened them for their antimicrobial potential by employing the cup plate method against *Candida albicans*, *Aspergillus flavus*, *Aspergillus fumigatus*, and *Trichophyton rubrum*. Derivative **S141** (Fig. 5.49) demonstrated the strongest antifungal effect (22–26 mm diameters of inhibition zones) among the tested compounds being also more potent than ciclopirox olamine (20–22 mm diameters), which was used as reference agent.

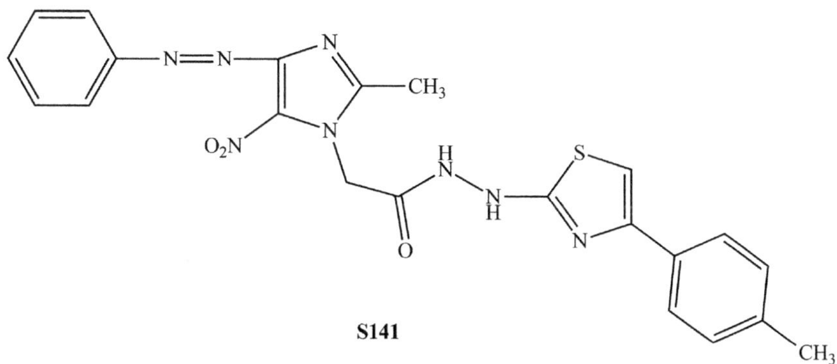

S141

FIG. 5.49 Chemical structure of compound **S141**.

5.3 Thiazolidinones

A series of venlafaxine derivatives containing thiazolidin-4-one moiety were synthesized by Kavitha et al. (2006). All the compounds were evaluated for their in vitro antifungal potential by using the disk diffusion and microdilution methods against *Aspergillus niger*, *A. flavus*, *Fusarium oxysporum*, *Trichoderma* species, and *F. moniliforme*, having nystatin as positive control. From the obtained results the authors observed that derivatives **S142**, **S143**, **S144**, **S145**, and **S146** (Fig. 5.50) showed good inhibitory activity (MICs of 12–26 μM) compared to nystatin (MICs of 29–36 μM).

Several thiazole-based thiazolidinones and azetidinones were prepared by Patel and Mehta (2006) in order to be tested for their antifungal effects against *Penicillium expansum*, *Botryodiplodia theobromae*, *Nigrospora* sp., and *Trichothecium* sp. As reference antifungal was used sulfanilamide. From the evaluation of the antifungal potential of the synthesized compounds was found that derivatives **S147**, **S148**, **S149**, **S150**, **S151**, and **S152** (Fig. 5.51) exhibited antifungal effect similar to that of the reference compound.

Ahmed et al. (2006) studied the antifungal activity of some newly synthesized succinimido(2-aryl-4-oxo-3-{[(quinolin-8-yloxy)acetyl]amino}-1,3-thiazolidin-5-yl)acetates against *C. albicans* and *A. fumigatus*. For the in vitro evaluation of the activity of the compounds the disk diffusion method was employed, having fluconazole as reference drug. Derivatives **S153** (zone of inhibi tion 24–25 mm at 300 μg/mL) and **S154** (zone of inhibition 23–24 mm at 300 μg/mL) (Fig. 5.52) demonstrated the strongest antifungal effects among the synthesized compounds, but still being less potent than fluconazole (zone of inhibition 25–27 mm at 300 μg/mL).

Bondock et al. (2007) described the synthesis of some thiazole, thiazolidinone, and thiazoline derivatives in order to elucidate their antifungal activity against *Aspergillus niger* and *A. oryzae*. For the biological evaluation of the novel compounds was employed the agar diffusion method, while fluconazole

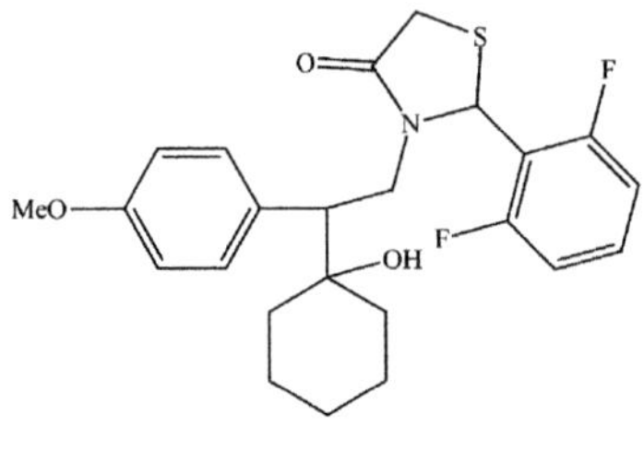

S142

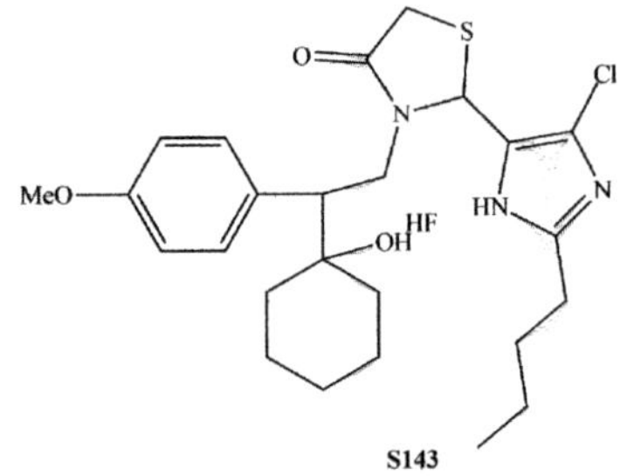

S143

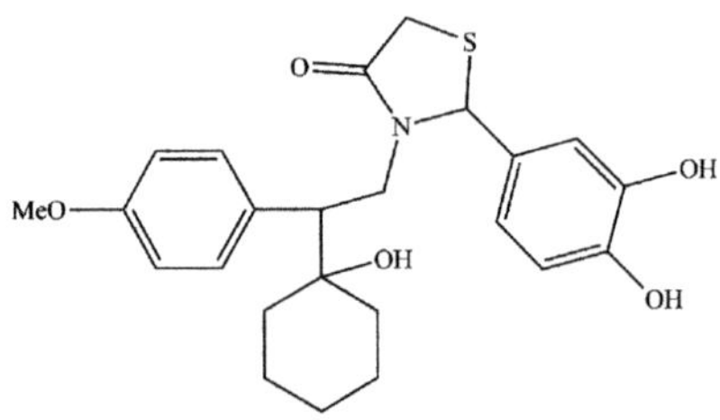

S144

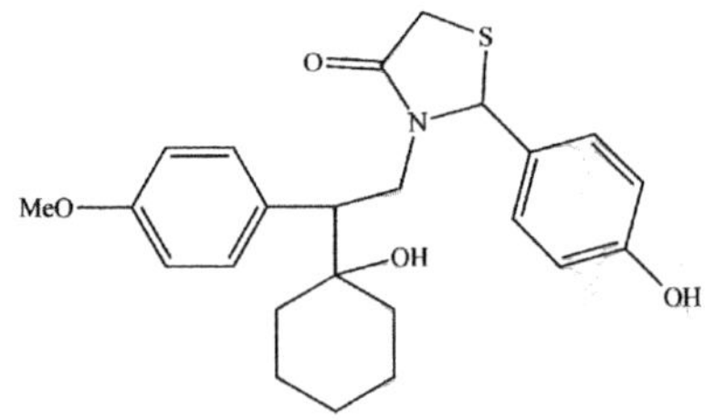

S145

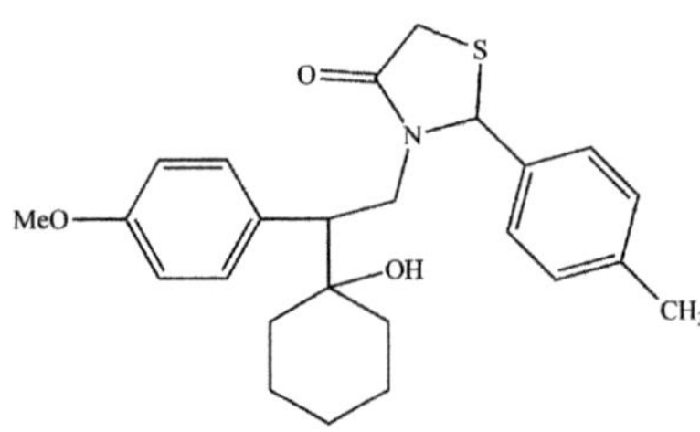

S146

FIG. 5.50 Chemical structures of compounds **S142–S146**.

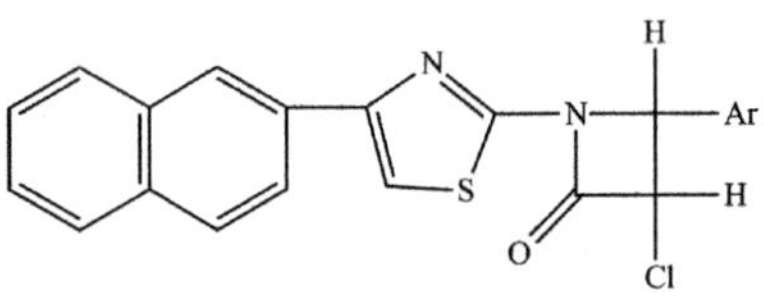

S147: Ar = 4-OH-C_6H_4

S148: Ar = 2-OH-C_6H_4

S149: Ar = 4-OCH_3-C_6H_4

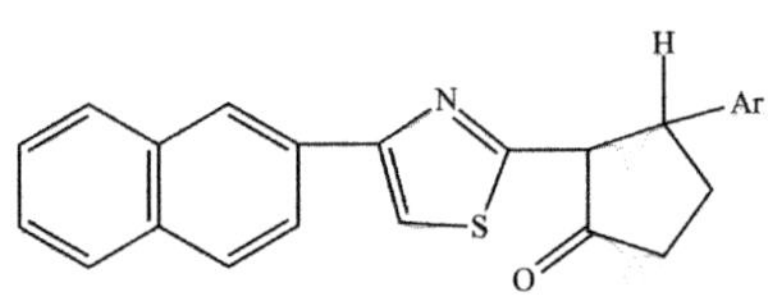

S150: Ar = 4-OCH_3-C_6H_4

S151: Ar = 2-OH-C_6H_4

S152: Ar = 3,4-methylenedioxyphenyl

FIG. 5.51 Chemical structures of compounds **S147–S152**.

was chosen as reference drug. The strongest antifungal effect was exerted by thiazole derivative **S155**, followed by compounds **S156**, **S157**, **S158**, and **S159** (Fig. 5.53). It should be mentioned that in this series, thiazole and thiazolidine derivatives were in general more potent than the thiazolidinone ones.

A series of 4-thiazolidinones were synthesized by de Aquino et al. (2008) and were subsequently evaluated in vitro for their antifungal activity against

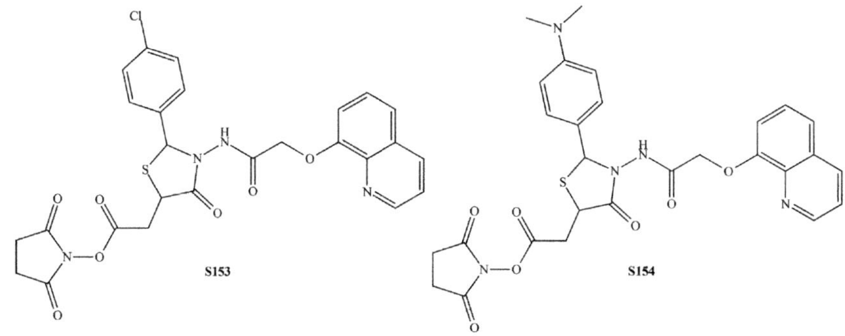

FIG. 5.52 Chemical structures of compounds **S153** and **S154**.

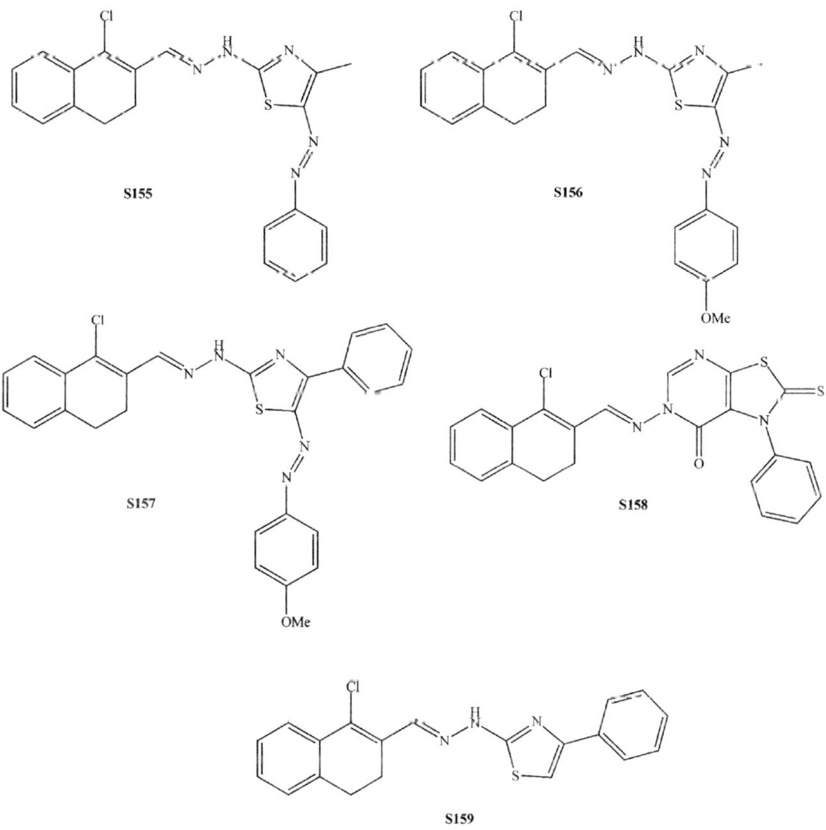

FIG. 5.53 Chemical structures of compounds **S155–S159**.

some fungal strains by using the disk diffusion method. After determining MICs and MFCs, the authors observed that the best antifungal potential was exerted by derivatives **S160**, **S161**, **S162**, and **S163** (Fig. 5.54) against *Candida* sp. (IMUR 4249) with MIC = 80 μg/mL and MFC = 70 μg/mL, being also equipotent with nystatin that was used as reference agent.

A series of novel methylene-bis-pyrimidinyl-spiro-4-thiazolidinones were synthesized by Srinivas et al. (2008) and their antifungal activity was investigated against *Aspergillus niger*, *A. flavus*, *Candida albicans*, and *Rhizopus oryzae* by employing the cup plate method. Derivative **S164** (Fig. 5.55) was found to be highly active against all the tested fungal strains, being also more potent than griseofulvin, which was used as reference compound.

Several thiazolidinones of the general structure shown in Fig. 5.56 were synthesized by Ramachandran et al. (2009) to be tested for their potential to act as antifungals against a collection of five strains (*Cr. neoformans*, *C. albicans*, *Rhizopus* sp., *A. niger*, and *A. flavus*). As reference antifungal was chosen the commercial drug amphotericin B. The SAR studies revealed that introduction of fluorine substituent at 3′ and 4′ positions of phenyl ring gave derivatives **S165** and **S166** (MICs = 6.25 μg/mL) (Fig. 5.56) with activity stronger than that of

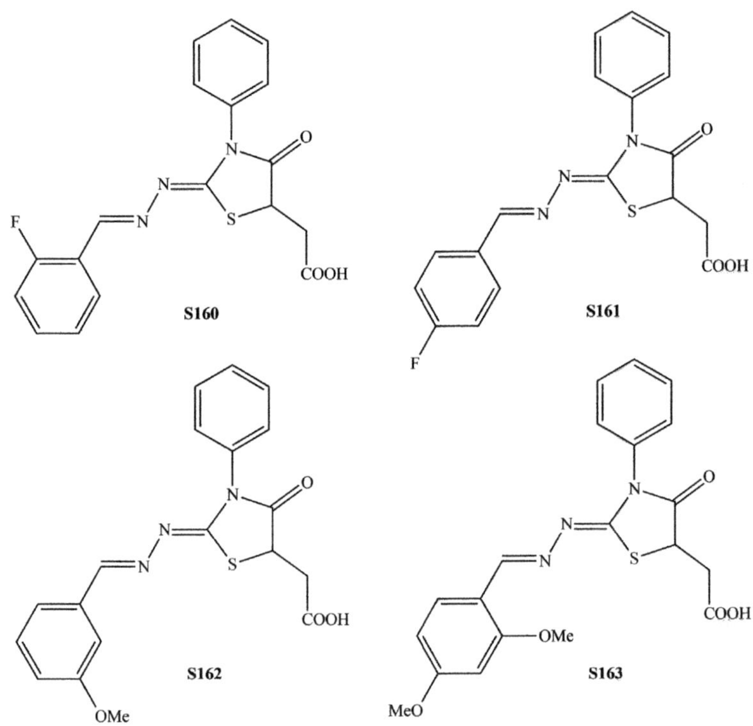

FIG. 5.54 Chemical structures of compounds **S160**–**S163**.

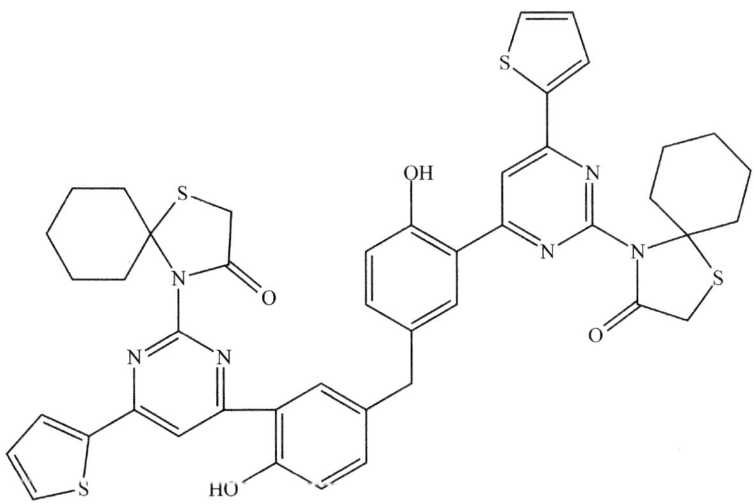

S164

FIG. 5.55 Chemical structure of compound **S164**.

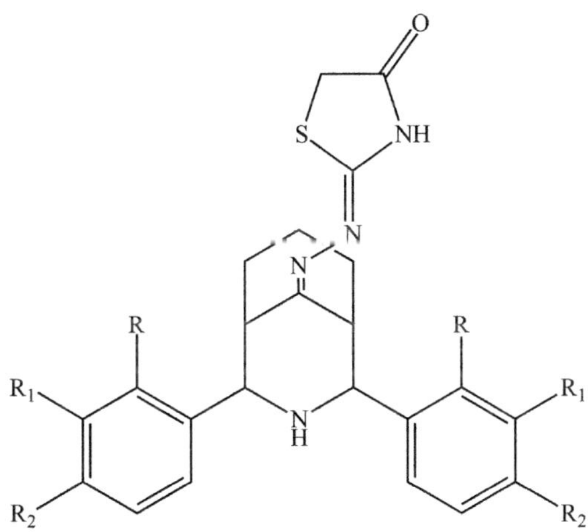

75

S165: $R = R_1 = H,\ R_2 = F$

S166: $R = R_2 = H,\ R_1 = F$

FIG. 5.56 Chemical structures of compounds **S165** and **S166**.

the reference drug (MIC = 25.0 μg/mL) regarding *C. albicans*. It should be also pointed out that compound **S166** was found to be more potent than amphotericin B also against *Rhizopus* sp. Furthermore, it could be concluded that for this series of thiazolidinones, fluorine, and chlorine substituents at the phenyl ring favored the activity against the tested strains.

Kouatli et al. (2010) designed and synthesized some thiazole-based thiazolidinones containing adamantane moiety in order to screen them for their antifungal activity against a collection of eight fungal strains (*A. flavus*, *A. versicolor*, *A. niger*, *T. viride*, *F. fulvum*, *P. funiculosum*, and *P. ochrochloron*), having as reference drugs ketoconazole and bifonazole. Initially was synthesized compound **S167** (Fig. 5.57), with MICs ranging at $0.60–2.38\times10^{-2}$ μmoL/mL and MFC values of $2.38–3.35\times10^{-2}$ μmoL/mL, being more potent than ketoconazole and bifonazole against all the tested fungi. Subsequently, this compound was condensed with various aromatic aldehydes to give derivatives **S168-S177** (Fig. 5.57), which were all found to exhibit stronger antifungal activity than **S167** and ketoconazole with MICs of $0.52–2.38\times10^{-2}$ μmoL/mL and MFCs of $1.05–3.35\times10^{-2}$ μmoL/mL. Thiazolidinone derivative **S176** exerted the strongest antifungal effect against the chosen strains with MIC values of $0.52–1.57\times10^{-2}$ μmoL/mL and MFCs of $1.05–2.31\times10^{-2}$ μmoL/mL. It was observed by the authors that the most sensitive strain was *F. fulvum*, while *A. versicolor* was the most resistant one.

A series of N-[(4-oxo-2-substituted aryl-1,3-thiazolidine)-acetamidyl]-5-nitroindazoles were designed and synthesized by Upadhyay et al. (2010). The novel derivatives were tested in vitro for their antifungal effect against *Aspergillus flavus*, *Penicillium citrinum*, and *Fusarium oxysporum* by using the paper disk method. By studying the results of the assay, the authors stated that the strongest antifungal activity was exhibited by compounds **S178** (MICs of 8–14 μg/mL) and **S179** (MICs of 9–15 μg/mL) (Fig. 5.58) being, nevertheless, less potent than griseofulvin (MICs of 6–11 μg/mL) that was used as reference drug.

Taking into consideration the variety of pharmacological activities of coumarins, Patel et al. (2010) presented the synthesis of several novel coumarin-based thiazolidinones, which were later tested for their antifungal potential

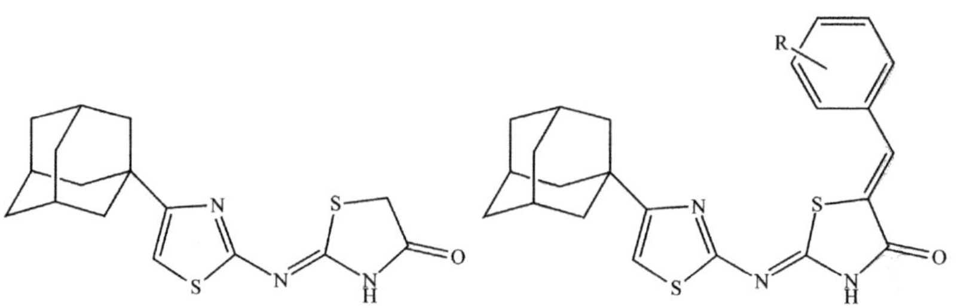

S167

S168: R = 4-Cl **S173:** R = 2-NO_2
S169: R = 3-Cl **S174:** R = 4-OH
S170: R = 2-Cl **S175:** R = 4-OH, 3-OMe
S171: R = 4-NO_2 **S176:** R = 4-OH, 3,5-OMe
S172: R = 3-NO_2 **S177:** R = 4-OMe

FIG. 5.57 Chemical structures of compounds **S167–S177**.

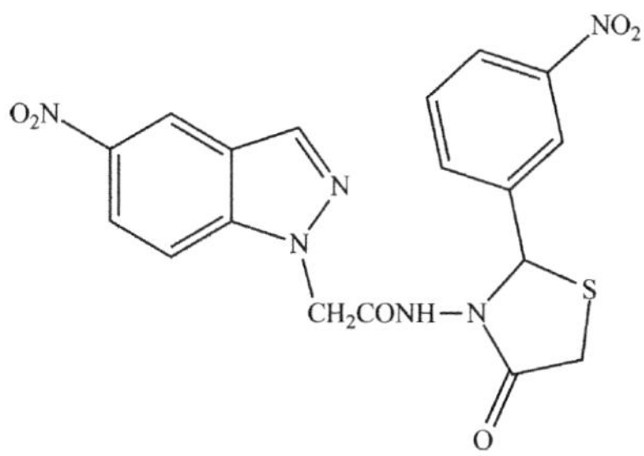

S178

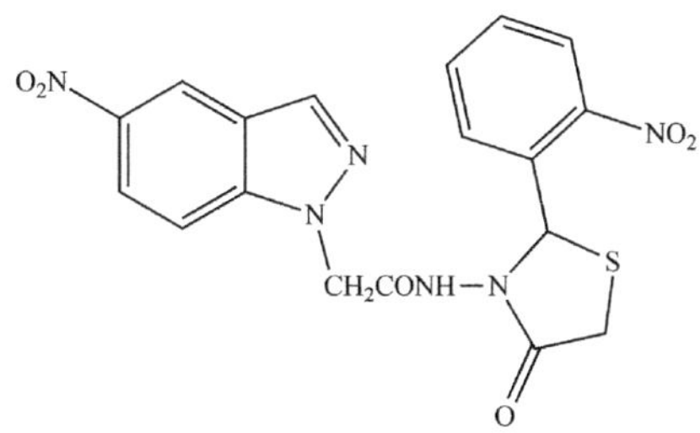

S179

FIG. 5.58 Chemical structures of compounds **S178** and **S179**.

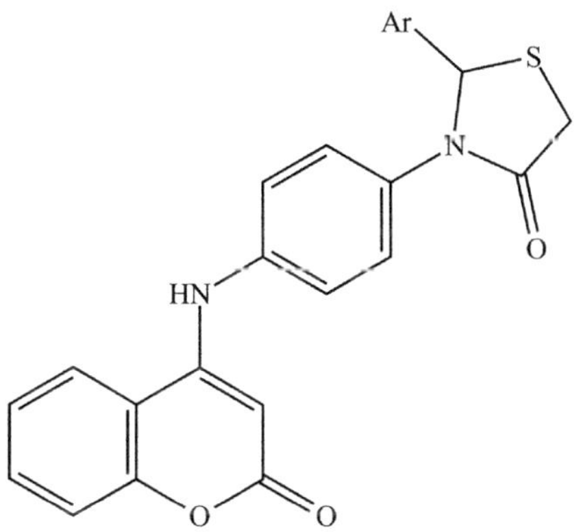

S180: Ar = 4-nitro-phenyl

S181: Ar = 4-methyl-phenyl

S182: Ar = phenyl-2-carboxaldehyde

S183: Ar = naphthyl

S184: Ar = 2-choloro-quinonyl

FIG. 5.59 Chemical structures of compounds **S180–S184**.

against *Candida albicans* and *Aspergillus niger* by using the broth dilution method. From the results of the assay was observed that derivatives **S180** and **S181–S184** (Fig. 5.59) were effective against *C. albicans* with MICs of 200–250 mg/mL, being much more potent than griseofulvin but less potent than nystatin. Interesting is the fact that the synthesized thiazolidinones were inactive against *A. niger* and in all of the cases less effective than fluconazole.

Another series of thiazolidinones, this time containing in the molecular scaffold N-ethylpiperazine moiety, were synthesized by Patel et al. (2012) in order to be tested for their antifungal potential with the paper disc diffusion method against some fungal species (*C. albicans*, *A. niger*, *A. fumgatus*, and *A. clavatus*), having as reference drug ketoconazole. The general structure of the synthesized thiazolidinones is given in Fig. 5.60. The compounds exhibited, in general, good-to-moderate activity with MICs ranging between 6.25 and 100 mg/mL. It was observed that electron-withdrawing substituents (such as –Cl and –F) at the phenyl ring favored the antifungal potential against the tested strains.

Mishra et al. (2012) designed and synthesized nine novel thiazolidinones containing thiophene moiety with the general structure that is shown in Fig. 5.61

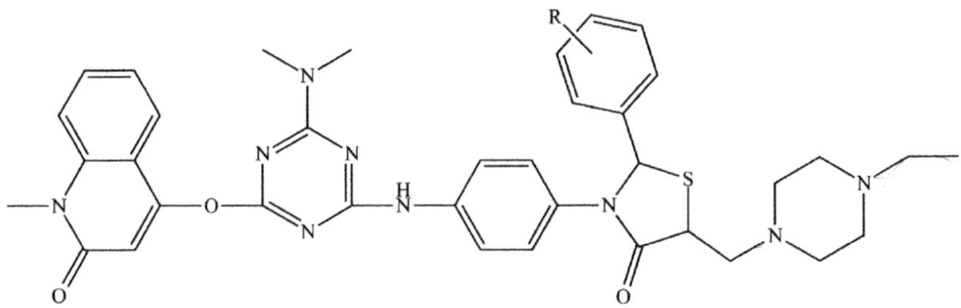

FIG. 5.60 General structure of the synthesized thiazolidinones.

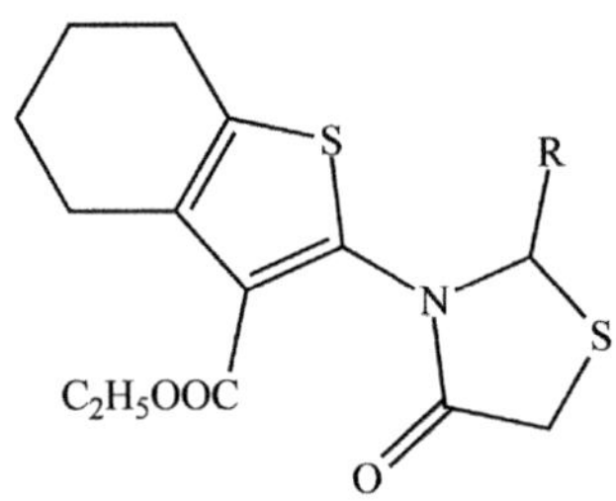

FIG. 5.61 General structure of the synthesized thiazolidinones.

and examined their antifungal potential against *C. albicans* by using the agar disc diffusion assay. Taking into account the results of the assay (inhibition zone of 9–15 mm), the authors came to the conclusion that this group of heterocycles reveal moderate-to-low antifungal effect against the tested strain.

A group of 2,5-disubstituted 4-thiazolidinone analogues were synthesized by Chawla et al. (2012) and were evaluated for their in vitro antifungal effect against *C. albicans*, having as reference drug fluconazole. From the obtained results was revealed that four of the novel thiazolidinones, namely, **S185**, **S186**, **S187**, and **S188** (Fig. 5.62) were very potent (inhibition zone of 10–26 mm), exhibiting activity even stronger than fluconazole (inhibition zone of 35 mm).

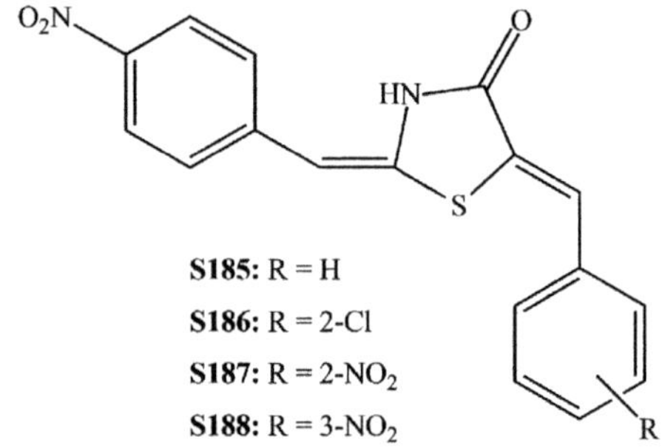

S185: R = H

S186: R = 2-Cl

S187: R = 2-NO_2

S188: R = 3-NO_2

FIG. 5.62 Chemical structures of compounds **S185–S188**.

Shrivastava et al. (2012) synthesized some pyridine-based thiazolidinones and evaluated their antifungal activity against *Botrytis fabae*, *Fusarium oxysporum*, and *Trichoderma viride*. For the antifungal screening was used the agar diffusion method and as reference drug was chosen fluconazole. Most of the compounds were found to exhibit moderate-to-low inhibitory activity against the used strains. Nevertheless, thiazolidinone **S189** was the most potent compound against *B. fabae* and *T. viride*, while **S190** and **S191** (Fig. 5.63) were effective against *F. oxysporum* with MIC = 6.25 μg/mL but still all of them were found to be less potent than fluconazole (MIC = 3.125 μg/mL).

A series of thiazole-based thiazolidinones were designed, synthesized, and evaluated in vitro by Apostolidis et al. (2013) for their antimicrobial potential against several fungal and bacterial species. Regarding their antifungal activity, the novel compounds were found to be very potent with MICs ranging at $0.73–23.2 \times 10^{-2}$ μmoL/mL and MFCs at $1.52–38.5 \times 10^{-2}$ μmoL/mL. It would be interesting to point out that the tested derivatives exhibited stronger activity than the reference drugs bifonazole (15–28 fold) and ketoconazole (15–250 fold). From the obtained results was revealed that compound **S192** (Fig. 5.64)

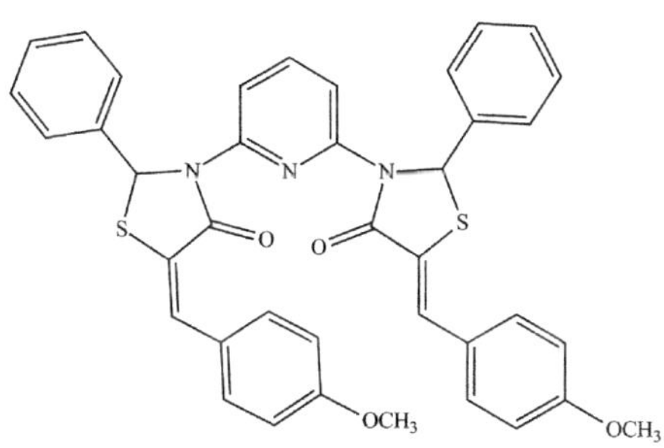

S189

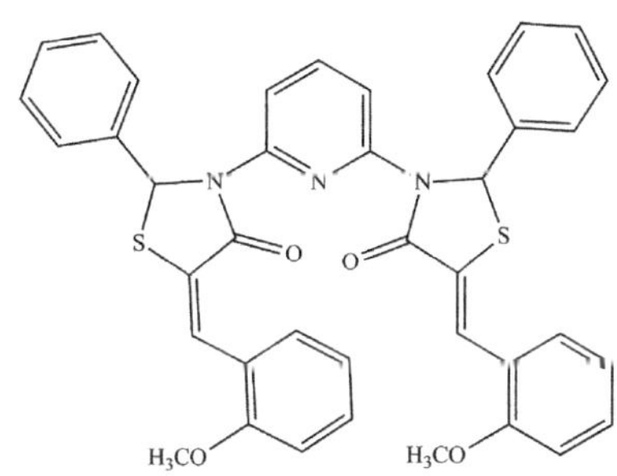

S190

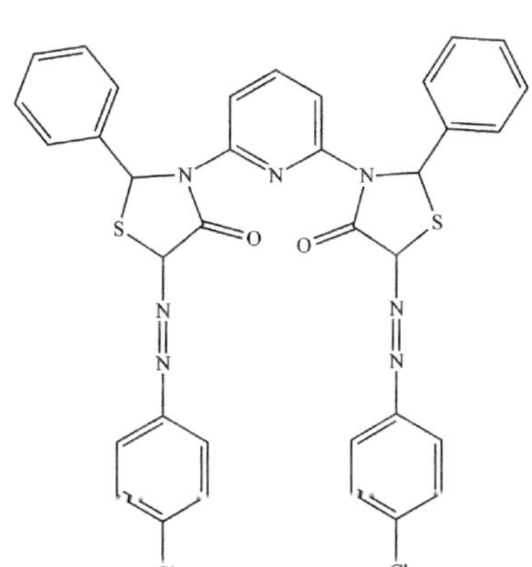

S191

FIG. 5.63 Chemical structures of compounds **S189–S191**.

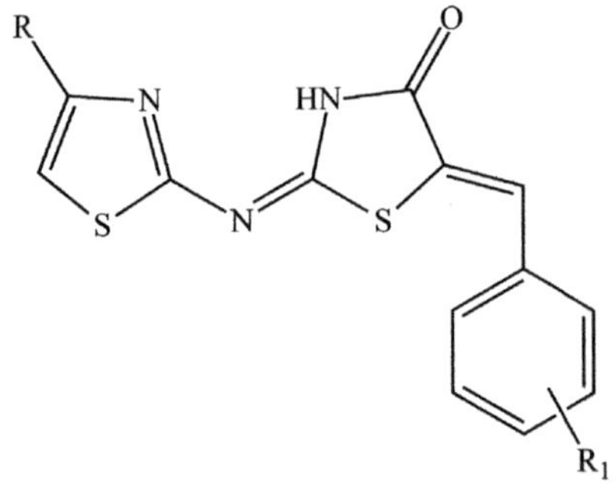

	R	R_1
S192:	H	4-N(CH$_3$)$_2$
S193:	Ph	2-Cl
S194:	Ph	4-Cl
S195:	Ph	3-Cl
S196:	CH$_3$	4-Cl

FIG. 5.64 Chemical structures of compounds **S192–S196**.

was the most potent with MIC values of $0.76–1.52 \times 10^{-2}$ µmoL/mL and MFC values at 1.52×10^{-2} µmoL/mL. Concerning the fungi, *Trichoderma viride* and *Penicillium ochrochloron* were found to be the most sensitive species, while *Aspergillus flavus* was the most resistant one. From the SAR studies was observed that, in general, bulky substituents have positive effect on the antifungal potential. Another interesting fact is that regarding 5-(chlorobenzylidene)-2-(4-phenyl-1,3-thiazol-2-ylimino)-1,3-thiazolidin-4-ones, the position of chloro-substituent at the phenyl ring is significant for the activity. Thus 2-Cl derivative (**S193**) was the most potent one, followed by 4-Cl (**S194**) and 3-Cl (**S195**) ones. It should be mentioned that the nature of substituent of thiazolyl moiety had also effect on the antifungal activity. For example, replacing 4-phenyl group of compound **S194** with 4-CH$_3$ (**S196**) had as a result 1.6-fold increase in the activity.

A series of azetidone and thiazolidinone derivatives containing quinoline moiety were prepared by Mistry and Jauhari (2013) and screened for their antifungal effects against *A. niger* and *C. albicans* by using the agar streak dilution assay. As reference compound was used the commercial drug ketoconazole. Azetidones **S197**, **S198**, and **S199** were effective against *A. niger* with MIC = 12.5 µg/mL, while thiazolidinones **S200** and **S201** (Fig. 5.65) were also potent against the same species with MIC = 12.5 µg/mL. As far as *C. albicans* is concerned, derivative **S200** exhibited the same potential with MIC = 12.5 µg/mL, while compound **S201** was less effective (half-fold) with MIC = 25.0 µg/mL. It is noteworthy though, that all the novel compounds of this series were less effective than ketoconazole against the two selected strains.

Pitta et al. (2015) reported the microwave-assisted synthesis of ten novel compounds (**S202-S211**, Fig. 5.66) that contain thiazole, adamantane, and 4-thiazolidinone moieties. These derivatives were evaluated for their antibacterial and antifungal potential by using the microdilution method. The compounds were found to be, in general, very potent as antifungals (MICs of 0.021–87.60 µmoL/mL $\times 10^{-2}$ and MFCs of 0.045–109.50 µmoL/mL $\times 10^{-2}$) compared to reference drugs bifonazole (MICs of 32.0–64.0 µmoL/mL $\times 10^{-2}$ and MFCs of 64.0–80.0 µmoL/mL $\times 10^{-2}$) and ketoconazole (MICs of 38.0–475.0 µmoL/mL $\times 10^{-2}$ and MFCs of 95.0–570 µmoL/mL $\times 10^{-2}$). Thiazolidinone deriva-

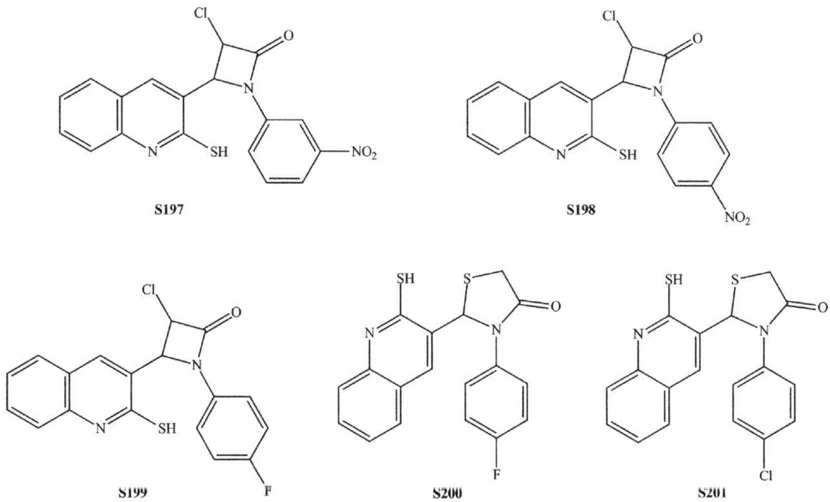

FIG. 5.65 Chemical structures of compounds **S197–S201**.

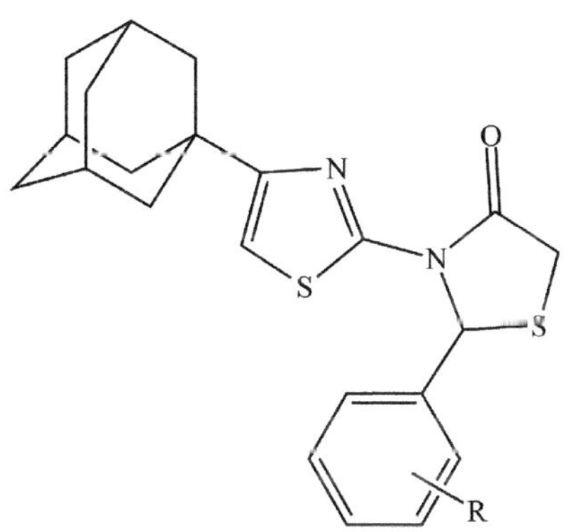

R

S202: 2-Cl
S203: 3-Cl
S204: 4-Cl
S205: 2,3-diCl
S206: 2,6-diCl
S207: 3-F
S208: 4-Br
S209: 4-NO_2
S210: 4-OMe
S211: 2,5-diOCH_3

FIG. 5.66 Chemical structures of compounds **S202–S211**.

tive **S205** demonstrated the best antifungal activity with MIC values ranging from 0.021 to 0.53 μmoL/mL × 10^{-2} and MFC values from 0.06 to 1.075 μmoL/mL × 10^{-2}. Furthermore, SAR studies revealed that introduction of 2,3-diCl, 3-Cl, and 4-Cl substituents at the phenyl ring enhances the antifungal effect. The most sensitive species was found to be *Aspergillus ochraceus*, while *Candida albicans* was the most resistant one.

As a continuation on their research efforts regarding the synthesis of novel antimicrobials, Haroun et al. (2016) reported the design, synthesis, and evaluation of antifungal activity of a series of fourteen thiazole-based thiazolidinones. All the novel compounds were assayed against eight fungi by using microdilution method having as reference drugs bifonazole and ketoconazole.

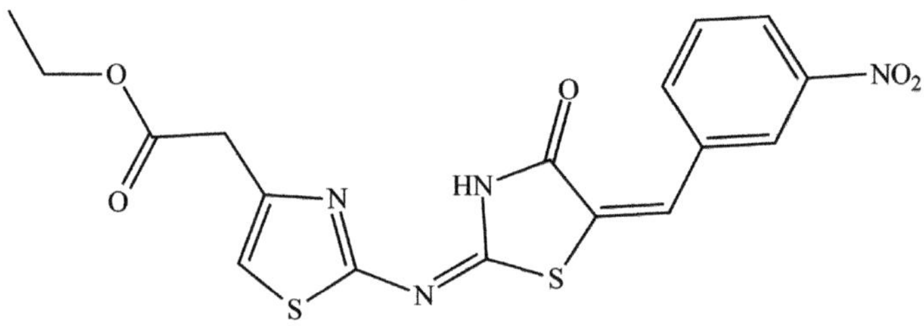

S212

FIG. 5.67 Chemical structure of compound **S212**.

The vast majority of the synthesized thiazolidinones were found to be more potent than the two previously mentioned antimycotics with MIC range of 0.3–38.6 μmol × 10^{-2}/mL and MFC range of 0.6–77.2 μmol × 10^{-2}/mL. It should be mentioned that *Trichoderma viride* was the most sensitive species against the effect of the compounds and *Candida albicans* the most resistant one. The 3-NO_2 derivative **S212** (Fig. 5.67) exhibited the strongest antifungal potential among all fourteen thiazole-based thiazolidinones.

Six novel arylidene substituted 4-thiazolidinone derivatives were synthesized and screened for their antimicrobial activity by Gupta et al. (2016). For the evaluation of the potential of the compounds by cup-plate method, *Aspergillus niger* and *Penicillium chrysogenum* were used. Furthermore, MICs of synthesized thiazolidinones were determined by tube dilution method and were found to be between 100 and 400 μg/mL. Commercial drug fluconazole was used as reference with MICs of 1.25 μg/mL and 2.5 μg/mL for *P. chrysogenum* and *A. niger*, respectively. Although less potent than fluconazole, derivatives **S213, S214, and S215** (Fig. 5.68) achieved the highest inhibitory effect (MIC = 100 μg/mL) against the two tested fungal species.

A group of twenty-three novel thiazolin-4-ones were synthesized by Stana et al. (2017) as potential lanosterol 14α-demethylase inhibitors. The antifungal activity of the compounds was evaluated in vitro against a panel of *Candida*

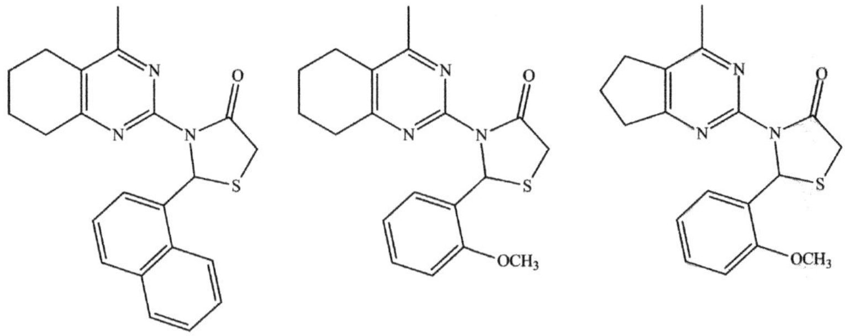

S213 S214 S215

FIG. 5.68 Chemical structures of compounds **S213–S215**.

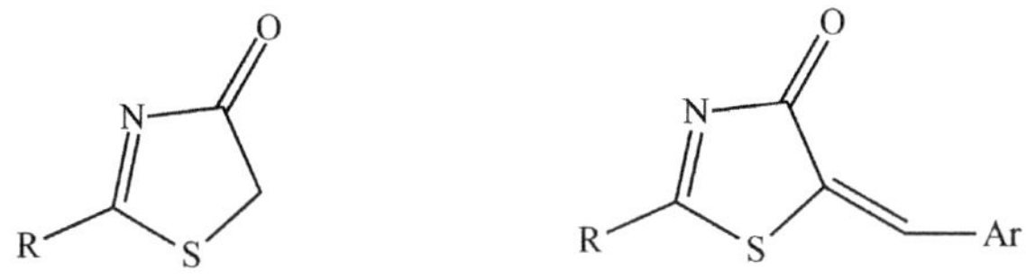

S217, S218, S219, S221 **S216, S220**

S216: R = a-$C_{10}H_7$-NH-	Ar = 2,4-diCl-phenyl
S217: R = CH_3-NH	Ar = -
S218: R = C_6H_5-NH	Ar = -
S219: R = a-$C_{10}H_7$-NH-	Ar = -
S220: R = C_6H_5-NH	Ar = 4-NO_2-phenyl
S221: R = C_6H_5-	Ar = -

FIG. 5.69 Chemical structures of compounds **S216–S221**.

strains and their MICs and MFCs were determined by broth microdilution method. All the novel thiazolidinones were found to be efficient against the tested fungi, with derivatives **S216** (MIC=0.015 μg/mL and MFCs of 0.015–0.03 μg/mL) and **S217** (MIC=0.015 μg/mL and MFCs of 0.015–0.03 μg/mL) (Fig. 5.69) being over 500-fold more potent than fluconazole (MICs of 7.81–15.62 μg/mL and MFCs of 15.62–31.25 μg/mL), which was used as positive standard. Taking also into account the results of docking and ADME–Tox prediction studies, which were also performed, the authors came to the conclusion that the synthesized compounds are potential anti-Candida agents, which might act by inhibiting 14α-demethylase and thus could be subjects of further investigation. It should be mentioned that thiazolidinones **S218** and **S219** (Fig. 5.69) were identified as "hits" by the authors, being both free of UMSs (Undesirable Moieties and Substructures) in toxicity issues, covalent inhibitors, and PAINs (Pan Assay Interference Compounds) but PPI (Protein-Protein Interactions) unfriendly, alongside with derivatives **S220**, which is the only one of the lead compounds that is PPI friendly, and **S221**.

A series of twelve novel benzothiazole-based thiazolidinones were designed and synthesized by Haroun et al. (2018) in order to be evaluated for their antifungal and antibacterial potential. The antifungal activity of the compounds was tested by using the microdilution method and from the obtained results can be observed that all the synthesized thiazolidinones are potent as antifungal agents with MIC values of $0.55–28.8 \times 10^{-2}$ μmoL/mL and MFC values of $1.49–38.4 \times 10^{-2}$ μmoL/mL. Derivative **S222** (Fig. 5.70) exhibited the best activity against the tested fungi with MICs ranging between 1.10 and 5.94×10^{-2} μmoL/mL and MFCs between 1.49 and 11.88×10^{-2} μmoL/mL. It should be pointed out that compounds **S222** and **S223** demonstrated good activity against both the selected fungal and bacterial species. *Trichoderma viride* was found to be the most sensitive fungus against the synthesized thiazolidinones, while the most

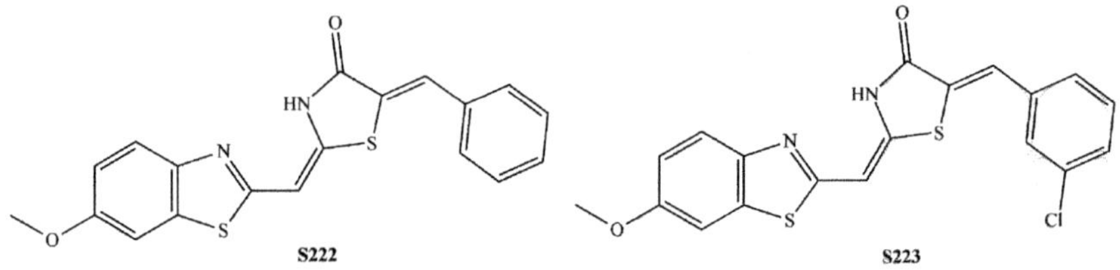

FIG. 5.70 Chemical structures of compounds **S222** and **S223**.

resistant species was *Candida albicans*. Regarding the SAR studies, it is noteworthy that introduction of substituents at the phenyl ring had as consequence a decrease in the antifungal effect. From the obtained results concerning the prediction of potential cytotoxicity of the synthesized molecules (OPEN TOX and PROTOX programs), authors characterized them as inactive for carcinogenesis and mutagenesis and with LD50 values ranging between 350 and 1000 mg/kg can be considered as safe for further studies.

5.4 Echinocandins

Echinocandins are antifungal agents that have the ability to inhibit the fungal cell wall β-(1,3)-D-glucan synthase in a similar way with that of penicillin against bacteria (Denning, 2002; Hossain and Ghannoum, 2000). In that way the synthesis of the essential fungal cell wall polysaccharide β-(1,3)-D-glucan is being disrupted, inhibiting cell wall growth. The most known derivatives of this class are caspofungin, micafungin, and anidulafungin (Douglas et al., 1997). Although these three representatives exhibit a promising antifungal profile (Table 5.1), there are available only parenteral formulations for use in clinical practice (Nett and Andes, 2016). As far as the metabolism is concerned, although echinocandins are not processed by CYP450, they are metabolized in the liver and this is a variable that should be taken into account for patients with hepatic conditions (Wagner et al., 2006).

In general, echinocandins are effective against *Aspergillus* spp. and *Candida* spp. such as *Candida albicans*, *C. glabrata*, and *C. tropicalis*, but exhibit limited antifungal effect against *Cryptococcus neoformans* due to minimal or no presence of β-(1,3)-D-glucan synthase (Arathoon, 2001; Bachmann et al., 2002; Denning, 1997). It should be also noted that they are not effective against Mucorales, endemic dimorphic fungi, *Scedosporium* spp., and *Fusarium* spp. (Cuenca-Estrella et al., 1999; Diekema et al., 2003; Nett and Andes, 2016).

The first semisynthetic derivative of this group that was approved by FDA was **caspofungin** (2001) (Fig. 5.71) for patients with invasive aspergillosis who are not responding or are unable to tolerate other therapeutic schemes. It should

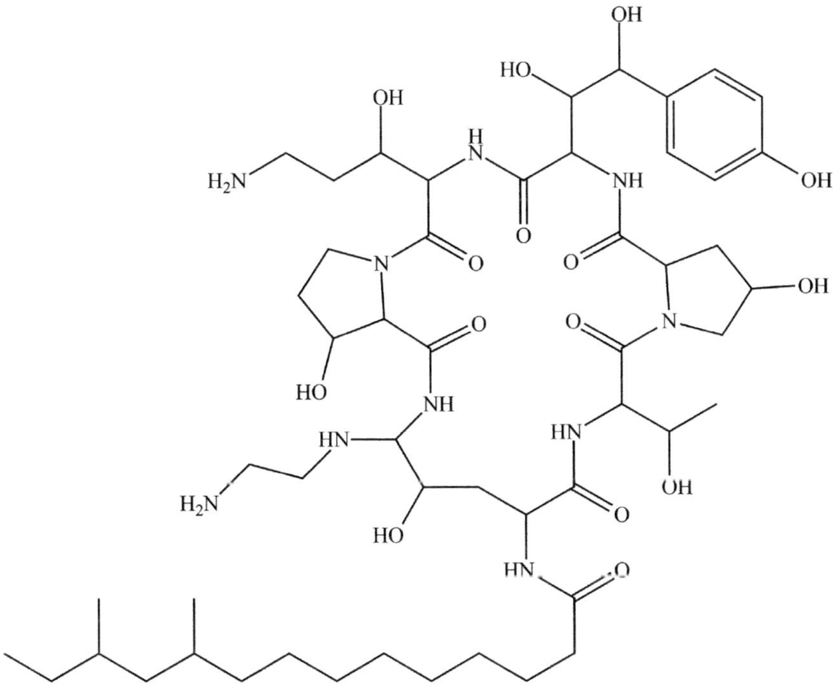

FIG. 5.71 Chemical structure of **caspofungin**.

be mentioned that caspofungin is potent in cases with esophageal and pharyngeal candidiasis, while having fewer side effects compared to amphotericin B. Nevertheless, some of the more frequent side effects of caspofungin are fever, headache, rash, histamine-related symptoms, and thrombophlebitis (Arathoon et al., 2002; Keating and Jarvis, 2001; Villanueva et al., 2001).

Anidulafungin (Fig. 5.72) is a semisynthetic cyclic lipopeptide that exhibits antifungal activity against all *Candida* species, including strains that have become resistant to azole drugs and amphotericin B. This antifungal agent is also effective against *Aspergillus* spp. and should be mentioned that it doesn't interact with any enzymes of cytochrome P450 (Ostrosky-Zeichner et al., 2003).

Micafungin (Fig. 5.73) is another semisynthetic echinocandin antifungal drug that can be administered for the treatment of invasive fungal infection (e.g., aspergillosis). It was found to achieve better response rates concerning antifungal prophylaxis, compared to fluconazole for patients who received hematopoietic stem-cell transplant (Kohno et al., 2001; Ratanatharathorn et al., 2002; Van Burik et al., 2002).

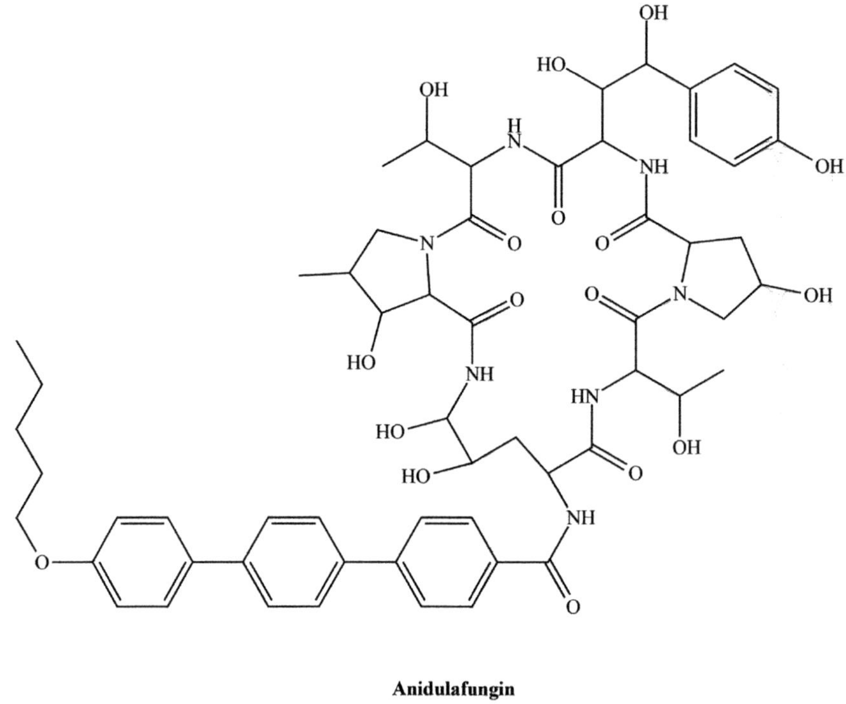

FIG. 5.72 Chemical structure of **anidulafungin**.

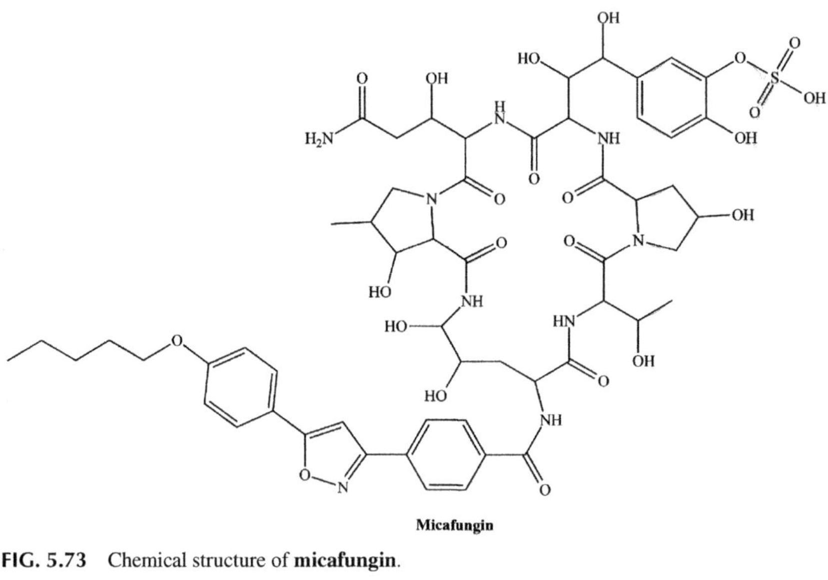

FIG. 5.73 Chemical structure of **micafungin**.

5.5 Other synthetic antifungals

Other synthetic analogues of natural compounds

Sordarins

Sordarins are a group of compounds with antifungal properties, which contain a tetracyclic diterpene core that includes a norbornene system. They have a unique mode of action, since they act by inhibiting protein synthesis in fungi by stabilizing the ribosome/Elongation Factor 2 (EF2) complex. This diverse mechanism of action is the reason that turns sordarins into a promising class of antifungal agents and spurs the scientific interest for further research efforts (Liang, 2008) (Fig. 5.74).

Sordarin was isolated in 1969 from the fungus *Sordaria araneosa* and by treating compound with concentrated aqueous HCl in acetone scientists were able to degrade it to the aglycone **sordaricin** (Hauser and Sigg, 1971; Liang, 2008; Sigg and Stoll, 1969). It should be mentioned that for many years sordarin wasn't considered as antifungal agent, but later was found that it can inhibit, in vitro, the protein synthesis in *Candida albicans*. Sordarin acts on EF2, which catalyzes the translocation of the ribosome along mRNA during elongation of the polypeptide chain that is under formation. As it was mentioned, sordarin has the ability to stabilize the ribosome/EF2 complex. Except the previously mentioned effectiveness of this promising agent against *Candida albicans*, sordarin exhibits strong antifungal effect also toward *Saccharomyces cerevisiae* (Anderson et al., 2003; Domínguez and Martín, 1998; Liang, 2008; Liu et al., 2006; Odds et al., 2003; Shastry et al., 2001).

Synthetic analogues of sordarin

In order to expand the therapeutic spectrum of sordarins to cover different kinds of fungal species, scientists from leading pharmaceutical laboratories have introduced several modifications to both the glycosyl moiety and diterpene core of the natural isolated compound.

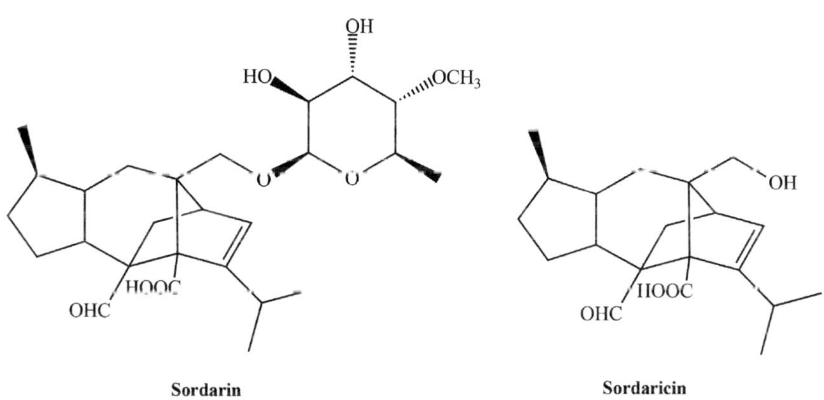

FIG. 5.74 Chemical structures of **sordarin** and **sordaricin**.

Glycosyl analogues

The isolation of sordarin analogue **GR135402** (Fig. 5.75) from *Graphium putredinis,* which is a selective and potent inhibitor of *Candida albicans* protein synthesis inspired scientists to develop novel derivatives of this promising natural antifungal.

A series of representative glycosyl analogues that have been synthesized by some of the leading pharmaceutical laboratories and scientific groups are presented in Figs. 5.76 and 5.77 (Dominguez et al., 1998; Liang, 2008).

It should be noted that, in general, **GM** compounds (Fig. 5.76) showed activity against a series of pathogenic fungi, including *Candida, Aspergillus,* and *Pneumocystis* spp., while **GW** group (azasordarins, Fig. 5.77) exhibited similar to superior activity against the same pathogens. It would be interesting to point out that synthesized glycosyl analogues demonstrated lower in vitro toxicity than amphotericin B on mammalian cell lines (Herreros et al., 1998, 2001).

Dominguez et al. (1998) tested a group of previously mentioned **GM** compounds, in comparison with **sordarin** and **GR135402**, for their ability to inhibit in vitro protein synthesis and cell growth of five pathogenic fungal strains: *Candida albicans, C. glabrata, C. parapsilosis, C. krusei,* and *Cryptococcus neoformans.* Moreover, they evaluated their inhibitory effect on protein synthesis in mammalian systems (rabbit reticulocyte). All the results are presented in Table 5.2. Regarding IC_{50} values, all tested compounds demonstrated good activity against *C. albicans*, *C. glabrata*, and *Cr. neoformans*, while *C. krusei* and *C. parapsilosis* were found to be resistant against all tested derivatives in both of the assays. This resistance effect, alongside with the strong activity of compounds against *C. albicans,* could indicate that these derivatives bind spe-

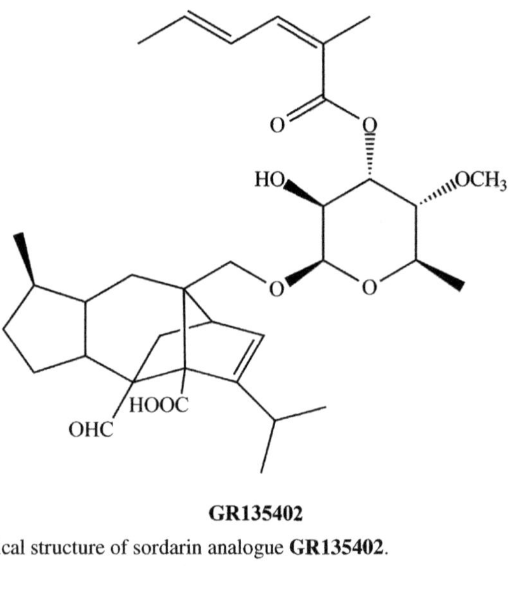

GR135402

FIG. 5.75 Chemical structure of sordarin analogue **GR135402**.

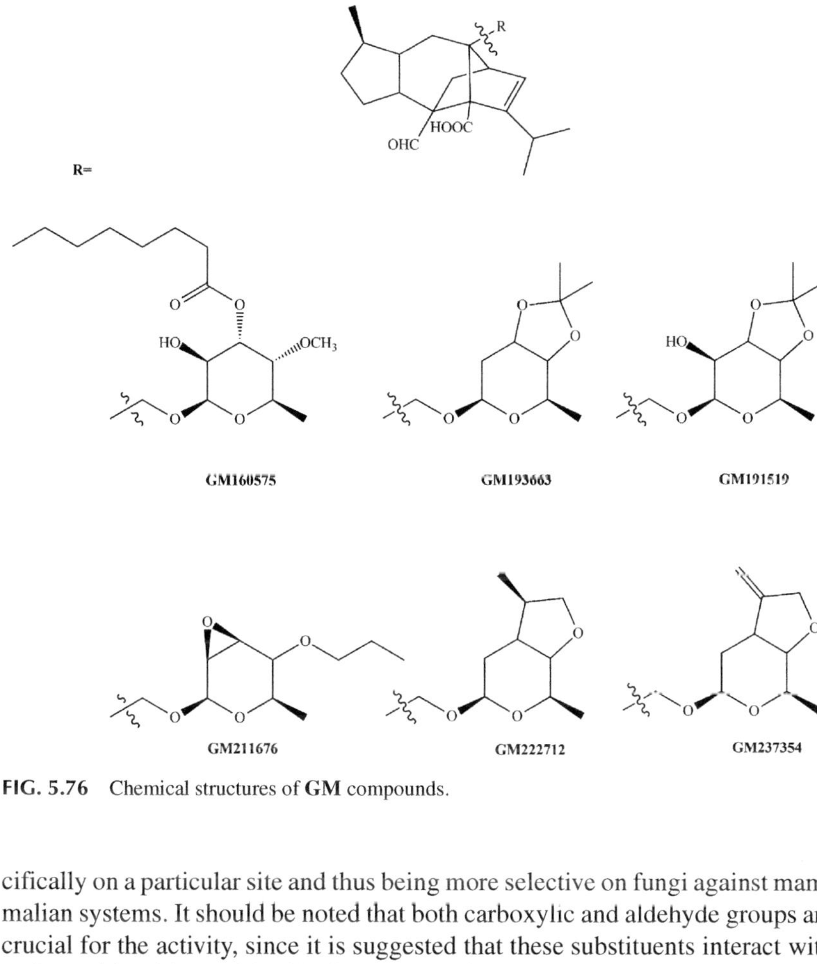

FIG. 5.76 Chemical structures of **GM** compounds.

cifically on a particular site and thus being more selective on fungi against mammalian systems. It should be noted that both carboxylic and aldehyde groups are crucial for the activity, since it is suggested that these substituents interact with amino acid residues of the binding pocket of the fungi. Further tests took place on compound **GM160575**, which was considered a representative of this group, in order to confirm the hypothesis that the antifungal effect of these analogues is due to inhibition of protein synthesis. In particular, the experiments that were used to elucidate the effect of these sordarin analogues on de novo protein and RNA synthesis confirmed the mode of action of sordarins as specific inhibitors of protein synthesis and more specifically as inhibitors of the elongation cycle.

Oxime **S224**, isoxazoline **S225**, isoxazole **S226**, and oxazepane **S227** (Fig. 5.78) were found to be less potent than the compounds of **GM** and **GW** series. Nevertheless, the strongest activity among these derivatives was observed for compounds that were substituted at the nitrogen position (Arai et al., 2002; Kaneko et al., 2002; Serrano-Wu et al., 2002a,b).

Regarding activity against the yeast *Saccharomyces cerevisiae*, it should be pointed out that sordarin exhibits weak potential, while sordaricin exhibits no

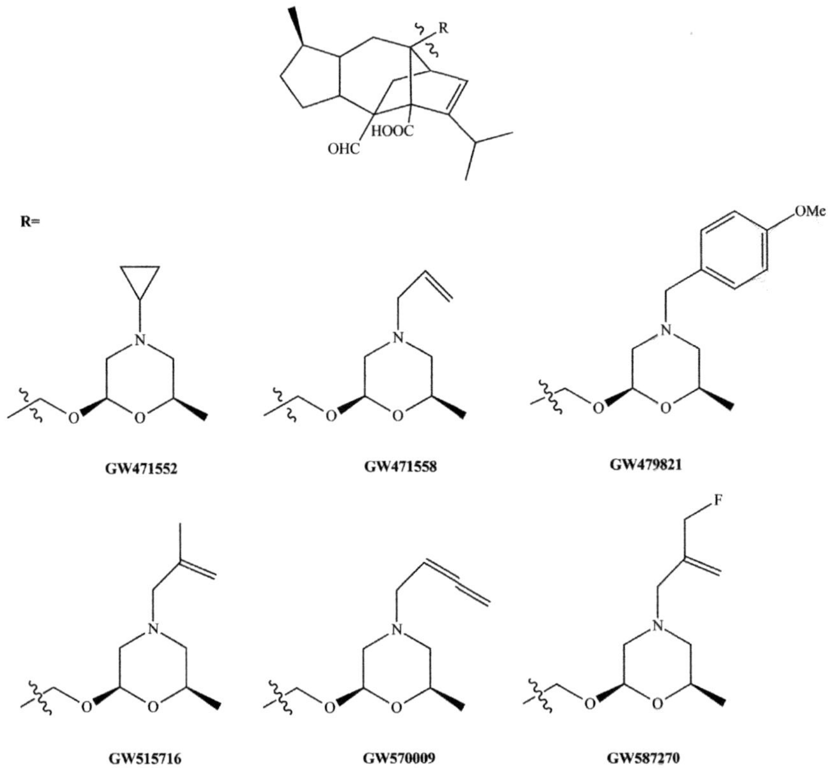

FIG. 5.77 Chemical structures of **GW** compounds.

activity at all. Taking into account the promising results concerning the antifungal activity of isobutyl ether derivative **L-493,422** (Fig. 5.79), Tse et al. (1998) reported the synthesis of a series of aliphatic side-chain derivatives of sordarin.

Structures of synthesized analogues **S228–S238** are shown in Fig. 5.80. The compounds were tested for their antifungal potential against *Saccharomyces cerevisiae* and data are presented in Table 5.3 as IC_{50} values (μg/ml), which were the concentrations required to inhibit 50% of growth, compared to untreated control.

From the obtained results it is clear that replacement of glycosyl moiety of sordarin with aliphatic chain had as a result an increase in the activity against *S. cerevisiae*. As far as the nitrile derivatives are concerned, the antifungal potential increased for C1 to C9 analogues and decreased for compounds with longer carbon chains, indicating that lipohilicity of side chains of sordarins is crucial for the antifungal activity observed. The best activity was observed for derivatives with C5 chain and thus, further tests regarding the activity of unsaturated C5 analogues were performed. Results showed that *cis*-pent-2-enyl derivative

TABLE 5.2 Effect of some glycosyl analogues of sordarin on the cell-free protein synthesis assay (IC_{50}s) and on cell growth (MICs).

Compound	*C. albicans* IC_{50} (μg/mL)	*C. albicans* MIC (μg/mL)	*C. glabrata* IC_{50} (μg/mL)	*C. glabrata* MIC (μg/mL)	*C. parapsilosis* IC_{50} (μg/mL)	*C. parapsilosis* MIC (μg/mL)	*C. krusei* IC_{50} (μg/mL)	*C. krusei* MIC (μg/mL)	*C. neoformans* IC_{50} (μg/mL)	*C. neoformans* MIC (μg/mL)	IC_{50} (μg/mL) rabbit retic.
Sordarin	0.01	8.0	0.2	>125	>100	>125	>100	>125	0.06	>125	>100
GR135402	0.2	0.015	0.8	>125	>100	>125	>100	>125	0.2	0.25	>100
GM160575	0.08	<0.001	0.4	>125	>100	>125	>100	>125	0.01	0.25	>100
GM191519	<0.005	0.12	0.5	31.0	100	>125	100	>125	0.005	125	>100
GM193663	<0.005	<0.001	0.02	31.0	100	>125	>100	>125	0.2	8.0	>100
GM211676	<0.005	0.001	0.01	8.0	100	>125	100	>125	0.12	1.0	>100

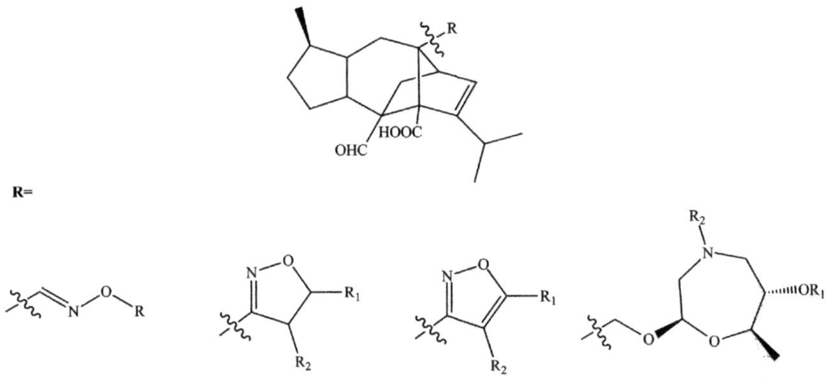

S224 S225 S226 S227

FIG. 5.78 Chemical structures of compounds **S224–S227**.

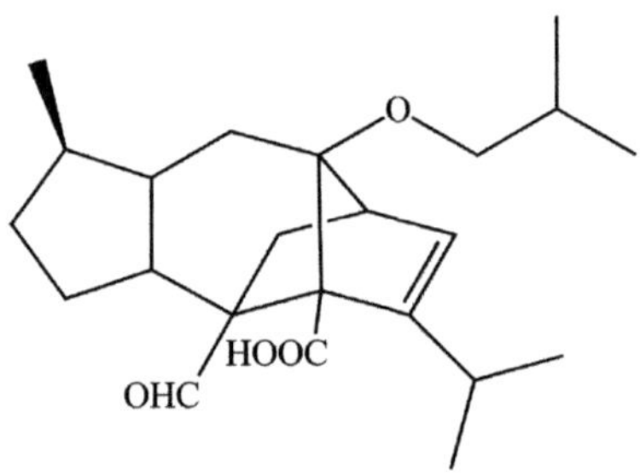

L-493,422

FIG. 5.79 Chemical structure of compound **L-493,422**.

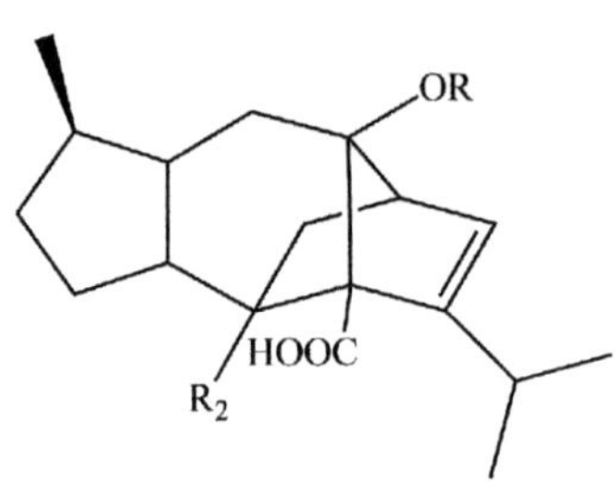

S228: R = Me, R_2 = CHO or CN

S229: R = Et, R_2 = CHO or CN

S230: R = n-Pr, R_2 = CHO or CN

S231: R = n-Bu, R_2 = CHO or CN

S232: R = iso-Bu, R_2 = CHO or CN

S233: R = n-Pentyl, R_2 = CHO or CN

S234: R = iso-Pentyl, R_2 = CHO or CN

S235: R = n-Hexyl, R_2 = CHO or CN

S236: R = n-Heptyl, R_2 = CHO or CN

S237: R = n-Octyl, R_2 = CHO or CN

S238: R = n-Nonyl, R_2 = CHO or CN

FIG. 5.80 Chemical structures of compounds **S228–S238**.

TABLE 5.3 Antifungal activity results of compounds S228–S238 against *Saccharomyces cerevisiae* as IC_{50} values (μg/mL).

	R_2=CHO	R_2=CN
Sordarin	20	2
Sordaricin	>250	20
S228: R=Me	0.1	1
S229: R=Et	0.06	0.5
S230: R=*n*-Pr	0.06	0.09
S231: R=*n*-Bu	0.0001	0.03
S232: R=isoBu	0.0001	0.03
S233: R=*n*-Pentyl	0.00001	0.00001
S234: R=isoPentyl	0.00001	0.00001
S235: R=*n*-Hexyl	0.015	0.0003
S236: R=*n*-Heptyl	0.3	0.1
S237: R=*n*-Octyl	0.4	0.2
S238: R=*n*-Nonyl	50	1

was the most potent (IC_{50}=0.000001 μg/mL), followed by *trans*-pent-2-enyl (IC_{50}=0.01 μg/mL) and pent-2-ynyl (IC_{50}=0.025 μg/mL) sordarin analogues.

Having in mind the possible geometrical constraints, alongside with the biological activity results, it was suggested that the potent derivatives of this series (e.g., *cis*-alkene derivatives) of sordarin analogues have the ability of side-chain folding.

Diterpene core analogues

Although the vast majority of SAR studies of sordarin derivatives focused on the glycosyl unit, it has been confirmed that aldehyde group can be replaced by a nitrile substituent without loss of antifungal activity (Tse et al., 1998).

Cuevas and coworkers (Cuevas and Martos, 1998; Cuevas et al., 1999), after performing modeling studies, accomplished to reduce the tetracyclic moiety of sordarin to the simple cyclopentane analogue **S239** (Fig. 5.81).

This compound contains aldehyde, carboxylic acid, and hydroxymethyl substituents that are important for the pharmacological activity. It should be noted at this point that in terms of stereochemistry, the dihedral angles of CHO-CH_2-CH_2-COOH backbone in this derivative are very similar to those of the carbon chain in sordaricin. A series of analogues containing this cyclopentane moiety

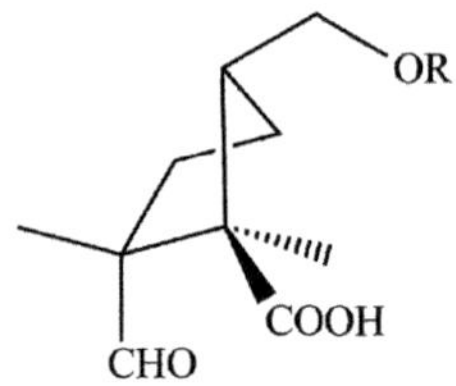

S239

FIG. 5.81 Chemical structure of compound **S239**.

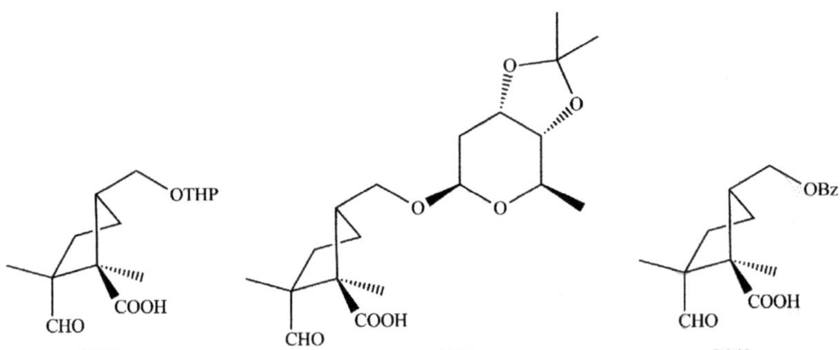

S240 **S241** **S242**

FIG. 5.82 Chemical structures of compounds **S240–S242**.

were synthesized and were tested for their antifungal activity against *Candida albicans* (2005E assay). Derivatives **S240**, **S241**, and **S242** (Fig. 5.82) exhibited stronger antifungal effect than sordaricin, having 3.7–6.7 times lower MICs, indicating that sordarin skeleton can be simplified without loss of activity.

Tricyclic analogues

The ongoing research efforts of scientists in the field of sordarin derivatives led to the synthesis of novel compounds with further structural modifications, including a series of tricyclic analogues of sordarin. The general structure of these tricyclic analogues is shown in Fig. 5.83. Compounds **S243** and **S244** were chosen to be further studied in order to elucidate their biological activity (Liang et al., 2007; Liang, 2008).

Pradimicins

Pradimicin A (Fig. 5.84) is a natural antibiotic that is produced by *Actinomadura hibisca* P157–2 alongside with other members of the pradimicin family. It demonstrates broad-spectrum antifungal activity against different fungal strains, including *Candida albicans*, *Aspergillus fumigatus,* and *Cryptococcus neoformans*. From SAR studies has been observed that replacement of D-alanine at C-15 by glycine or D-serine didn't reduce antifungal potential. On the other

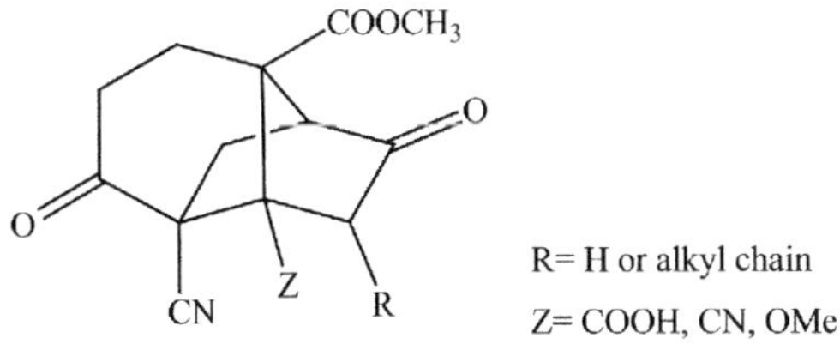

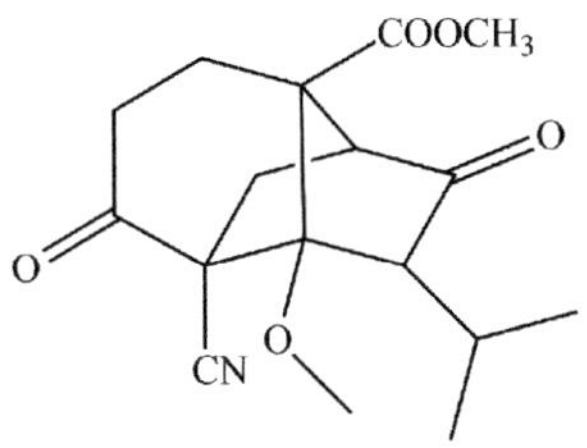

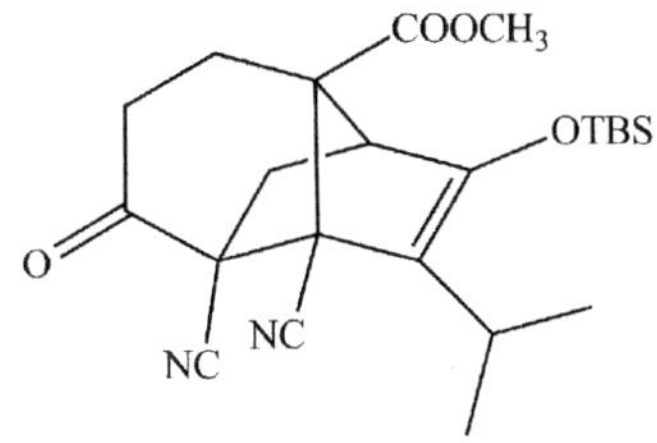

S243 **S244**

FIG. 5.83 General structure of tricyclic analogues of sordarin and chemical structures of compounds **S243** and **S244**.

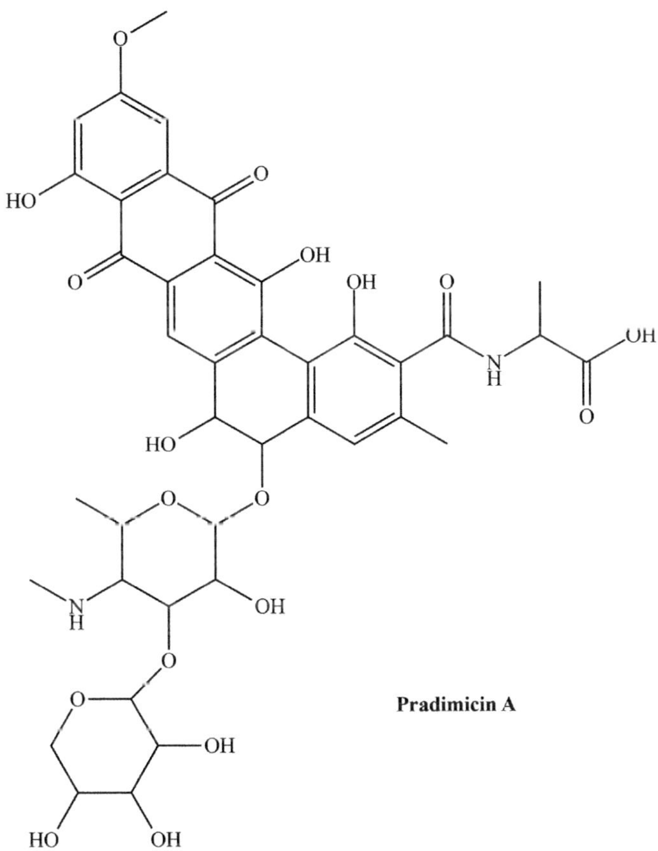

FIG. 5.84 Chemical structure of **pradimicin A**.

hand, change to L-alanine had as outcome the complete loss of antifungal effect. These observations led to the assumption that presence of amino acid at C-15 of pradimicin A is crucial for antifungal activity (Kakushima et al., 1990; Oki et al., 1988; Sawada et al., 1990).

Okuyama et al. (1993) reported the synthesis and evaluation of antifungal activity of a series of alanine-exchanged derivatives of pradimicin A. In vitro activity of synthesized compounds was determined against *Candida albicans, Aspergillus fumigatus,* and *Cryptococcus neoformans* by the two-fold agar dilution method on yeast morphology. From the obtained results was observed that the nature of C-17 substituent in D-α-amino acid derivatives led, in general, to changes in the antifungal potential. Neutral or basic substituents did not change the potential against *C. neoformans* and *A. fumigatus*, while acidic groups exerted weaker inhibitory effect against *A. fumigatus*. Compound **S245** was found to be two-fold less potent than pradimicin A against *C. albicans* and *A. fumigatus*, while derivative **S246** exhibited good activity against *C. neoformans* and *A. fumigatus*, but was less potent against *C. albicans*. It should be mentioned that derivative **S247** (D-aspartic acid analogue) demonstrated very weak potential against *A. fumigatus*. The most potent derivative was found to be the D-diaminopropionic acid analogue **S248**, which demonstrated almost the same effect with pradimicin A against *C. neoformans* and *A. fumigatus*. It should be noted that substitution on *N*-16 had as result the loss of activity, leading to the suggestion that the amide proton is important for the antifungal potential. It would be interesting to point out that the aminomethyltetrazole derivative **S249** retained the activity. Compounds **S247** and **S248** were also evaluated for their in vivo antifungal activity against a systemic *C. albicans* infection in mice. From the obtained results can be confirmed that derivative **S248** was in general as potent as pradimicin A. The structures and biological activity results of mentioned compounds are presented in Fig. 5.85 and Table 5.4, respectively.

Undecylenic acid

Undecylenic acid (Fig. 5.86) is an unsaturated fatty acid that is usually administered topically to treat fungal infections of the skin as well as symptoms such as irritation, burning, and itching that are connected with such conditions. This compound is also used in dentistry against *Candida albicans*. It exhibits mainly fungistatic activity, although in high concentrations and long-term use fungicidal effect might also be observed. Regarding its mode of action it is proposed that it interacts with components of the fungal cell membrane. Undecylenic acid is considered relatively safe as far as adverse effects are concerned, although skin irritation and burning sensation have been reported as consequence of its use. There are limitations in its use due to its oily nature and low solubility. (Kumar et al., 2014; Mazu et al., 2016; Petrovic et al., 2018; Shapiro and Rothman, 1945).

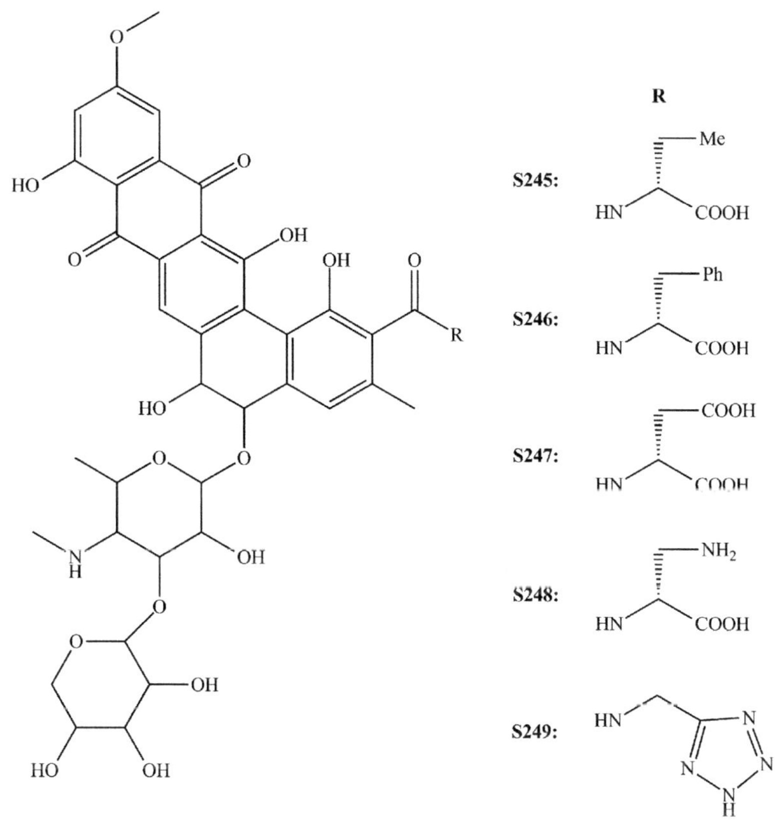

FIG. 5.85 Chemical structures of compounds **S245–S249**.

TABLE 5.4 In vitro (IC_{50}s) and in vivo (PD_{50}s) activities of pradimicine derivatives **S245–S249**.

Compound	*C. albicans* IC_{50} (μg/mL)	*C. neoformans* IC_{50} (μg/mL)	*A. fumigatus* IC_{50} (μg/mL)	*C. albicans* PD_{50} (mg/kg, iv)
S245	25	3.1	6.3	–
S246	>100	1.6	6.3	–
S247	12.5	6.3	25	27
S248	25	1.6	3.1	10
S249	6.3	3.1	3.1	–
Pradimicine A	12.5	0.8	3.1	8.9

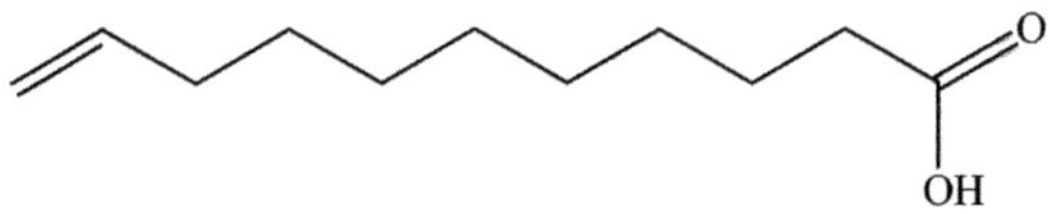

Undecylenic acid

FIG. 5.86 Chemical structure of **undecylenic acid**.

(−)Incrustoporine analogues

The synthesis and evaluation of antifungal activity of a series of (−)incrustoporine analogues was presented by Pour et al. (2000). The compounds were assayed in vitro with microdilution method against several human pathogenic fungal strains. While saturated lactones were found to be inactive, derivatives **S250–S259** (Fig. 5.87) exhibited antifungal effects. Furthermore, it was observed that phenyl substitution with halogens favored the antifungal activity, especially against filamentous fungi (MICs of 32.25–250 μmoL/L).

Usnic acid derivatives

Kartsev et al. (2018) designed and synthesized thirteen usnic acid derivatives that were selected after prediction of their potential biological activity with computer software PASS, in order to evaluate in vitro their antimicrobial properties. All the tested compounds were found to be very potent against the selected fungi, with activity 220–230 times higher than that of ketoconazole (MICs of 38–475×10^{-2} mmoL/mL and MFCs of 95–570×10^{-2} mmoL/mL) and 8–89 times higher than that of bifonazole (MICs of 48–64×10^{-2} mmoL/mL and MFCs of 64–80×10^{-2} mmoL/mL). Usnic acid derivative **S260** (Fig. 5.88) exhibited the strongest antifungal potential (MICs of 0.35–0.7×10^{-2} mmoL/mL and MFCs of 0.7–1.41×10^{-2} mmoL/mL), followed by compound **S261** and **S262**. As far as the fungi were concerned, *Trichoderma viride* was the most resistant species and *Penicillium verrucosum* var. *cyclopium* the most resistant one. It should be

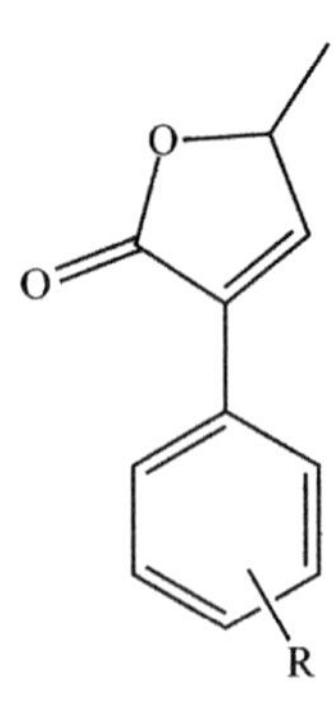

S250: R = H
S251: R = 4-Me
S252: R = 4-OMe
S253: R = 4-Cl
S254: R = 3,4-Cl
S255: R = 4-F
S256: R = 3-F
S257: R = 4-Br
S258: R = 3-Br
S259: R = 4-NO_2

FIG. 5.87 Chemical structures of compounds **S250–S259**.

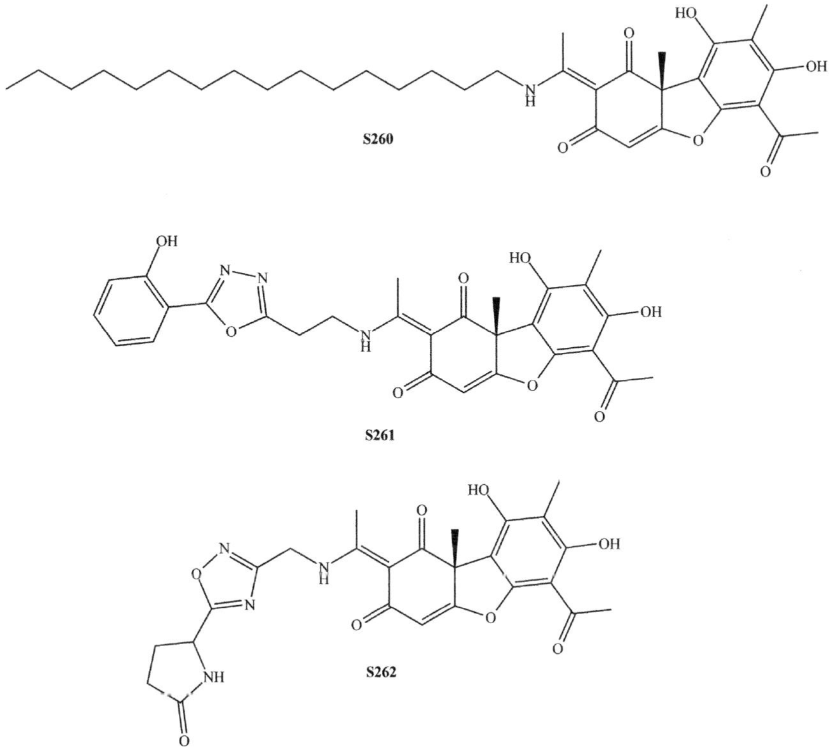

FIG. 5.88 Chemical structures of compounds **S260–S262**.

mentioned that the vast majority of the synthesized compounds were found to be more potent than usnic acid against several of the selected fungi. The docking studies that were performed were in agreement with the antifungal potential of the most active derivatives **S260**, **S261**, and **S262**.

Tobramycin analogues

A series of six tobramycin analogues (**S263–S268**, Fig. 5.89) were synthesized by Shrestha et al. (2015) and were evaluated for their in vitro antifungal activity, cytotoxicity, and membrane disruptive activity. The antifungal potential of the compounds was tested in vitro against a panel of yeasts and filamentous fungi having as reference drugs posaconazole, itraconazole, and fluconazol. The authors observed that the analogues with a linear alkyl chain at position 6′ in thioether linkage revealed chain length-dependent antifungal effects. It was also found the derivatives with C_{12} (MICs of 1.95–62.5 mg/L) and C_{14} (MICs of 1.95–7.8 mg/L) chains revealed a promising antifungal profile against the majority of the strains being more potent than reference drugs with the exception against *Cryptococcus neoformans* MYA-895, *Candida albicans* 1023 L,

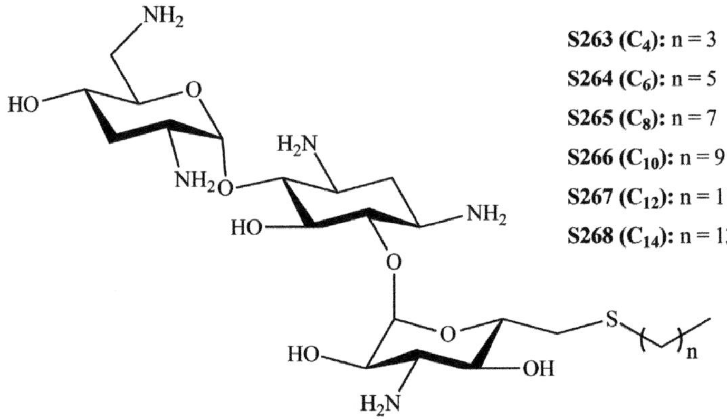

FIG. 5.89 Chemical structures of compounds **S263–S268**.

Aspergillus nidulans 38,163, and *Fusarium graminearum* 053. It was proposed by the authors that these two compounds appear to possess a novel mode of action, inhibiting fungi by inducing apoptosis and disrupting the fungal membrane. Furthermore, it should be mentioned that cytotoxicity evaluation in mammalian cell lines showed that the synthesized tobramycin analogues demonstrate minimal cell toxicity.

p-Coumaric acid derivatives

Khatkar et al. (2017) synthesized several p-coumaric acid derivatives, which were later tested in vitro with tube dilution method for their ability to inhibit *C. albicans* and *A. niger*, using fluconazole as reference drug. MICs were determined (μg/ml) and were transformed into pMIC values (− log MIC in μM/ml) for the subsequent QSAR studies that were performed by the authors. It was found that derivative **S269** (Fig. 5.90) was the post potent one with pMIC values of 1.67 and 1.97 μM/mL for *C. albicans* and *A. niger*, respectively, with fluconazole having pMIc of 2.64 for both fungal strains. The authors also observed that MBCs were 3-fold higher than the corresponding MFCs leading them to

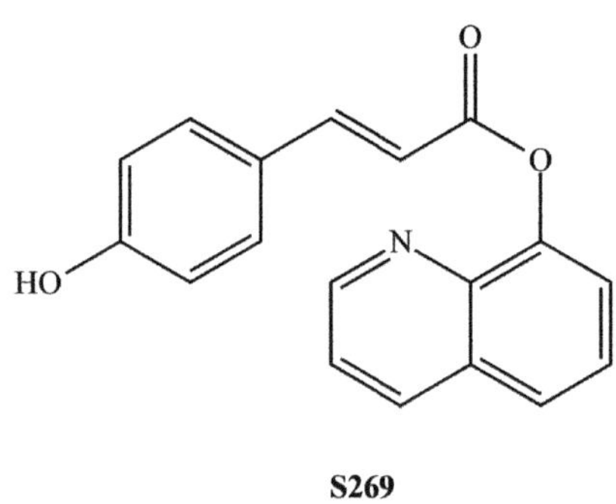

S269

FIG. 5.90 Chemical structure of compound **S269**.

the conclusion that the compounds are fungistatic. Furthermore, by studying the results of QSAR studies, it became clear that the antifungal activity (and antimicrobial in general) was greatly affected by the electronic parameter, electronic energy (Ee), topological parameters first-order molecular connectivity index ($^1\chi$), and Wiener index (*W*).

Miscellaneous synthetic antifungals

Ciclopirox

Ciclopirox (Fig. 5.91) is a hydroxypyridone derivative, which exerts its antifungal activity with a mode of action that has not been fully elucidated yet. It is believed to act by chelating polyvalent cations (e.g., Fe^{3+}, Al^{3+}), which leads to inhibition of a series of enzymes such as cytochromes. It is also thought that it induces modifications in fungal membrane and subsequent malfunctions of cell. It is usually administered topically for the treatment of superficial mycoses, revealing broad-spectrum fungistatic and fungicidal activities, while it is thought to possess also anti-inflammatory and antibacterial properties. It should be mentioned that burning, irritation, and redness are some of potential adverse effects of ciclopirox formulations when applied topically (Kumar et al., 2014; Leem et al., 2003).

Flucytosine

Flucytosine (Fig. 5.92) is an antimetabolite that enters fungal cells via cytosine permease, where it is converted intracellularly to fluorouracil (5-FU), which acts as a false nucleoside, inhibiting synthesis of both DNA and RNA (Harvey et al., 2009, Wu et al., 2004).

It is usually administered in combination with amphotericin B or fluconazole due to the occurrence of resistance when used as antifungal monotherapy (Pappas et al., 2009; Tassel and Madoff, 1968). It should be noted that mutations in the enzymes, which are responsible for the metabolism of flucytosine to the toxic molecules 5-fluorouracil and 5-fluorouridine monophosphate can

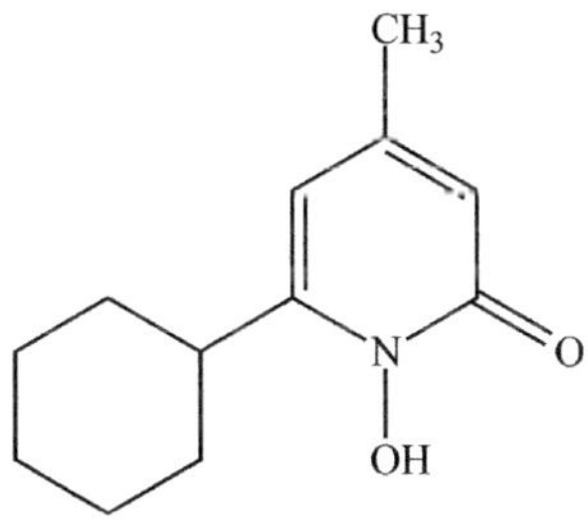

Ciclopirox

FIG. 5.91 Chemical structure of **ciclopirox**.

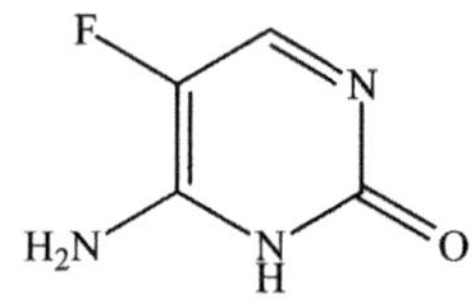

Flucytosine

FIG. 5.92 Chemical structure of **flucytosine**.

also promote resistance to this antifungal drug (Nett and Andes, 2016). It can be used for the treatment of severe disseminated *Candida* infections, cryptococcal meningitis, and also cryptococcosis in AIDS patients. Although its therapeutic spectrum is considered to be limited, there are reported cases regarding aspergillosis and chromoblastomycosis where fluocytosine could be recommended for administration in clinical practice (Larsen et al., 1994; Polak et al., 1982).

Resistance in flucytosine can develop rapidly and that is why this drug is often administered in combination amphotericin B or azoles (e.g., fluconazole).

The main toxic effects of flucytosine include hepatic toxicity, myelosuppression, and gastrointestinal intolerance (Polak et al., 1982).

Pentamidine

Pentamidine (Fig. 5.93) is an aromatic diamidine that besides its antiprotozooal activity is also effective against the fungus *Pneumocystis carinii*. The exact mechanism of action of pentamidine is not completely elucidated, but it is believed that the activity of the drug is connected to inhibition of the synthesis of DNA, RNA, phospholipids, and proteins (Waller and Sampson, 2018). It is used to prevent and treat pneumocystis pneumonia, especially in HIV-infected patients but it can cause several side effects by lowering blood sugar, blood pressure, and leading to central nervous system damage (Bloom and Ryan, 2013).

Other synthetic derivatives

Taking into consideration the role of N-myristoyltransferase (Nmt) in the viability of fungi, Masubuchi et al. (2001) designed and synthesized a group of novel benzofurans as potential inhibitors of *C. albicans* Nmt. From the obtained results was discovered the lead derivative **S270** (Fig. 5.94) with IC_{50}=0.98 mM that was found to be highly selective over human Nmt but showed no in vitro potency toward *C. albicans*. Several structural modifications were performed by the authors that finally led to derivative **S271** with elongated side chain compared to **S270** and improved antifungal potential against *C. albicans* (53 μM). Further modifications led to compound **S272** (Fig. 5.94), which exhibited the strongest inhibitory activity toward *C. albicans* Nmt (IC_{50}=0.1 μM), being highly selective (>5000-fold) over the human enzyme. Derivative **S272** was also the most potent in vitro against *C. albicans* (IC_{50}=1.62 μM), exhibiting

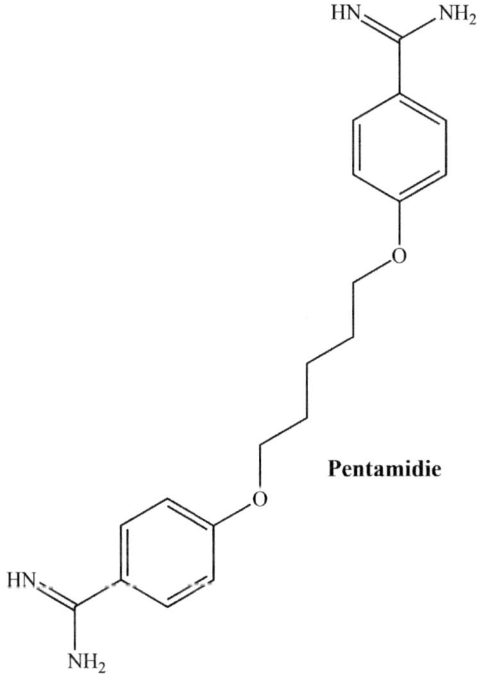

Pentamidie

FIG. 5.93 Chemical structure of **pentamidine**.

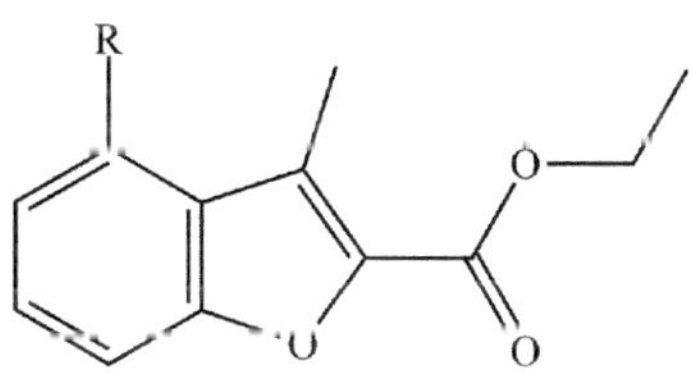

S270: R = $OCH_2CH(OH)CH_2NHCH(CH_3)_2$

S271: R = $O(CH_2)_5NHC(CH_3)_3$

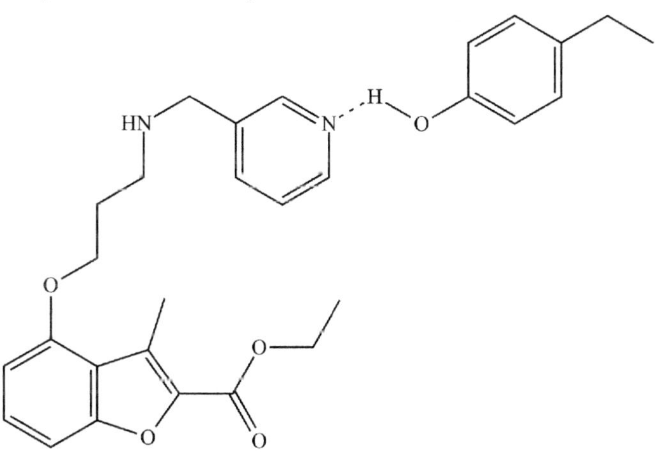

S272

FIG. 5.94 Chemical structures of compounds **S270–S272**.

similar effect with fluconazole (IC_{50}=0.72 μM) but showed no activity in a mouse systemic candidiasis model.

Kawasaki et al. (2003) presented the results of their scientific efforts regarding the design, synthesis and evaluation of antifungal activity of a series of acid-stable benzofurans as potential inhibitors of *C. albicans* Nmt. From the synthesized benzofurans, derivative **S274** (Fig. 5.95) was found to be a strong *C. albicans* Nmt inhibitor with an IC_{50} of 0.0035 μM when its isomer **S273** was 180 times less potent, revealing the significance of nitrogen atom position at the pyridine ring.

Subsequently, the authors proceeded to structural modifications of compound **S274** realizing that heteroaromatic derivatives containing nitrogen were in general strong Nmt inhibitors. Moreover, it was found that introduction of methyl substituents in thiazole and pyridine derivatives increased their inhibitory potency. It should be pointed out that most of the derivatives that exhibited strong inhibitory activity against *C. albicans* Nmt were also highly selective against human Nmt, while none of the compounds revealed a strong inhibitory profile toward *A. fumigatus* Nmt. Concerning the in vitro antifungal screening, the strong *C. albicans* Nmt inhibitors **S274–S279** and **S281–S284** (Fig. 5.96) were also very potent against *C. albicans* CY1002. Compounds **S275** and **S284** were assayed also by using a murine systemic candidiasis model, exhibiting a clear antifungal effect.

A collection of forty-eight novel cyclic b-amino acids were synthesized by Mittendorf et al. (2003) and screened in vitro for their ability to act as antifungals against *C. albicans*. From the obtained results it was revealed that compounds **S285** and **S286** (Fig. 5.97) were very effective with IC_{50} values of 0.03 and 0.13 mg/L, respectively, but the subsequent modifications in the structure of these two molecules had no positive results. Moreover, by performing SAR

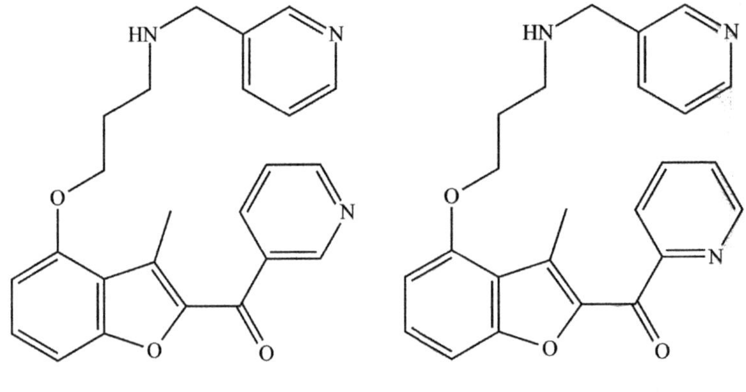

S273 **S274**

FIG. 5.95 Chemical structures of compounds **S273** and **S274**.

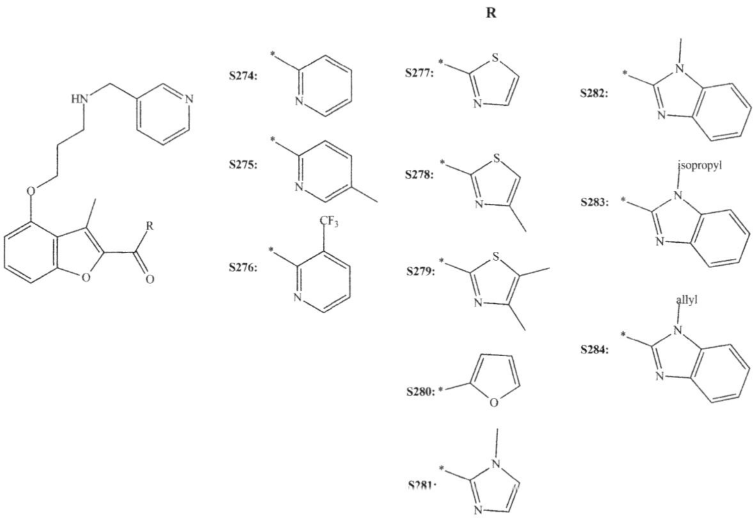

FIG. 5.96 Chemical structures of compounds **S274–S284**.

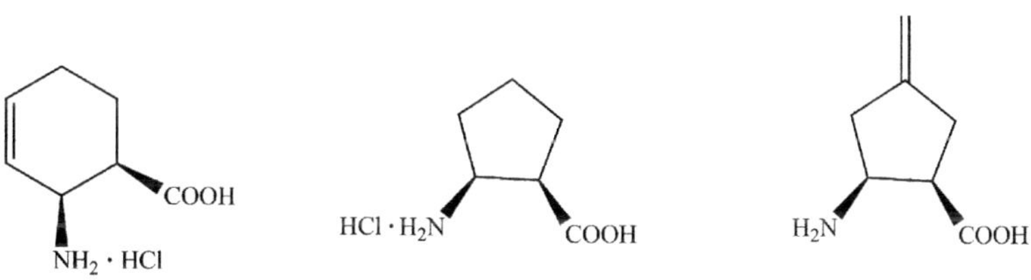

S285 **S286** **S287**

FIG. 5.97 Chemical structures of compounds **S285–S287**.

studies, the authors chose derivative **S287** (Fig. 5.97) with $IC_{50}=0.13\,mg/L$, which exhibited the same activity with **S286** to be further modified but ultimately without being able to improve its potential. It should be pointed out though that compound **S287** managed to enter clinical trials for the oral treatment of yeast infections.

Haitao et al. (2003) after investigating the fungal lanosterol 14α-demethylase by performing molecular studies on the *Candida albicans* enzyme, synthesized a series of nonazole derivatives in order to evaluate them as lead antifungals against a selection of pathogenic strains (*Candida albicans*, *Sporothrix schenckii*, *Cryptococcus neoformans*, *Fonsecaea pedrosoi*, *Aspergillus fumigatus*, *Microsporum canis*, and *Trichophyton rubrum*). Although less potent than fluconazole, compounds **S288–S307** (Fig. 5.98) revealed the best antifungal profile against the selected strains. SAR studies were also performed by the authors, coming to the conclusion that the 2-hydrophobic side chain, the 4-hydrogen bond donor/acceptor, as well as the 7-hydrophilic hydrogen bond donor

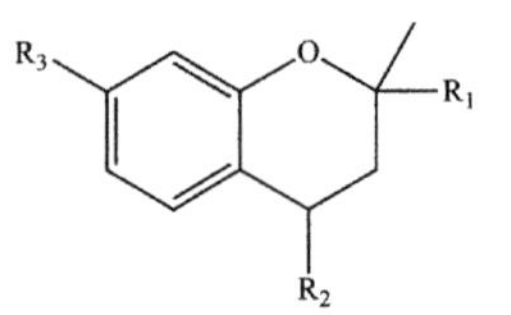

S288-304

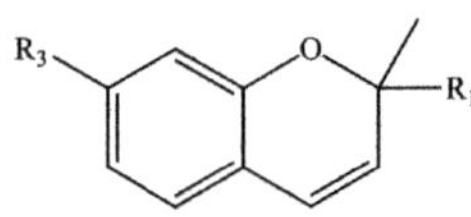

S305-307

S288: $R_1 = CH_2(CH_2)_3CH_3$, $R_2 = OH$, $R_3 = 7\text{-}OH$
S289: $R_1 = CH_2(CH_2)_5CH_3$, $R_2 = OH$, $R_3 = 7\text{-}OH$
S290: $R_1 = CH_2CH_2CH{=}C(CH_3)_2$, $R_2 = OH$, $R_3 = 7\text{-}OH$
S291: $R_1 = CH_2(CH_2)_3CH_3$, $R_2 = {=}NOH$, $R_3 = 7\text{-}OH$
S292: $R_1 = CH_2(CH_2)_5CH_3$, $R_2 = {=}NOH$, $R_3 = 7\text{-}OH$
S293: $R_1 = CH_2CH_2CH{=}C(CH_3)_2$, $R_2 = {=}NOH$, $R_3 = 7\text{-}OH$
S294: $R_1 = CH_2(CH_2)_3CH_3$, $R_2 = {=}O$, $R_3 = 7\text{-}OH$
S295: $R_1 = CH_2(CH_2)_5CH_3$, $R_2 = {=}O$, $R_3 = 7\text{-}OH$
S296: $R_1 = CH_2CH_2CH{=}C(CH_3)_2$, $R_2 = {=}O$, $R_3 = 7\text{-}OH$
S297: $R_1 = CH_2(CH_2)_7CH_3$, $R_2 = {=}O$, $R_3 = 7\text{-}OH$
S298: $R_1 = CH_2CH_2(CH_3)_2$, $R_2 = {=}O$, $R_3 = 7\text{-}OH$
S299: $R_1 = CH_2(CH_2)_3CH_3$, $R_2 = {=}O$, $R_3 = 7\text{-}OCH_3$
S300: $R_1 = CH_2(CH_2)_5CH_3$, $R_2 = {=}O$, $R_3 = 7\text{-}OCH_3$
S301: $R_1 = CH_2CH_2CH{=}C(CH_3)_2$, $R_2 = {=}O$, $R_3 = 7\text{-}OCH_3$
S302: $R_1 = CH_2(CH_2)_3CH_3$, $R_2 = OH$, $R_3 = 7\text{-}OCH_3$
S303: $R_1 = CH_2(CH_2)_5CH_3$, $R_2 = OH$, $R_3 = 7\text{-}OCH_3$
S304: $R_1 = CH_2CH_2CH{=}C(CH_3)_2$, $R_2 = OH$, $R_3 = 7\text{-}OCH_3$
S305: $R_1 = CH_2(CH_2)_3CH_3$, $R_3 = 7\text{-}OCH_3$
S306: $R_1 = CH_2(CH_2)_5CH_3$, $R_3 = 7\text{-}OCH_3$
S307: $R_1 = CH_2CH_2CH{=}C(CH_3)_2$, $R_3 = 7\text{-}OCH_3$

FIG. 5.98 Chemical structures of compounds **S288–S307**.

are crucial for the potential of the derivatives of this series, facts which are in agreement with the molecular model analysis.

The synthesis of a series of novel 1-[2-(benzoxazol-2-yl)ethoxy]-2,6-diarylpiperidin-4-ones (compounds **S308–S316**, Fig. 5.99) was reported by Ramalingan et al. (2004). All the compounds were tested for their inhibitory activity against *Candida-6, Candida albicans, Candida-51, Aspergillus niger*, and *Aspergillus flavus* having as reference drug amphotericin-B. The authors observed that derivative **S311** (MICs of 12.5–100.0 μg/mL) exhibited in general stronger antifungal activity compared to parent compound **S308** (MICs

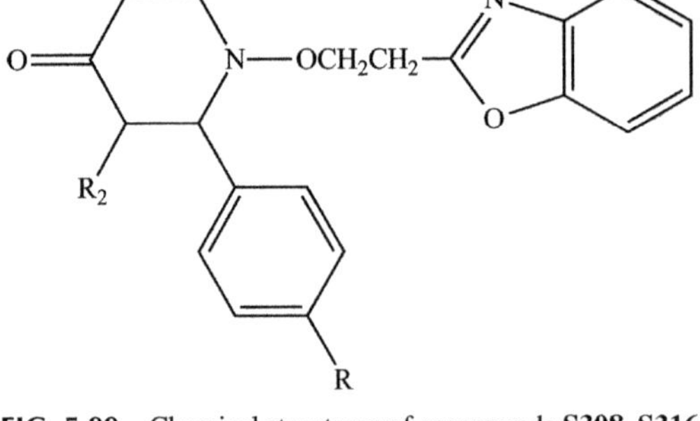

S308: $R = R_1 = R_2 = H$
S309: $R = R_2 = H$, $R_1 = CH_3$
S310: $R = R_1 = CH_3$, $R_2 = H$
S311: $R = R_1 = H$, $R_2 = Cl$
S312: $R = H$, $R_1 = CH_3$, $R_2 = Cl$
S313: $R = R_1 = CH_3$, $R_2 = Cl$
S314: $R = R_1 = H$, $R_2 = OCH_3$
S315: $R = H$, $R_1 = CH_3$, $R_2 = OCH_3$
S316: $R = R_1 = CH_3$, $R_2 = OCH_3$

FIG. 5.99 Chemical structures of compounds **S308–S316**.

of 50.0–200.0 μg/mL), being also more potent than the reference drug against *A. niger* and equipotent toward *Candida-51* and *A. fumigatus*. Introduction of methyl substituent at position 3′ of 2,6-diarylpiperidin-4-one ring of **S311** led to the synthesis of **S312**, which exhibited stronger antifungal effect against *C. albicans* and *Candida-6* but was less potent against *Candida-51*, *A. niger*, and *A. flavus*. Placing a second methyl group at position 5′ of the same ring gave derivative **S313** with stronger potential than **S311** against *A. flavus* but decreased against *C. albicans* and *Candida-6*. When the chloro atom of **S311** was replaced by a methoxy substituent (**S314**) a stronger antifungal effect against *Candida-6* and *Candida-51* was exerted, while the potency toward *C. albicans* and *A. flavus* was suppressed. It should also be noted that derivatives **S315** and **S316** were found to be stronger as antifungals against all tested strains compared to parent compound **S308** and also toward *Candida-51* and *Candida-6* in comparison with derivative **S311**.

Taking into consideration the antifungal potential of 1,4-naphthoquinones, Tandon et al. (2004a) synthesized a series of novel 1,4-naphthoquinones containing a sulfur atom and screened them in vitro to reveal their inhibitory effect against several pathogenic fungal strains. From the obtained results should be pointed out that derivatives **S317** and **S318** (Fig. 5.100) were more potent

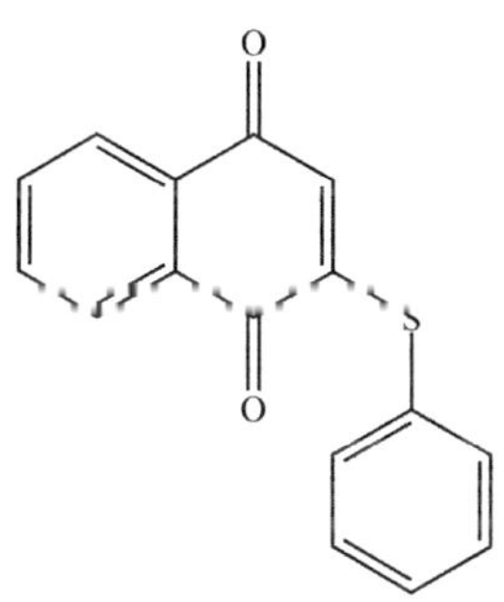

S317

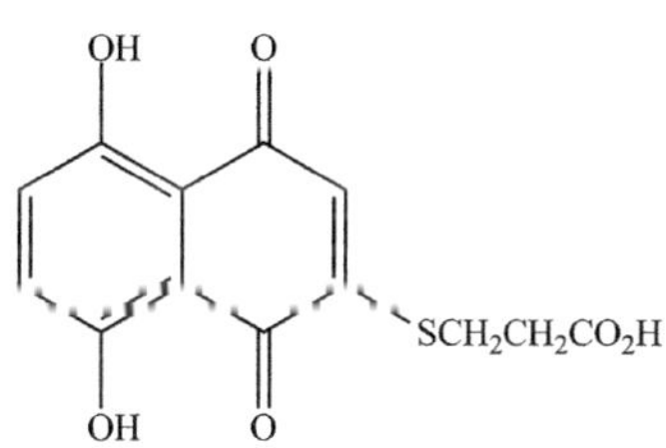

S318

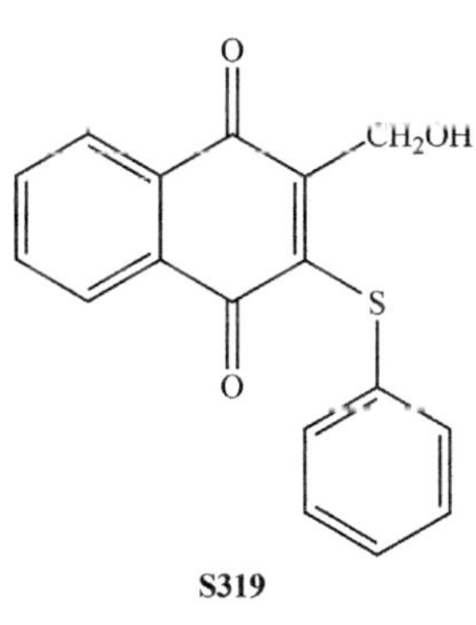

S319

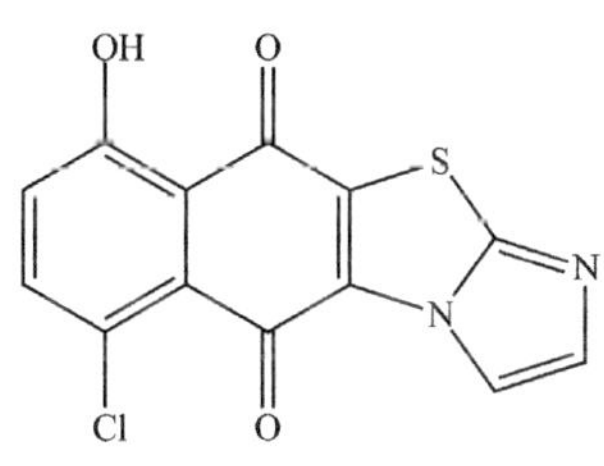

S320

FIG. 5.100 Chemical structures of compounds **S317–S320**.

than miconazole (MIC = 25.0 μg/mL) against *C. albicans* with MIC = 12.5 μg/mL and MIC = 3.125 μg/mL respectively being at the same time potent against *Cr. neoformans* (MIC = 1.56 μg/mL). It is noteworthy that compound **S317** exhibited good activity also against *Tr. mentagrophytes* and *M. cannis* with MICs < 0.78 and < 0.98 μg/mL, respectively. It should be mentioned that naphthoquinone **S319** exerted strong antifungal effect toward *Cr. neoformans* with MIC = 1.56 μg/mL. Except the derivatives that contained sulfur atom, the authors observed that also naphthoquinones, which contained imidazole moiety were active in vitro against many of the selected strains. In particular, derivative **S320** was found to be very effective against *Cr. neoformans* (MIC = 6.25 μg/mL) and *A. fumigatus* (MIC = 3.1 μg/mL) and more potent than reference drugs ketoconazole and miconazole, respectively.

Later in the same year, Tandon et al. (2004b) reported the synthesis of another group of 1,4-naphthoquinone derivatives that contained sulfur atom and assayed them in vitro against a collection of pathogenic fungal strains to elucidate their potential to act as antifungals. From the novel naphthoquinones, **S321** and **S322** (Fig. 5.101) exhibited, in general, the best antifungal activity against all tested fungi. In particular, derivative **S321** exhibited stronger antifungal effect against *C. albicans* (MIC = 3.125 μg/mL) and *Cr. neoformans* (MIC = 1.56 μg/mL) than miconazole (MIC = 25.0 μg/mL and MIC = 12.5 μg/mL, respectively) that was used as reference drug. Similar antifungal effect against the two previously mentioned strains showed compound **S322** with MICs of 12.5 and 1.56 μg/mL, respectively.

In the next year, Tandon et al. (2005) inspired by their previous results regarding the antifungal potential of 1,4-naphthoquinones, proceeded to the synthesis and evaluation of antifungal activity of thirteen novel 1,4-naphthoquinones that contained sulfur and/or nitrogen atoms at positions 2′ or 3′ of the side chain or ring. The in vitro assay was performed by using microdilution method. From the synthesized derivatives, compound **S323** (Fig. 5.102) was the most potent, even better than miconazole, against *C. albicans* and *A. fumigatus* with MICs = 12.5 μg/mL.

The synthesis of a series of novel 8-methyl-7,9-diaryl-1,2,4,8-tetraazaspiro[4.5]decan-3-thiones as potential antifungal agents was described by Balasubramanian et al. (2005). The general structure of the compounds is

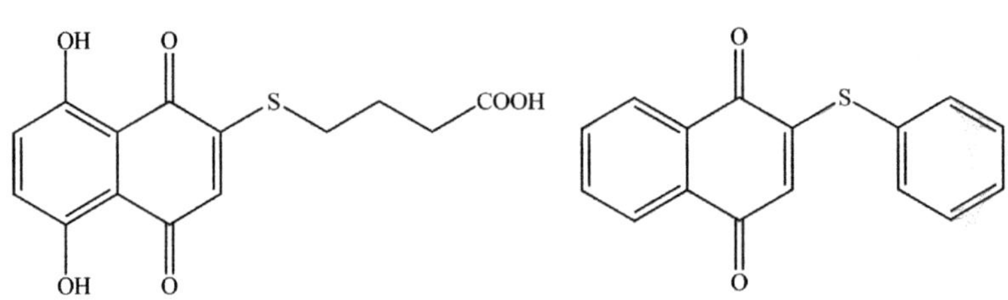

S321 S322

FIG. 5.101 Chemical structures of compounds **S321** and **S322**.

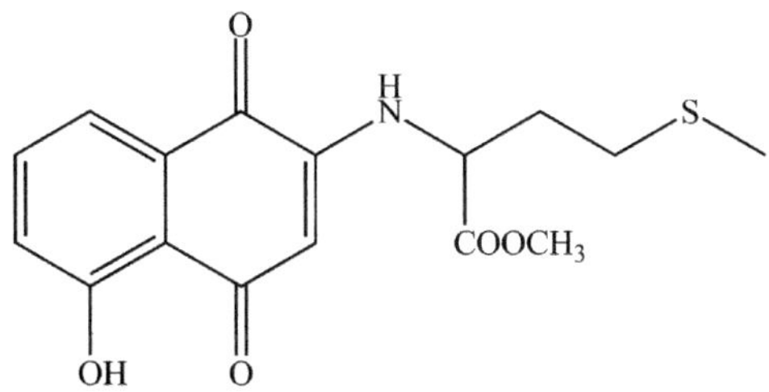

S323

FIG. 5.102 Chemical structure of compound **S323**.

presented in Fig. 5.103 (compounds **S324–S331**). All the new derivatives were screened in vitro for their inhibitory activity against *C. albicans*, *Candida-6*, *Candida-51*, *A. niger*, and *A. flavus* having amphotericin B as reference drug. It was observed that that this series of derivatives exhibited relatively moderate antifungal potential with compounds **S329-S331** being equipotent with reference drug.

Chhillar et al. (2006) reported the synthesis of eleven 4-aryl-1,4-dihydropyridines, including the intermediate epoxide and three 4-aryl-1,2,3,4-tetrahydropyrimidine derivatives. The compounds were evaluated for their antifungal effects against pathogenic strains of *Aspergillus fumigatus* and *Candida albicans* by using disc diffusion (DDA), micro broth dilution (MDA) and percentage spore germination methods (PSGI). Derivative diethyl 4-(4-methoxyphenyl)-2,6-dimethyl-1,4-dihydropyridin-3,5-dicarboxylate **S332** (Fig. 5.104) was found to exhibit the strongest antifungal potential (DDA MIC = 2.92 μg/disc, MDA and PSGI MIC = 15.62 μg/mL), while being sixteen times less toxic than amphotericin B, revealing that dihydropyrimidine scaffold could be further studied for designing safer antifungal agents.

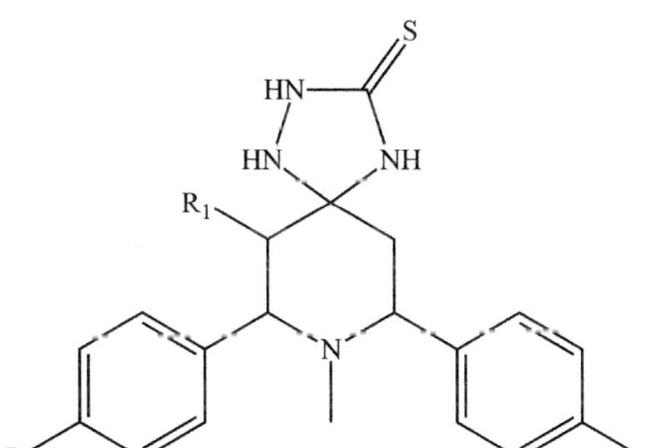

S324: R = R_1 = H	
S325: R = H,	$R_1 = CH_3$
S326: R = CH_3,	R_1 = H
S327: R = $R_1 = CH_3$	
S328: R = Cl,	R_1 = H
S329: R = Cl,	$R_1 = CH_3$
S330: R = OCH_3,	R_1 = H
S331: R = OCH_3,	$R_1 = CH_3$

FIG. 5.103 Chemical structures of compounds **S324–S331**.

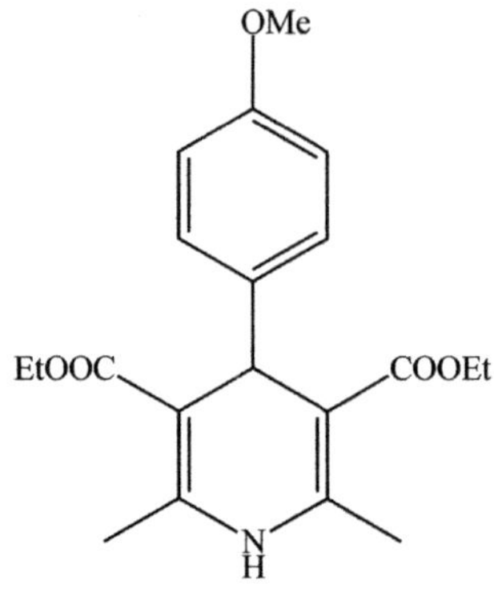

S332

FIG. 5.104 Chemical structure of compound **S332**.

Inspired from the results mentioned above, Zhu et al. (2006) reported the synthesis and evaluation of antifungal activity of several tetrahydroisoquinoline derivatives by using the same method as in their previous publication (Haitao et al., 2003). Based on their SAR studies, they inserted long lipophilic alkyl side chains at position 2′ (for interaction with hydrophobic S3 subsite) and phenolic hydroxyls at positions 6′ and 7′ of the tetrahydroisoquinoline moiety (for hydrogen bonds formation with the S4 subsite residue). It was observed that all the lead compounds were effective against the selected fungal strains. Derivatives **S333** and **S334** (MICs of 3.125 to 50.0 μg/mL) (Fig. 5.105) exhibited the same or stronger activity than fluconazole (MICs of 6.25 to > 200.0 μg/mL) against the majority of the used strains, with the exception of *Candida albicans*. Additionally, the authors reported that too short or too long side chain at position 2′ has a negative impact in the antifungal potential of this class of compounds.

As a continuation of their ongoing project, Bin et al. (2007) replaced the tetrahydroisoquinoline moiety with 2-aminotetralin and synthesized a series of novel compounds of the general structure given in Fig. 5.106, taking into account the SAR data and molecular studies results of their previous work (Zhu et al., 2006; Haitao et al., 2003). The synthesized derivatives were evaluated for their antifungal potential using, in general, the same methodology as in

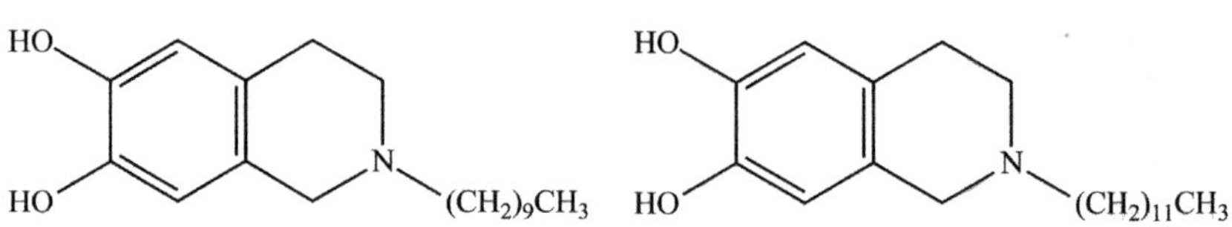

S333 S334

FIG. 5.105 Chemical structures of compounds **S333** and **S334**.

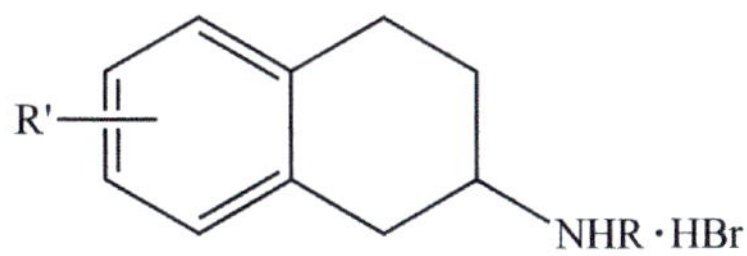

S335: R' = 6-OH, R = $(CH_2)_8CH_3$

S336: R' = 6-OH, R = $(CH_2)_9CH_3$

S337: R' = 5-OCH_3, R = $(CH_2)_{11}CH_3$

S338: R' = 7-OCH_3, R = $(CH_2)_{11}CH_3$

S339: R' = 6-OH, R = $(CH_2)_{11}CH_3$

S340: R' = 6-OCH_3, R = $(CH_2)_9CH_3$

FIG. 5.106 Chemical structures of compounds **S335–S340**.

their previously published research efforts, having as reference drug fluconazole. From the obtained results it was observed that derivatives containing N-substituted aromatic rings were less potent than the ones with alkyl substitution (7–13 carbon atoms). Moreover, authors stated that alkyl chains of 9–10 carbons atoms as side chain were preferable in terms of antifungal activity. Furthermore, it was found that regarding the tetralin aromatic ring, substitution with hydroxyl or methoxy groups at 5′ and 6′ positions are preferable compared to 7′ and 8′. As a next step, the authors chose the most potent compounds against *C. albicans* (MICs <0.125 μmoL/L) to be screened against four fluconazole-resistant strains of this pathogenic fungus. Derivatives **S335**, **S336**, **S337**, **S338**, and **S339** exhibited the same or greater potency than terbinafine or ketoconazole. It should be pointed out that derivative **S340** appeared to exhibit the strongest antifungal effect (MICs of 0.0625 to 0.125 μmoL/L) against the fluconazole-resistant strains, even when compared to amphotericin B (MICs of 0.25 to >64.0 μmoL/L).

A series of 2-(arylidene/5-arylfurfurylidene)-5-(4-methylthiophenyl)-6-carbethoxy-7-methyl-5H-thiazolo[2,3-*b*]pyrimidin-3(1H)-ones as potential antifungal agents have been synthesized by Ashok et al. (2007). The novel compounds (Fig. 5.107) were investigated for their antifungal activity against four strains (*Aspergillus flavus*, *A. fumigatus*, *Candida albicans*, and *Penicillium marneffei*) by serial dilution assay. In general, all the synthesized derivatives exerted moderate-to-good potential compared to ciclopirox olamine that was used as reference drug (MICs of 10–20 μg/mL).

A series of novel benzamides (compounds **S341–S348**, Fig. 5.108A and B) were prepared by Kumar et al. (2007) and all the compounds were tested in vitro for their ability to inhibit two *Aspergillus* strains (*Aspergillus ficcum* and *A. parasiticus*). It was revealed by the assay results that all the synthesized derivatives demonstrate appreciable potential against the two tested fungi with –log(1/MIC in μM/mL) values of 1.36–1.78 and 1.20–1.83 for *A. ficcum* and *A. parasiticus*,

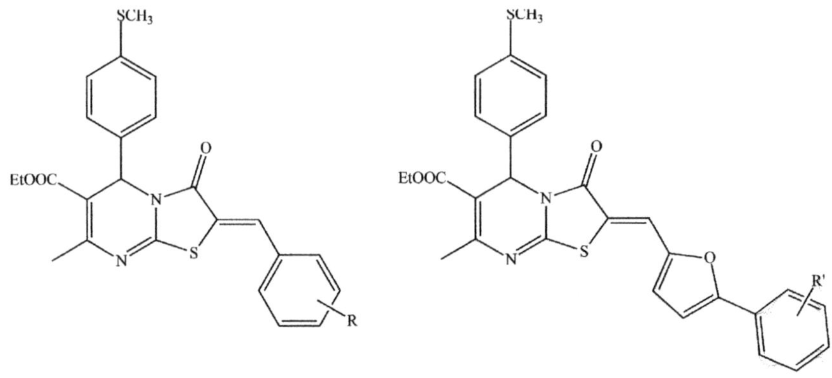

R = 4-SCH$_3$, 4-OCH$_3$, 4-Cl, 4-F, 4-OH, 2,4-Cl$_2$, 4-OH-3-OCH$_3$, 4-F-3-OPh, 3,4-OCH$_2$O

R' = 4-NO$_2$, 4-Cl, 4-Br, 2,4-Cl$_2$

FIG. 5.107 Chemical structures of 2-(arylidene/5-arylfurfurylidene)-5-(4-methylthiophenyl)-6-carbethoxy-7-methyl-5H-thiazolo[2,3-*b*]pyrimidin-3(1H)-ones.

respectively. It should be pointed out that compound **S345** (Fig. 5.108) exhibited the strongest antifungal effect against the two selected strains with –log(1/MIC in μM/ml) values of 1.78 (*A. ficcum*) and 1.83 (*A. parasiticus*). Furthermore, the authors stated that the presence of –OCH$_2$Ph, –COOCH$_3$, and –OH groups in the structure of derivative **S345** could be responsible for its enhanced antimicrobial potential.

The synthesis and evaluation of anti-infective activity of a series of cryptoleptine derivatives has been reported by Zhu et al. (2007). Parent compound **S349** (Fig. 5.109) was found to be very potent as antifungal (MICs for *C. albicans* and *C. neoformans*, 1.0 and 2.5 μg/mL, respectively) and with no cytotoxicity at 23.8 μg/mL. It would be interesting to mention that the 2-substituted derivative of **S349** did not reveal any increase in the activity. Nevertheless, the data concerning **S349** seem to be promising for further lead optimization efforts.

As part of their studies on dihydropyrimidone derivatives as potential antifungal agents, Singh et al. (2008) presented the synthesis and in vitro evaluation of antifungal activity of compounds **S350–S355** (Fig. 5.110) against *Trichoderma hamatum*, *T. koningii*, and *Aspergillus niger*. These 1,4-dihydropyridines were found to be potent against the previously mentioned strains with MICs in the range 0.025–0.35 μg/mL.

Taking into account the wide spectrum of biological activities of salicylanilides, Skala et al. (2009) presented the synthesis of a series of analogues of the general structures included in Fig. 5.111 in order to screen them for their antifungal effect against eight fungal species with fluconazole as reference drug. The authors observed that the vast majority of the compounds (except two) exhibited antifungal activity stronger or same with that of fluconazole (MIC > 125.0 μmoL/L) against *A. fumigatus* and *A. corymbifera* with MIC values of 31.25–125.0 μmoL/L. It should be pointed out that dihydroxy analogue **S356**

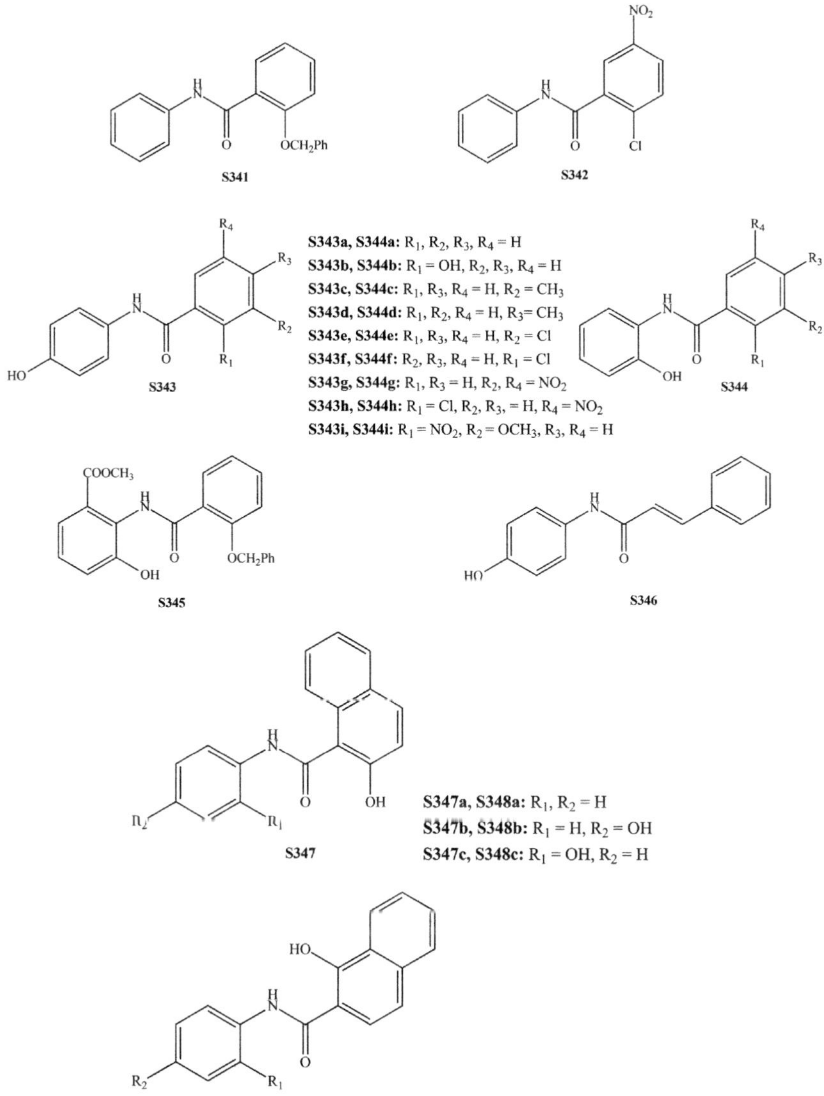

FIG. 5.108 (A) Chemical structures of compounds **S341–S346**. (B) Chemical structures of compounds **S347–S348**.

revealed in general very good antifungal potential, being more potent than reference drug with MIC = 1.95 μmoL/L after 72/120 h. Cyclic derivatives revealed reduced activity, except **S357**, which was found to be equipotent with fluconazole after 48 h (MIC = 3.91 μmoL/L) against *C. glabrata*. Authors observed

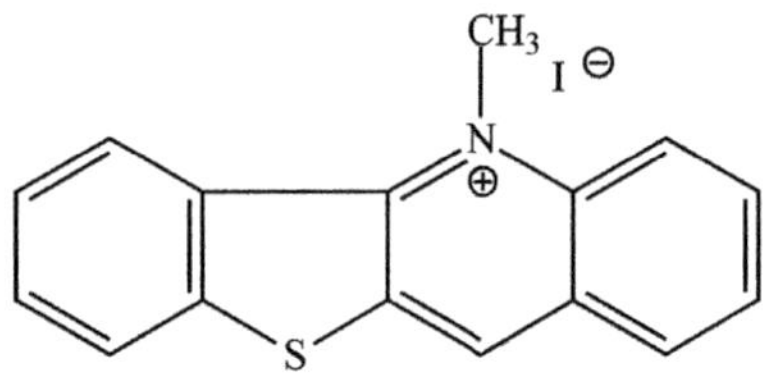

S349

FIG. 5.109 Chemical structure of compound **S349**.

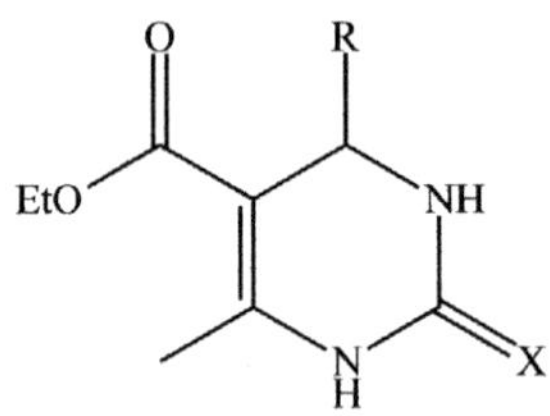

	R	X
S350:	C_6H_5	O
S351:	C_6H_5	S
S352:	2-HOC_6H_4	O
S353:	2-HOC_6H_4	S
S354:	4-$Me_2NC_6H_4$	S
S355:	4-$Me_2NC_6H_4$	O

FIG. 5.110 Chemical structures of compounds **S350–S355**.

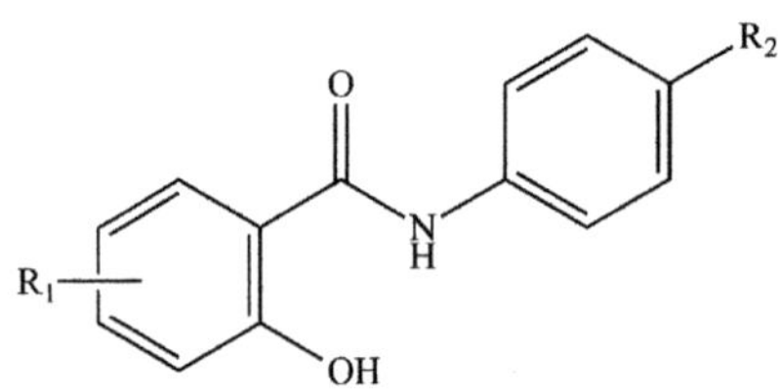

S356: R_1 = 6-OH, R_2 = 4-Cl

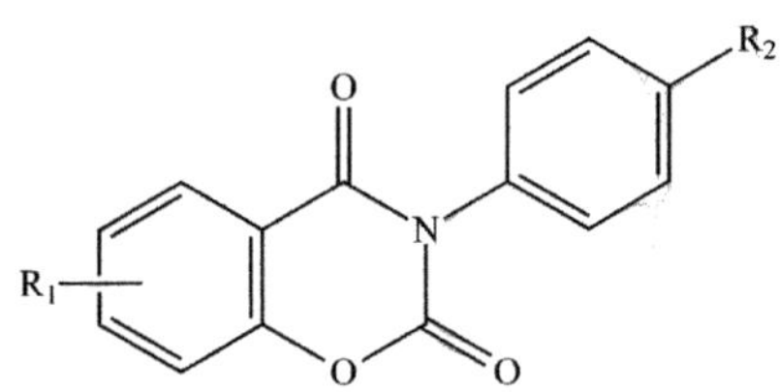

S357: R_1 = 8-OH, R_2 = 4-Cl

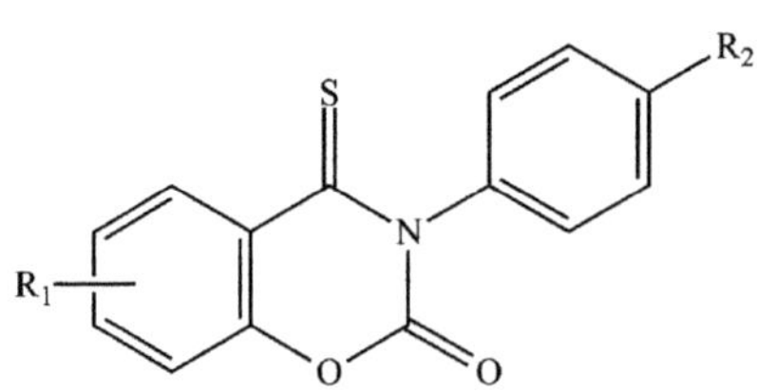

S358

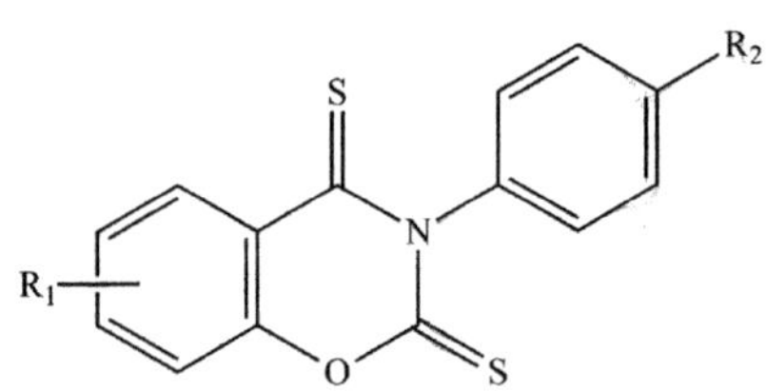

S359

FIG. 5.111 General structures of salicylanilides, their thioanalogues, and compounds **S356** and **S357**.

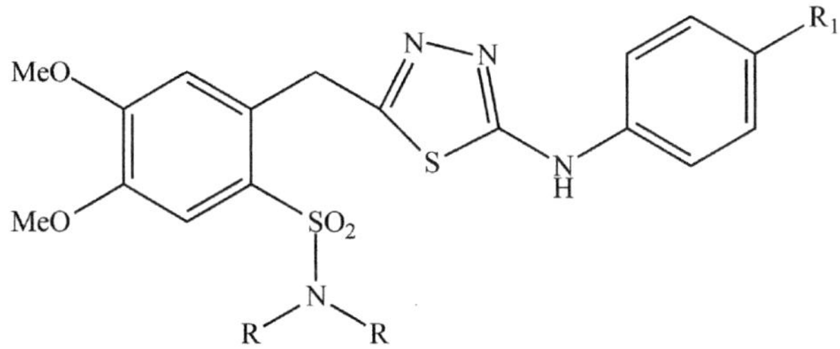

S360: $N(R)_2$ = methylpiperidine, $R_1 = CF_3$

FIG. 5.112 General structure of synthesized sulfonamides and chemical structure of compound S360.

that thione derivatives showed no antifungal effect against the selected fungal strains that were used in this assay.

Camoutsis et al. (2010) presented the synthesis of several sulfonamides (Fig. 5.112) that contained 1,3,4-thiadiazole moiety, aiming to evaluate their antifungal potential against a series of eight fungal strains, having as reference drugs bifonazole and ketoconazole. All synthesized compounds were found to exhibit strong antifungal effect with MICs of 0.89–2.30 $\times 10^{-2}$ μmoL/mL and MFCs of 1.79–8.52 $\times 10^{-2}$ μmoL/mL, being more potent than bifonazole and ketoconazole with MICs of 32.0–64.0 and 38.0–475.0 $\times 10^{-2}$ μmoL/mL, respectively. The strongest antifungal effect was exerted by derivative **S360** with MICs of 0.89–1.79 $\times 10^{-2}$ μmoL/mL and MFCs of 1.79–3.58 $\times 10^{-2}$ μmoL/mL. It should be additionally noted that *T. viride* was the most sensitive strain, while *A. versicolor* was the most resistant one.

From the SAR studies was proposed by the authors that, in general, derivatives containing $N(R)_2$ = pyrrolidine group were more potent than the rest. Conformational studies were also performed for some of the derivatives by using NMR and molecular docking methods. In particular, homonuclear double quantum filtered correlation spectroscopy (DQF-COSY) and rotating frame overhauser enhancement spectroscopy (ROESY) procedures were used, coming in the conclusion that the conformation of the molecules' scaffold is not influenced by the substitutions that were performed.

A group of amino acid esters of salicylanilides were synthesized by Kratky et al. (2010) to be assayed in vitro for their inhibitory potential against a collection of microorganisms, including eight fungal strains (*C. krusei*, *C. glabrata*, *C. albicans*, *C. tropicalis*, *T. beigelii*, *A. fumigatus*, *T. mentagrophytes*, and *A. corymbifera*). The biological activity screening was performed by using the broth microdilution method, having as reference drug fluconazole. By studying the obtained results, the authors observed that the synthesized esters were in general potent as antifungal agents. Derivatives **S361** (MIC = 3.9 μmoL/L),

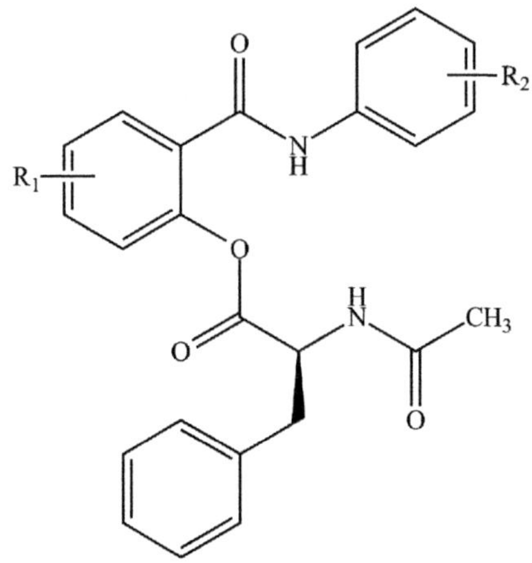

S361: $R_1 = 4\text{-Cl}, R_2 = 4\text{-Cl}$
S362: $R_1 = 5\text{-Cl}, R_2 = 4\text{-Cl}$
S363: $R_1 = 4\text{-Br}, R_2 = 4\text{-CF}_3$
S364: $R_1 = 5\text{-Cl}, R_2 = 4\text{-Br}$

FIG. 5.113 Chemical structures of compounds **S361–S364**.

S362 (MIC = 7.81 μmoL/L), and **S363** (MIC = 7.81 μmoL/L) exhibited stronger activity than fluconazole (MIC > 125 μmoL/L) against *A. corymbifera.* Esters **S364** and **S363** (Fig. 5.113) were very effective toward *T. mentagrophytes* with MICs of 3.9 and 7.81 μmoL/L, respectively. The authors concluded that *A. corymbifera* and *T. mentagrophytes* were the most sensitive species, while *C. glabrata* was the most resistant one.

Taking into account their previous work in the field, Kratky and Vinsova (2012) prepared a series of novel salicylanilide esters of the structures that are given in Fig. 5.114A and B.

Subsequently, the authors evaluated the antifungal potential of both parent (**S365a-t**) and final compounds (**S366a-s**) by performing the same methods that they reported in their earlier work (Kratky et al., 2010). It was found that derivatives **S365j** and **S366j** were effective against *C. albicans* with MICs of 7.81 and 15.62 μmoL/L (after 24 h), respectively, but also were found to exert strong activity against *C. krusei* with MICs of 7.81 and 31.25 μnol/l, respectively, being more potent than fluconazole (MIC > 50.0 μmoL/L), which was used as reference antifungal. *T. asahii* was found to be the most sensitive yeast strain with compounds **S365f**, **S365h**, **S365i**, **S365j**, **S365r**, and **S365s** being effective against it with MICs ranging between 1.95 and 3.9 μmoL/L, being also more potent than fluconazole (MIC = 4.0 μmoL/L). Moreover, the authors observed that ester derivative **S366e** was active against *C. tropicalis* and *C. krusei* with MICs of 7.81 and 31.25 μmoL/L, respectively, being, nevertheless, less potent than reference drug. Furthermore, it was found that only two esters were potent against *A. fumigatus*, while derivatives **S365a**, **S365f**, **S365g**, **S365i**, **S365j**, **S366j**, **S365o**, **S365r**, **S366r**, and **S365s** with MICs of 3.9–15.62 μmoL/L were more effective than fluconazole (MIC > 50.0 μmoL/L) against *A. corymbifera.* It could be stated that parent compound **S365t** was in general the most potent

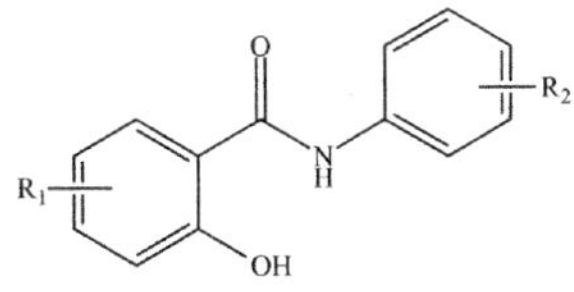

S365

S365a: R_1 = 5-Cl, R_2 = 3-Cl
S365b: R_1 = 4-Cl, R_2 = 3-Cl
S365c: R_1 = 5-Cl, R_2 = 4-Cl
S365d: R_1 = 4-Cl, R_2 = 4-Cl
S365e: R_1 = 5-Cl, R_2 = 3,4-diCl
S365f: R_1 = 4-Cl, R_2 = 3,4-diCl
S365g: R_1 = 5-Cl, R_2 = 3-Br
S365h: R_1 = 4-Cl, R_2 = 3-Br
S365i: R_1 = 5-Cl, R_2 = 4-Br
S365j: R_1 = 4-Cl, R_2 = 4-Br
S365k: R_1 = 5-Cl, R_2 = 3-F
S365l: R_1 = 4-Cl, R_2 = 3-F
S365m: R_1 = 5-Cl, R_2 = 4-F
S365n: R_1 = 4-Cl, R_2 = 4-F
S365o: R_1 = 5-Cl, R_2 = 3-CF_3
S365p: R_1 = 4-Cl, R_2 = 3-CF_3
S365q: R_1 = 5-Cl, R_2 = 4-CF_3
S365r: R_1 = 4-Cl, R_2 = 4-CF_3
S365s: R_1 = 5-Br, R_2 = 4-CF_3
S365t: R_1 = H, R_2 = H

(A)

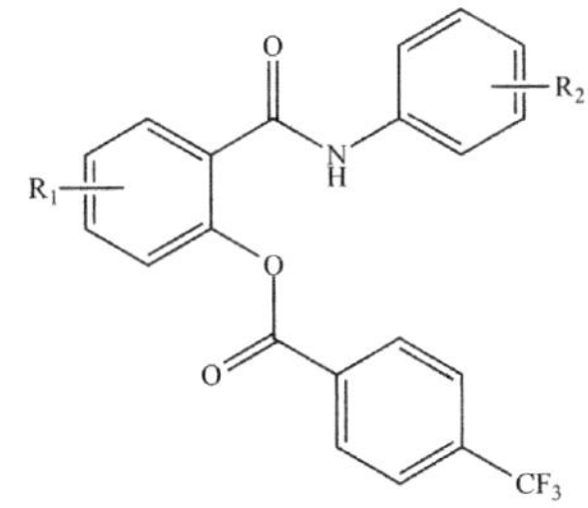

S366

S366a: R_1 = 4-Cl, R_2 = 3-Cl
S366b: R_1 = 5-Cl, R_2 = 3-Cl
S366c: R_1 = 4-Cl, R_2 = 4-Cl
S366d: R_1 = 5-Cl, R_2 = 4-Cl
S366e: R_1 = 4-Cl, R_2 = 3,4-diCl
S366f: R_1 = 5-Cl, R_2 = 3,4-diCl
S366g: R_1 = 4-Cl, R_2 = 3-Br
S366h: R_1 = 5-Cl, R_2 = 3-Br
S366i: R_1 = 4-Cl, R_2 = 4-Br
S366j: R_1 = 5-Cl, R_2 = 4-Br
S366k: R_1 = 4-Cl, R_2 = 3-F
S366l: R_1 = 5-Cl, R_2 = 3-F
S366m: R_1 = 4-Cl, R_2 = 4-F
S366n: R_1 = 5-Cl, R_2 = 4-F
S366o: R_1 = 4-Cl, R_2 = 3-CF_3
S366p: R_1 = 5-Cl, R_2 = 3-CF_3
S366q: R_1 = 4-Cl, R_2 = 4-CF_3
S366r: R_1 = 5-Cl, R_2 = 4-CF_3
S366s: R_1 = 4-Br, R_2 = 4-CF_3

(B)

FIG. 5.114 Chemical structures of compounds (A) **S365a-t** and (B) **S366a-s**.

of the tested molecules, while *T. mentagrophytes* was the most sensitive fungus with twenty-two compounds being more potent than fluconazole against it.

Taking into consideration the interesting properties of sugar esters, Shen et al. (2012) synthesized two groups of monosaccharide esters and screened them in vitro to determine their antifungal effect against *Aspergillus flavus*, *A. fumigatus*, *A. niger*, and *Geotrichum candidum*. The biological assay was performed under the NCCLS M27-A standard and clotrimazole was chosen as reference drug (National Committee For Clinical Laboratory Standards, 1997). From the assay results was observed that derivative **S367** (Fig. 5.115) was the most potent against the tested strains (MICs of 2.0–4.0 μg/mL) with activity stronger than that of clotrimazol (MICs of 4.0–16.0 μg/mL), followed by ester **S368** (MICs of 4.0–8.0 μg/mL).

Scotti et al. (2012) evaluated the antifungal potential of fifty 2-aminothiophene derivatives, which were previously designed and synthesized, with the general structure that is given in Fig. 5.116 (Mendonça Junior et al., 2011; de Oliveira, 2010; Souza, 2011).

The majority of the derivatives (thirty out of fifty) exerted antifungal effects with MICs of 1.25–5.0 μg/mL. The strongest potential against *Cryptococcus neoformans* possessed compounds **S370**, **S371**, **S372**, **S373**, **S374**, **S376**, **S377**, **S378**, and **S379** with MIC = 1.25 μg/mL. Regarding *Candida krusei*, the best activity was exerted by derivatives **S369**, **S370**, **S375**, and **S379** with MIC = 1.25 μg/mL. SAR studies were also performed and it was observed that the antifungal effect depends on the size of the ring, which is condensed with

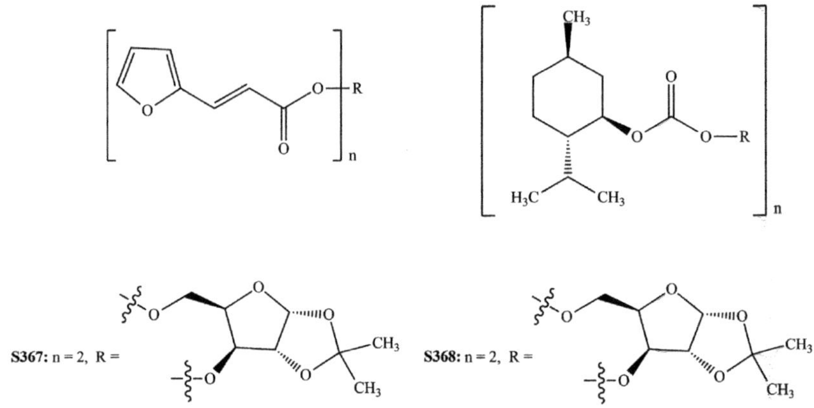

FIG. 5.115 Chemical structures of compounds **S367** and **S368**.

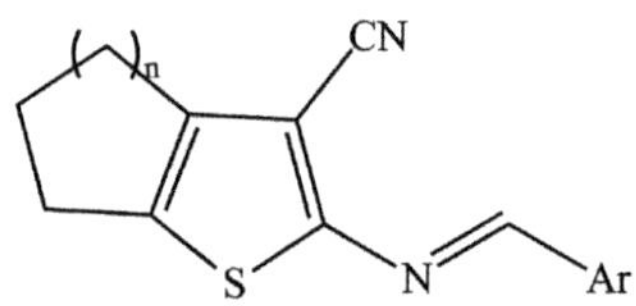

S369-S372: n = 1

S373-S375: n = 2

S376-S379: n = 3

S369: Ar = Ph

S370: Ar = 4-NO_2-Ph

S371: Ar = 2-indole

S372: Ar = 4-OMe-Ph

S373: Ar = 4-NO_2-Ph

S374: Ar = 3,4-Cl-Ph

S375: Ar = 3,4-OBz-Ph

S376: Ar = 4-iPr-Ph

S377: Ar = 4-Cl-Ph

S378: Ar = 3,4-Cl-Ph

S379: Ar = 3,4,5-OMe-Ph

FIG. 5.116 Chemical structures of compounds **S369–S379**.

thiophene moiety as well as on the type and position of phenyl ring (Ar) substituents. The in vitro assay followed computational studies, including molecular modeling, use of chemometric tools and QSAR-3D studies for better understanding of the impact of structural modifications to the activity observed.

Prabhaker and Saundane (2014) presented the synthesis of some novel indole derivatives as potential antimicrobial agents. The compounds were assayed in vitro for their antifungal activity against four *Aspergillus* species by using the cup-plate method, with fluconazole as reference antimycotic (20–22 mm inhibition zone). The new indole derivatives exhibited moderate activity, lower than that of fluconazole. Nevertheless, it should be mentioned that compound **S380** (Fig. 5.117) demonstrated the best antifungal profile with inhibition zone in the range of 18–19 mm.

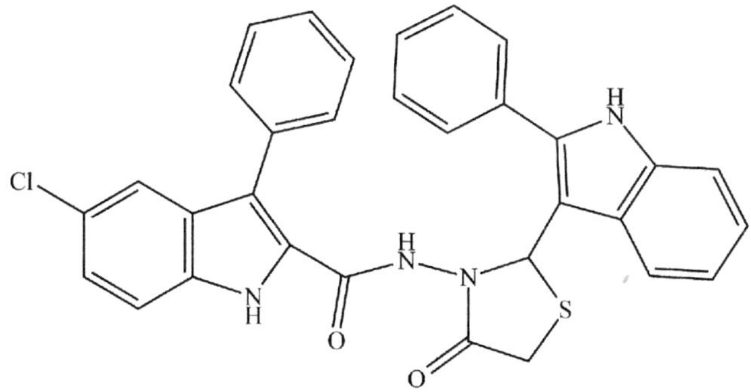

S380

FIG. 5.117 Chemical structure of compound **S380**.

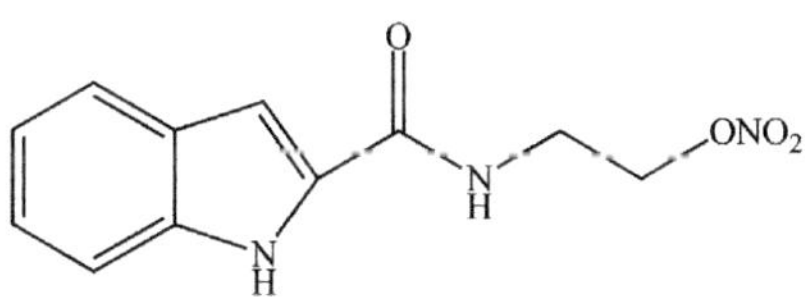

S381

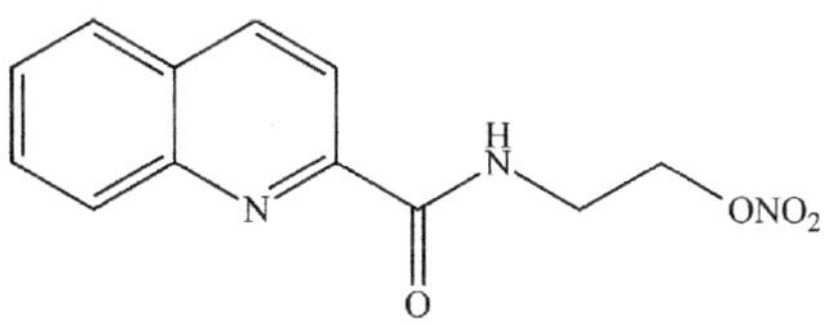

S382

FIG. 5.118 Chemical structures of compounds **S381** and **S382**.

Smiljkovic et al. (2018) selected four nitrate esters of heteroaromatic compounds, which were previously synthesized (Fotopoulou et al., 2008; Koufaki et al., 2012), in order to evaluate their activity in vitro against four *Candida albicans* strains.

All the tested nitrate esters exerted moderate-to-good antifungal potential (average MIC=0.0028 mg/mL, average MFC=0.0056 mg/mL) but lower than that of ketoconazole, which was used as reference drug. Nevertheless, compounds **S381** and **S382** (Fig. 5.118) possessed good inhibitory effect on biofilm formation and thus they could be characterized as novel anti-quorum-sensing agents. Moreover, it was proved that only nitrate ester **S381** binds to CYP51 of *Candida albicans* via interaction with heme Fe. In addition, NMR studies and in silico docking and molecular dynamic simulations for compound **S381** showed that its possible binding pose is via a metal coordination complex that is absent in the case of the other esters. It was also suggested by the authors that compound **S381** is selective toward *C. albicans* (K_D=4.19 μM) and doesn't bind to human CYP51. This indicated that the previously mentioned nitrate ester could lack hepatotoxic side effects, which was confirmed by performing cytotoxicity test on PLP2 cells from pig's liver (GI_{50} values >400 μg/mL).

References

Ahmed, M., Sharma, R., Nagda, D.P., Jat, J.L., Talesara, G.L., 2006. Synthesis and antimicrobial activity of succinimido(2-aryl-4-oxo-3-{[(quinolin-8-yloxy)acetyl]amino}-1,3-thiazolidin-5-yl) acetates. ARKIVOC 11, 66–75.

Althagafi, I., El-Metwaly, N., Farghaly, T.A., 2019. New series of thiazole derivatives: synthesis, structural elucidation, antimicrobial activity, molecular modeling and MOE docking. Molecules 24 (9). https://doi.org/10.3390/molecules24091741.

Anderson, G.R., Nissen, P., Nyborg, J., 2003. Elongation factors in protein biosynthesis. Trends Biochem. Sci. 28, 434–441.

Apostolidis, I., Liaras, K., Geronikaki, A., Hadjipavlou-Litina, D., Gavalas, A., Sokovic, M., Glamoclija, J., Ciric, A., 2013. Synthesis and biological evaluation of some 5-arylidene-2-(1,3-thiazol-2-ylimino)-1,3-thiazolidin-4-ones as dual anti-inflammatory/antimicrobial agents. Bioorg. Med. Chem. 21 (2), 532–539.

de Aquino, T.M., Liesen, A.P., da Silva, R.E., Lima, V.T., Carvalho, C.S., de Faria, A.R., de Araújo, J.M., de Lima, J.G., Alves, A.J., de Melo, E.J., Góes, A.J., 2008. Synthesis, anti-*Toxoplasma gondii* and antimicrobial activities of benzaldehyde 4-phenyl-3-thiosemicarbazones and 2-[(phenylmethylene)hydrazono]-4-oxo-3-phenyl-5-thiazolidineacetic acids. Bioorg. Med. Chem. 16 (1), 446–456.

Arai, M., Harasaki, T., Fukuoka, T., Kaneko, S., Konosu, T., 2002. Bioorg. Med. Chem. Lett. 12, 2733–2736.

Arathoon, E.G., 2001. Clinical efficacy of echinocandin antifungals. Curr. Opin. Infect. Dis. 14, 685–691.

Arathoon, E.G., Gotuzzo, E., Noriega, L.M., Berman, R.S., Di Nubile, M.J., Sable, C.A., 2002. Randomized, double-blind, multicenter study of caspofungin versus amphotericin B for treatment of oropharyngeal and esophageal candidiasis. Antimicrob. Agents Chemother. 46, 451–457.

Ashok, M., Holla, B.S., Kumari, N.S., 2007. Convenient one pot synthesis of some novel derivatives of thiazolo[2,3-b]dihydropyrimidinone possessing 4-methylthiophenyl moiety and evaluation of their antibacterial and antifungal activities. Eur. J. Med. Chem. 42 (3), 380–385.

Bachmann, S.P., Patterson, T.F., Lopez-Ribot, J.L., 2002. *In vitro* activity of caspofungin (MK-0991) against *Candida albicans* clinical isolates displaying different mechanisms of azole resistance. J. Clin. Microbiol. 40, 2228–2230.

Balasubramanian, S., Ramalingan, C., Aridoss, G., Kabilan, S., 2005. Synthesis and study of antibacterial and antifungal activities of novel 8-methyl-7,9-diaryl-1,2,4,8-tetraazaspiro[4.5] decan-3- thiones. Eur. J. Med. Chem. 40, 694–700.

Ballington, D.A., Laughlin, M.M., 2008. Pharmacology, third ed. CBS Publishers & Distributors, New Delhi, pp. 97–100.

Barar, F.S.K., 2006. Essentials of Pharmacotherapeutics, third ed. S. Chand & Company Ltd, pp. 501–507.

Barchiesi, F., Arzeni, D., Camiletti, V., Simonetti, O., Cellini, A., Offidani, A., Scalise, G., 2001. *In vitro* activity of posaconazole against clinical isolates of dermatophytes. J. Clin. Microbiol. 39 (11), 4208–4209.

Bartlett, M.S., Edlind, T.D., Durkin, M.M., Shaw, M.M., Queener, S.F., Smith, J.W., 1992. Antimicrotubule benzimidazoles inhibit *in vitro* growth of *Pneumocystis carinii*. Antimicrob. Agents Chemother. 36, 779–782.

Bartlett, M.S., Edlind, T.D., Lee, C.H., Dean, R., Queener, S.F., Shaw, M.M., Smith, J.W., 1994. Albendazole inhibits *Pneumocystis carinii* proliferation in inoculated immunosuppressed mice. Antimicrob. Agents Chemother. 38 (8), 1834–1837.

Bennet, J.E., 2006. Antimicrobial agents: antifungal agents. In: Goodman and Gilman's, The Pharmacological Basis of Therapeutics, eleventh ed. McGraw-Hill Publication, pp. 1029–1057.

Bin, Y., Haitao, J., Yongbin, C., Zhou, H., Zhu, J., Lu, J., Li, Y., Chen, J., Zheng, C., Jiang, Y., Liang, R., Tang, H., 2007. Synthesis and antifungal activities of novel 2-aminotetralin derivatives. J. Med. Chem. 50, 5293–5300.

Bloom, A.K., Ryan, E.T., 2013. Hunter's Tropical Medicine and Emerging Infectious Disease, Ninth Edition.

Bondock, S., Khalifa, W., Fadda, A.A., 2007. Synthesis and antimicrobial evaluation of some new thiazole, thiazolidinone and thiazoline derivatives starting from 1-chloro-3,4-dihydronaphthalene-2- carboxaldehyde. Eur. J. Med. Chem. 42, 948–954.

Bondock, S., Fadaly, W., Metwally, M.A., 2009. Enaminonitrile in heterocyclic synthesis: synthesis and antimicrobial evaluation of some new pyrazole, isoxazole and pyrimidine derivatives incorporating a benzothiazole moiety. Eur. J. Med. Chem. 44, 4813–4818.

Bondock, S., Fadaly, W., Metwally, M.A., 2010. Synthesis and antimicrobial activity of some new thiazole, thiophene and pyrazole derivatives containing benzothiazole moiety. Eur. J. Med. Chem. 45, 3692–3701.

Bondock, S., Naser, T., Ammar, Y.A., 2013. Synthesis of some new 2-(3 pyridyl) 4,5 disubstituted thiazoles as potent antimicrobial agents. Eur. J. Med. Chem. 62, 270–279.

Borelli, C., Schaller, M., Niewerth, M., Nocker, K., Baasner, B., Berg, D., Tiemann, R., Tietjen, K., Fugmann, B., Lang-Fugmann, S., Korting, H.C., 2008. Modes of action of the new arylguanidine abafungin beyond interference with ergosterol biosynthesis and *in vitro* activity against medically important fungi. Chemotherapy 54 (4), 245–259.

Botta, M., Corelli, F., Manetti, F., Mugnaini, C., Tafi, A., 2001. Rational design and synthesis of homochiral azole antifungal agents. Pure Appl. Chem. 73 (9), 1477–1485.

Camoutsis, C., Geronikaki, A., Ciric, A., Sokovic, M., Zoumpoulakis, P., Zervou, M., 2010. Sulfonamide-1,2,4-thiadiazole derivatives as antifungal and antibacterial agents: Synthesis, biological evaluation, lipophilicity and conformational studies. Chem. Pharm. Bull. 58 (2), 160–167.

Chawla, P., Singh, R., Saraf, S.K., 2012. Syntheses and evaluation of 2,5-disubstituted 4-thiazolidinone analogues as antimicrobial agents. Med. Chem. Res. 21, 2064–2071.

Chandrika, N.T., Shrestha, S.K., Ngo, H.X., Garneau-Tsodikova, S., 2016. Synthesis and investigation of novel benzimidazole derivatives as antifungal agents. Bioorg. Med. Chem. 24 (16), 3680–3686.

Chhillar, A.K., Arya, P., Mukherjee, C., Kumar, P., Yadav, Y., Sharma, A.K., Yadav, V., Gupta, J., Dabur, R., Jha, H.N., Watterson, A.C., Parmar, V.S., Prasad, A.K., Sharma, G.L., 2006. Microwave-assisted synthesis of antimicrobial dihydropyridines and tetrahydropyrimidin-2-ones: novel compounds against aspergillosis. Bioorg. Med. Chem. 14 (4), 973–981.

Como, J.A., Dismukes, W.E., 1994. Oral azole drugs as systemic antifungal therapy. N. Engl. J. Med. 330 (4), 263–272.

Cuenca-Estrella, M., Ruiz-Díez, B., Martínez-Suárez, J.V., Monzón, A., Rodríguez-Tudela, J.L., 1999. Comparative *in vitro* activity of voriconazole (UK-109,496) and six other antifungal agents against clinical isolates of *Scedosporium prolificans* and *Scedosporium apiospermum*. J. Antimicrob. Chemother. 43 (1), 149–151.

Cuevas, J.C., Martos, J.L., 1998. Synthesis of a monocyclic core of the antifungal sordarins. Tetrahedron Lett. 39, 8553–8556.

Cuevas, J.C., Lavandera, J.L., Martos, J.L., 1999. Design and synthesis of simplified sordaricin derivatives as inhibitors of fungal protein synthesis. Bioorg. Med. Chem. Lett. 9, 103–108.

Denning, D.W., 1997. Echinocandins and pneumocandins – a new antifungal class with a novel mode of action. J. Antimicrob. Chemother. 40, 611–614.

Denning, D.W., Ribaud, P., Milpied, N., Caillot, D., Herbrecht, R., Thiel, E., 2002. Efficacy and safety of voriconazole in the treatment of acute invasive aspergillosis. Clin. Infect. Dis. 34, 563–571.

Denning, D.W., 2002. Echinocandins: a new class of antifungal. J. Antimicrob. Chemother. 49 (6), 889–891.

Diekema, D.J., Messer, S.A., Hollis, R.J., Jones, R.N., Pfaller, M.A., 2003. Activities of caspofungin, itraconazole, posaconazole, ravuconazole, voriconazole, and amphotericin B against 448 recent clinical isolates of filamentous fungi. J. Clin. Microbiol. 41 (8), 3623–3626.

Dominguez, J.M., Kelly, V.A., Kinsman, O.S., Marriot, M.S., de las Heras, F.G., Martin, J.J., 1998. Sordarins: a new class of antifungals with selective inhibition of the protein synthesis elongation cycle in yeasts. Antimicrob. Agents Chemother. 42, 2274–2278.

Domínguez, J.M., Martín, J.J., 1998. Identification of elongation factor 2 as the essential protein targeted by sordarins in Candida albicans. Antimicrob. Agents Chemother. 42, 2279–2283.

Douglas, C.M., D'Ippolito, J.A., Shei, G.J., Meinz, M., Onishi, J., Marrinan, J.A., Li, W., Abruzzo, G.K., Flattery, A., Bartizal, K., Mitchell, A., Kurtz, M.B., 1997. Identification of the FKS1 gene of *Candida albicans* as the essential target of 1,3-beta-D-glucan synthase inhibitors. Antimicrob. Agents Chemother. 41 (11), 2471–2479.

Emami, S., Foroumadi, A., Falahati, M., Lotfali, E., Rajabalian, S., Ebrahimi, S.A., Farahyar, S., Shafiee, A., 2008. 2-Hydroxyphenacyl azoles and related azolium derivatives as antifungal agents. Bioorg. Med. Chem. Lett. 18 (1), 141–146.

Ezabadi, I.R., Camoutsis, C., Zoumpoulakis, P., Geronikaki, A., Sokovic, M., Glamoclija, J., Ciric, A., 2008. Sulfonamide-1,2,4-triazole derivatives as antifungal and antibacterial agents: synthesis, biological evaluation, lipophilicity, and conformational studies. Bioorg. Med. Chem. 16, 1150–1161.

Fotopoulou, T., Iliodromitis, E.K., Koufaki, M., Tsotinis, A., Zoga, A., Gizas, V., Pyriochou, A., Papapetropoulos, A., Andreadou, I., Kremastinos, D.T., 2008. Design and synthesis of nitrate esters of aromatic heterocyclic compounds as pharmacological preconditioning agents. Bioorg. Med. Chem. 16, 4523–4531.

Fringuelli, R., Schiaffella, F., Guarraci, A., Perito, S., Bistoni, F., Vecchiarelli, A., 2002. Anti-*Candida albicans* properties of novel benzoxazine analogues. Bioorg. Med. Chem. 10 (6), 1681–1686.

Galgiani, J.N., Catanzaro, A., Cloud, G.A., Higgs, J., Friedman, B.A., Larsen, R.A., Graybill, J.R., 1993. Fluconazole therapy for coccidioidal meningitis. The NIAID-mycoses study group. Ann. Intern. Med. 119 (1), 28–35.

Godefroi, E.F., Van Cutsem, J., Van der Eycken, C.A.M., Janssen, P.A.J., 1967. 1-(1-Indanyl)- and 1-(1-tetralyl)imidazole-5-carboxylate esters, a novel type of antifungal agents. J. Med. Chem. 10, 1160–1161.

Godefroi, E.F., Heeres, J., Van Cutsem, J., Janssen, P.A.J., 1969. Preparation and antimycotic properties of derivatives of 1-phenethylimidazole. J. Med. Chem. 12, 784–791.

Gupta, A., Singh, R., Sonar, P.K., Saraf, S.K., 2016. Novel 4-thiazolidinone derivatives as anti-infective agents: synthesis, characterization, and antimicrobial evaluation. Biochem. Res. Int. 2016. https://doi.org/10.1155/2016/8086762.

Haitao, J., Zhang, W., Zhang, M., Kudo, M., Aoyama, Y., Yoshida, Y., Sheng, C., Song, Y., Yang, S., Zhou, Y., Lu, J., Zhu, J., 2003. Structure-based de novo design, synthesis and biological evaluation of non-azole inhibitors specific for lanosterol 14r-demethylase of fungi. J. Med. Chem. 46, 474–485.

Haroun, M., Tratrat, C., Tsolaki, E., Geronikaki, A., 2016. Thiazole-Based thiazolidinones as potent antimicrobial agents. Design, synthesis and biological evaluation. Comb. Chem. High Throughput Screen. 19 (1), 51–57.

Haroun, M., Tratrat, C., Kositzi, K., Tsolaki, E., Petrou, A., Aldhubiaba, B., Attimarad, M., Harsha, S., Geronikaki, A., Venugopala, K.N., Elsewedya, H.S., Sokovic, M., Glamoclija, J., Ciric, A., 2018. New benzothiazole-based thiazolidinones as potent antimicrobial agents. Design, synthesis and biological evaluation. Curr. Top. Med. Chem. 18, 75–87.

Harvey, R.A., Champe, P.C., Finkel, R., 2009. Lippincott's Illustrated Reviews: Pharmacology, fourth edition. Lippincott Williams and Wilkins, New Delhi, pp. 407–415.

Hauser, D., Sigg, H.P., 1971. Isolierung und abbau von sordarin. Helv. Chim. Acta 54, 1178–1190.

Heeres, J., Backx, L.J.J., Van Cutsem, J., 1984. Antimycotic azoles. 7. Synthesis and antifungal properties of a series of novel triazol-3-ones. J. Med. Chem. 27, 894–900.

Herreros, E., Martinez, C.M., Almela, M.J., Marriott, M.S., Gomez De Las Heras, F., Gargallo-Viola, D., 1998. Sordarins: In vitro activities of new antifungal derivatives against pathogenic yeasts, *Pneumocystis carinii*, and filamentous fungi. Antimicrob. Agents Chemother. 42, 2863–2869.

Herreros, E., Almela, M.J., Lozano, S., Gomez De Las Heras, F., Gargallo-Viola, D., 2001. Antifungal activities and cytotoxicity studies of six new azasordarins. Antimicrob. Agents Chemother. 45, 3132–3139.

Herrling, S., Sous, H., Kruppe, W., Osterloh, G., Muckter, H., 1959. Experimentelle Untersuchungenuber eine neue gegen Pilze wirksame Verbindung. Arneim. Forsch. 9, 489–494.

Hitchcock, C.A., Dickinson, K., Brown, S.B., Evans, E.G.V., Adams, D.J., 1990. Interaction of azole antifungal antibiotics with cytochrome P-450-dependent 14R-sterol demethylase purified from *Candida albicans*. Biochem. J. 266, 475–480.

Hossain, M.A., Ghannoum, M.A., 2000. New investigational antifungal agents for treating invasive fungal infections. Expert Opin. Investig. Drugs 9, 1797–1813.

Jacobs, L.G., Skidmore, E.A., Freeman, K., Lipschultz, D., Fox, N., 1996. Oral fluconazole compared with bladder irrigation with amphotericin B for treatment of fungal urinary tract infections in elderly patients. Clin. Infect. Dis. 22 (1), 30–35

Kakushima, M., Nishio, M., Numata, K., Konishi, M., Oki, T., 1990. Effect of stereochemistry at the C-17 position on the antifungal activity of pradimicin A. J. Antibiotics 43, 1028–1030.

Kaneko, S., Arai, M., Uchida, T., Harasaki, T., Fukuoka, T., Konosu, T., 2002. Synthesis and evaluation of N-substituted 1,4-oxazepanyl sordaricins as selective fungal EF-2 inhibitors. Bioorg. Med. Chem. Lett. 12, 1705–1708.

Kankate, R.S., Parag, S.G., Deepak, P.B., 2015. Design, synthesis and antifungal evaluation of novel benzimidazole tertiary amine type of fluconazole analogues. Arab. J. Chem. 12, 2224–2235.

Kaplancikli, Z.A., Turan-Zitouni, G., Revial, G., Guven, K., 2004. Synthesis and study of antibacterial and antifungal activities of novel 2-[[(Benzoxazole/benzimidazole-2- yl)sulfanyl]acetylamino]thiazoles. Arch. Pharm. Res. 27 (11), 1081–1085.

Kartsev, V., Lichitsky, B., Geronikaki, A., Petrou, A., Smiljkovic, M., Kostic, M., Radanovic, O., Sokovic, M., 2018. Design, synthesis and antimicrobial activity of usnic acid derivatives. Med. Chem. Commun. 9, 870–882.

Kavitha, C.V., Basappa, Nanjunda Swamy, S., Mantelingu, K., Doreswamy, S., Sridhar, M.A., Shashidhara Prasad, J., Rangappa, K.S., 2006. Synthesis of new bioactive venlafaxine analogs: novel thiazolidin-4-ones as antimicrobials. Bioorg. Med. Chem. 14, 2290–2299.

Kawasaki, K., Masubuchi, M., Morikami, K., Sogabe, S., Aoyama, T., Ebiike, H., Niizuma, S., Hayase, M., Fujii, T., Sakata, K., Shindoh, H., Shiratori, Y., Aoki, Y., Ohtsuka, T., Shimma, N., 2003. Design and synthesis of novel benzofurans as a new class of antifungal agents targeting fungal N-myristoyltransferase. Part 3. Bioorg. Med. Chem. Lett. 13, 87–91.

Keating, G.M., Jarvis, B., 2001. Caspofungin. Drugs 61, 1121–1129.

Khabnadideh, S., Rezaei, Z., Khalafi-Nezhad, A., Pakshir, K., Roosta, A., Baratzadeh, Z., 2008. Design and synthesis of imidazole and benzimidazole derivatives as antifungal agents. Anti-Infect. Agents Med. Chem. 7, 215–218.

Khatkar, A., Nanda, A., Kumar, P., Narasimhan, B., 2017. Synthesis, antimicrobial evaluation and QSAR studies of p-coumaric acid derivatives. Arab. J. Chem. 10, S3804–S3815.

Kohno, S., Masaoka, T., Yamaguchi, H., 2001. A multicenter, open label clinical study of FK463 in patients with deep mycosis in Japan. In: Program and Abstracts of the 41st Interscience Conference on Antimicrobial Agents and Chemotherapy 2001. Chicago, IL. USA. 384. (Abstract J-834).

Kouatli, O., Geronikaki, A., Zoumpoulakis, P., Camoutsis, C., Sokovic, M., Ciric, A., Glamoclija, J., 2010. Novel 4-thiazolidinone derivatives as potential antifungal and antibacterial drugs. Bioorg. Med. Chem. 18, 426–432.

Koufaki, M., Fotopoulou, T., Iliodromitis, E.K., Bibli, S.I., Zoga, A., Kremastinos, D.T., Andreadou, I., 2012. Discovery of 6-[4-((6-nitroxyhexanoyl)piperazin-1-yl)]-9H-purine, as pharmacological post-conditioning agent. Bioorg. Med. Chem. 20, 5948–5956.

Kratky, M., Vinsova, J., Buchta, V., Horvati, K., Bösze, S., Stolaríková, J., 2010. New amino acid esters of salicylanilides active against MDR-TB and other microbes. Eur. J. Med. Chem. 45, 6106–6113.

Kratky, M., Vinsova, J., 2012. Antifungal activity of salicylanilides and their esters with 4-(trifluoromethyl)benzoic acid. Molecules 17, 9426–9442.

Kumar, A., Narasimhanb, B., Kumara, D., 2007. Synthesis, antimicrobial, and QSAR studies of substituted benzamides. Bioorg. Med. Chem. 15, 4113–4124.

Kumar, J.R., Muralidharan, S., Parasuraman, S.J., 2014. Antifungal agents: new approach for novel delivery systems. Pharm. Sci. Res. 6 (5), 229–235.

Lackner, T.E., Clissold, S.P., 1989. Bifonazole. A review of its antimicrobial activity and therapeutic use in superficial mycoses. Drugs 38 (2), 204–225.

Larsen, R.A., Bozzette, S.A., Jones, B.E., Leal, M.A., Forthal, D., Bauer, M., Tilles, J.G., McCutchan, J.A., Leedom, J.M., 1994. Fluconazole combined with flucytosine for treatment of cryptococcal meningitis in patients with AIDS. Clin. Infect. Dis. 19 (4), 741–745.

Leem, S.H., Park, J.E., Kim, I.S., Chae, J.Y., Sugino, A., Sunwoo, Y., 2003. The possible mechanism of action of ciclopirox olamine in the yeast *Saccharomyces cerevisiae*. Mol. Cell 15 (1), 55–61.

Liang, H., Schulé, A., Vors, J.P., Ciufolini, M.A., 2007. An avenue to the sordarin vore adaptable to analog synthesis. Org. Lett. 9 (21), 4119–4122.

Liang, H., 2008. Sordarin, an antifungal agent with a unique mode of action. Beilstein J. Org. Chem. 4, 31.

Liaras, K., Geronikaki, A., Glamočlija, J., Ćirić, A., Soković, M., 2011a. Thiazole-based chalcones as potent antimicrobial agents. Synthesis and biological evaluation. Bioorg. Med. Chem. 19 (10), 3135–3140.

Liaras, K., Geronikaki, A., Glamočlija, J., Ćirić, A., Soković, M., 2011b. Novel (E)-1-(4-methyl-2-(alkylamino)thiazol-5-yl)-3-arylprop-2-en-1-ones as potent antimicrobial agents. Bioorg. Med. Chem. 19 (24), 7349–7356.

Liaras, K., Geronikaki, A., Glamočlija, J., Ćirić, A., Soković, M., 2014. Thiazole-based aminopyrimidines and *N*-phenylpyrazolines as potent antimicrobial agents: synthesis and biological evaluation. Med. Chem. Commun. 5, 915–922.

Lima-Neto, G.R., Cavalcante, M.N., Srivastava, M.R., Mendonça Junior, J.F., Wanderley, G.A., Neves, P.R., Anjos, V.J., 2012. Synthesis of 1,2,3-triazole derivatives and in vitro antifungal evaluation on Candida strains. Molecules 17, 5882–5892.

Lino, C.I., Gonçalves de Souza, I., Borelli, B.M., Silvério Matos, T.T., Santos Teixeira, I.N., Ramos, J.P., Maria de Souza Fagundes, E., de Oliveira Fernandes, P., Maltarollo, V.G., Johann, S., de Oliveira, R.B., 2018. Synthesis, molecular modeling studies and evaluation of antifungal activity of a novel series of thiazole derivatives. Eur. J. Med. Chem. 151, 248–260.

Liu, M., Healy, M.D., Dougherty, B.A., Esposito, K.M., Maurice, T.C., Mazzucco, C.E., Bruccoleri, R.E., Davison, D.B., Frosco, M., Barrett, J.F., Wang, Y.K., 2006. Conserved fungal genes as potential targets for broad-spectrum antifungal drug discovery. Eukaryot. Cell 5, 638–649.

Mallikarjuna, B.P., Sastry, B.S., Suresh Kumar, G.V., Rajendraprasad, Y., Chandrashekar, S.M., Sathisha, K., 2009. Synthesis of new 4-isopropylthiazole hydrazide analogs and some derived clubbed triazole, oxadiazole ring systems – a novel class of potential antibacterial, antifungal and antitubercular agents. Eur. J. Med. Chem. 44, 4739–4746.

Masubuchi, M., Kawasaki, K., Ebiike, H., Ikeda, Y., Tsujii, S., Sogabe, S., Fujii, T., Sakata, K., Shiratori, Y., Aoki, Y., Ohtsuka, T., Shimma, N., 2001. Design and synthesis of novel benzofurans as a new class of antifungal agents targeting fungal N-myristoyltransferase. Part 1. Bioorg. Med. Chem. Lett. 11, 1833–1837.

Mazu, T.K., Bricker, B.A., Flores-Rozas, H., Ablordeppey, S.Y., 2016. The mechanistic targets of antifungal agents: an overview. Mini-Rev. Med. Chem. 16 (7), 555–578.

Meerpoel, L., Backx, L.J., Van der Veken, L.J., Heeres, J., Corens, D., De Groot, A., Odds, F.C., Van Gerven, F., Woestenborghs, F.A., Van Breda, A., Oris, M., van Dorsselaer, P., Willemsens, G.H., Vermuyten, K.J., Marichal, P., Vanden Bossche, H.F., Ausma, J., Borgers, M., 2005. Synthesis and in vitro and in vivo structure-activity relationships of novel antifungal triazoles for dermatology. J. Med. Chem. 48, 2184–2193.

Mendonça Junior, F.J.B., Lima-Neto, R.G., de Oliveira, T.B., Lima, M.C.A., Pitta, I.R., Galdino, S.L., da Cruz, R.M.D., de Araújo, R.S.A., Neves, R.P., 2011. Synthesis and evaluation of the antifungal activity of 2-(substituted-amino)-4,5-dialkyl-thiophene-3-carbonitrile derivatives. Lat. Am. J. Pharm. 30, 1492–1499.

Merck & Co. Inc, 2014. Noxafil (posaconazole) [package insert]. Merck & Co. Inc., Whitehouse Station, NJ.

Mishra, R., Tomar, I., Singhal, S., Jha, K., 2012. Facile synthesis of thiazolidinones bearing thiophene nucleus as antimicrobial agents. Der Pharma Chemica 4 (1), 489–496.

Mistry, B., Jauhari, S., 2013. Synthesis and in vitro antimicrobial and antitubercular evaluation of some quinoline-based azitidinone and thiazolidinone analogues. Med. Chem. Res. 22, 635–646.

Mittendorf, J., Kunisch, F., Matzke, M., Militzer, H.C., Schmidt, A., Schonfeld, W., 2003. Novel antifungal -amino acids: synthesis and activity against *Candida albicans*. Bioorg. Med. Chem. Lett. 13, 433–436.

National Committee For Clinical Laboratory Standards, 1997. Reference method for broth dilution antifungal susceptibility testing of yeasts. Approved standard M27-A, Villanova, PA, NCCLS.

Nett, J.E., Andes, D.R., 2016. Antifungal agents: spectrum of activity, pharmacology, and clinical Iindications. Infect. Dis. Clin. N. Am. 30 (1), 51–83.

Nussbaumer, P., Ryder, N.S., Stutz, A., 1991a. Allylamine antimycotics: recent trends in structure-activity relationships and syntheses. Pestic. Sci. 31, 437–455.

Nussbaumer, P., Petranyi, G., Stutz, A., 1991b. Synthesis and structure-activity relationships of benzo[*b*]thienylallylamine antimycotics. J. Med. Chem. 34, 65–73.

Nussbaumer, P., Dorfstiitter, G., Grassberger, M.A., Leitner, I., Meingassner, J.G., Thirring, K., Stiltz, A., 1993. Synthesis and structure-activity relationships of phenyl-substituted benzylamine antimycotics: a novel benzylbenzylamine antifungal agent for systemic treatment. J. Med. Chem. 36, 2115–2120.

Odds, F.C., Brown, A.J.P., Gow, N.A.R., 2003. Antifungal agents: Mechanisms of action. Trends Microbiol. 211, 272–279.

Odds, F., Ausma, J., Van Gerven, F., Woestenborghs, F., Meerpoel, L., Heeres, J., Vanden Bossche, H., Borgers, M., 2004. In vitro and in vivo activity of the novel azole antifungal agent R126638. Antimicrob. Agents Chemother. 48, 388–391.

Oki, T., Konishi, M., Tomatsu, K., Tomita, K., Saitoh, K., Tsunakawa, M., Nishio, M., Miyaki, T., Kawaguchi, H., 1988. Pradimicin, a novel class of potent antifungal antibiotics. J. Antibiotics 41, 1701–1704.

Okuyama, S., Kakushima, M., Kamachi, H., Konishi, M., Oki, T., 1993. Synthesis and antifungal activities of alanine-exchanged analogs of pradimicin A. J. Antibiot. (Tokyo) 46 (3), 500–506.

de Oliveira, T.B., 2010. Síntese, elucidação estrutural e avaliação biológica de novos derivados tiazolidínicos e cicloalquil-tiofênicos. M.Sc. Thesis, Graduate Program in Pharmaceutical Sciences. Federal University of Pernambuco, Recife, Brazil.

Ostrosky-Zeichner, L., Rex, J.H., Pappas, P.G., Hamill, R.J., Larsen, R.A., Horowitz, H.W., Powderly, W.G., Hyslop, N., Kauffman, C.A., Cleary, J., Mangino, J.E., Lee, J., 2003. Antifungal susceptibility survey of 2000 bloodstream Candida isolates in the United States. Antimicrob. Agents Chemother. 47 (10), 3149–3154.

Ozdemir, A., Turan-Zitouni, G., Kaplancikli, Z.A., Revial, G., Güven, K., 2007. Synthesis and antimicrobial activity of 1-(4-aryl-2-thiazolyl)-3-(2-thienyl)-5-aryl-2-pyrazoline derivatives. Eur. J. Med. Chem. 42 (3), 403–409.

Ozdemir, A., Turan-Zitouni, G., Kaplancıklı, Z.A., Revial, G., Demirci, F., İşcan, G., 2010. Preparation of some pyrazoline derivatives and evaluation of their antifungal activities. J. Enz. Inhib. Med. Chem. 25 (4), 565–571.

Pandeya, S.N., Sriram, D., Nath, G., De Clerq, E., 1999. Synthesis, antibacterial, antifungal and anti-HIV activities of Schiff and Mannich bases derived from isatin derivatives and N-[4-(49-chlorophenyl)thiazol-2-yl]. Eur. J. Pharm. Sci. 9, 25–31.

Paphitou, N.I., Ostrosky-Zeichner, L., Paetznick, V.L., Rodriguez, J.R., Chen, E., Rex, J.H., 2002. In vitro antifungal susceptibilities of *Trichosporon* species. Antimicrob. Agents Chemother. 46, 1144–1146.

Pappas, P.G., Kauffman, C.A., Andes, D., Benjamin Jr., D.K., Calandra, T.F., Edwards Jr., J.E., Filler, S.G., Fisher, J.F., Kullberg, B.J., Ostrosky-Zeichner, L., Reboli, A.C., Rex, J.H., Walsh, T.J., Sobel, J.D., 2009. Infectious Diseases Society of America. Clinical practice guidelines for the management of candidiasis: update by the Infectious Diseases Society of America. Clin. Infect. Dis. 48 (5), 503–535.

Patel, K.H., Mehta, A.G., 2006. Synthesis and antifungal activity of azetidinone and thiazolidinones derivatives of 2-Amino-6-(2- naphthalenyl)thiazolo[3,2-*d*]thiadiazole. J. Chem. 3, 267–273.

Patel, D., Kumari, P., Patel, N., 2010. Synthesis and characterization of some new thiazolidinones containing coumarin moiety and their antimicrobial study. Arch. Appl. Sci. Res. 2 (6), 68–75.

Patel, D., Kumari, P., Patel, N., 2012. Synthesis and biological evaluation of some thiazolidinones as antimicrobial agents. Eur. J. Med. Chem. 48, 354–362.

Petranyi, G., Meingassner, J.G., Mieth, H., 1987. Activity of terbinafine in experimental fungal infections of laboratory animals. Antimicrob. Agents Chemother. 31, 1558–1561.

Petrovic, M., Bonvin, D., Hofmann, H., Ebersold, M.M., 2018. Fungicidal PMMA-undecylenic acid composites. Int. J. Mol. Sci. 19, 184.

Pfaller, M.A., Messer, S.A., Hollis, R.J., Jones, R.N., 2002. Antifungal activities of posaconazole, ravuconazole, and voriconazole compared to those of itraconazole and amphotericin B against 239 clinical isolates of *Aspergillus* spp. & other filamentous fungi: report from SENTRY antimicrobial surveillance program, 2000. Antimicrob. Agents Chemother. 46, 1032–1037.

Pfaller, M.A., Diekema, D.J., Gibbs, D.L., Newell, V.A., Meis, J.F., Gould, I.M., Fu, W., Colombo, A.L., Rodriguez-Noriega, E., 2007. Results from the ARTEMIS DISK Global Antifungal Surveillance study, 1997 to 2005: an 8.5-year analysis of susceptibilities of Candida species and other yeast species to fluconazole and voriconazole determined by CLSI standardized disk diffusion testing. J. Clin. Microbiology 45 (6), 1735–1745.

Pereira de Sá, N., Lino, C.I., Fonseca, N.C., Borelli, B.M., Ramos, J.P., Souza-Fagundes, E.M., Rosa, C.A., Santos, D.A., Barbosa de Oliveira, R., Johann, S., 2015. Thiazole compounds with activity against *Cryptococcus gattii* and *Cryptococcus neoformans* in vitro. Eur. J. Med. Chem. 102, 233–242.

Pitta, E., Tsolaki, E., Geronikaki, A., Petrović, J., Glamočlija, J., Soković, M., Crespan, E., Maga, G., Bhunia, S.S., Saxena, A.K., 2015. 4-Thiazolidinone derivatives as potent antimicrobial agents: microwave-assisted synthesis, biological evaluation and docking studies. Med. Chem. Commun. 6, 319–326.

Plachta, D.A., Baranowski, A.M., Laudy, A.E., Kleps, J., 2007. Synthesis of 1-{4-[4-(adamant-1-yl)phenoxymethyl]-2-(4-bromophenyl)-1,3-dioxolan-2-ylmethyl}imidazole with expected antifungal and antibacterial activity. Acta Pol. Pharm. 64 (6), 535–541.

Polak, A., Schole, H.J., Wall, M., 1982. Combination therapy of experimental candidiasis, cryptococcosis and aspergillosis in mice. Chemotherapy 28 (6), 461–479.

Pour, M., Spulak, M., Balsanek, V., Kunes, J., Buchta, V., Waisser, K., 2000. Phenyl-5-methyl-2H,5H-furan-2-ones: tuning antifungal activity by varying substituents on the phenyl ring. Bioorg. Med. Chem. Lett. 10, 1893–1895.

Prabhaker, W., Saundane, A.R., 2014. Synthesis of novel indolyl-azetidinone and thiazolidinone derivatives as a potent antioxidant, antimicrobial and antitubercular agents. Der Pharma Chemica 6 (4), 70–79.

Prentice, A.G., Warnock, D.W., Johnson, S.A., Taylor, P.C., Oliver, D.A., 1995. Multiple dose pharmacokinetics of an oral solution of itraconazole in patients receiving chemotherapy for acute myeloid leukemia. J. Antimicrob. Chemother. 36 (4), 657–663.

Ram Shankar, U., Neelima, S., Sanjay, J., Nawal, K., Ramesh, C., Sudershan, K.A., 2004a. Optically active antifungal azoles: synthesis and antifungal activity of (2*R*,3*S*)-2-(2,4-difluorophenyl)-3-(5-{2-[4- aryl-piperazin-1-yl]-ethyl}-tetrazol-2-yl/1-yl)-1-[1,2,4]-triazol-1-yl-butan-2-ol. Bioorg. Med. Chem. 12, 2225–2238.

Ram Shankar, U., Sanjay, J., Neelima, S., Nawal, K., Ramesh, C., Sudershan, K.A., 2004b. Synthesis of novel substituted tetrazoles having antifungal activity. Eur. J. Med. Chem. 39, 579–592.

Ramachandran, R., Rani, M., Kabilan, S., 2009. Design, synthesis and biological evaluation of novel 2-[(2,4-diaryl-3-azabicyclo [3.3.1]nonan-9-ylidene)hydrazono]-1,3-thiazolidin-4-ones as a new class of antimicrobial agents. Bioorg. Med. Chem. Lett. 19, 2819–2823.

Ramalingan, C., Balasubramanian, S., Kabilan, S., Vasudevan, M., 2004. Synthesis and study of antibacterial and antifungal activities of novel 1-[2-(benzoxazol-2-yl)ethoxy]-2,6-diarylpiperidin-4-ones. Eur. J. Med. Chem. 39, 527–533.

Ratanatharathorn, V., Flynn, P., van Burik, J.A., McSweeney, P., Niederwieser, D., Kontoyiannis, D., 2002. Micafungin in combination with systemic antifungal agents in the treatment of refractory aspergillosis (RA) in bone marrow transplant (BMT) patients. Abstract 722. In: Proceedings of the 44th Annual American Society of Hematology Meeting 2002, Philadelphia, PA, USA.

Rezaei, Z., Khabnadideh, S., Pakshir, K., Hossaini, Z., Amiri, F., Assadpour, E., 2009. Design, synthesis, and antifungal activity of triazole and benzotriazole derivatives. Eur. J. Med. Chem. 44, 3064–3067.

Rezaei, Z., Khabnadideh, S., Zomorodian, K., Pakshir, K., Kashi, G., Sanagoei, N., Gholami, S., 2011. Design, synthesis and antifungal activity of some new imidazole and triazole derivatives. Arch. Pharm. Chem. Life Sci. 344, 658–665.

Ryder, N.S., 1987. Squalene epoxidase as the target of antifungal allylamines. Pestic. Sci. 21, 281–288.

Sabo, J.A., Abdel-Rahman, S.M., 2000. Voriconazole: a new triazole antifungal. Ann. Pharmacother. 34, 1032–1043.

Sawada, Y., Hatori, M., Yamamoto, H., Nishio, M., Miyaki, T., Oki, T., 1990. New antifungal antibiotics pradimicins FA-1 and FA-2: D-Serine analogs of pradimicins A and C. J. Antibiotics 43, 1223–1229.

Scotti, L., Tullius Scotti, M., De Oliveira Lima, E., Sobral da Silva, M., Do Carmo Alves de Lima, M., Da Rocha Pitta, I., Olímpio de Moura, R., Gonzaga Batista de Oliveira, J., Duarte da Cruz, R.M., Mendonça, B., Júnior, F.J., 2012. Experimental methodologies and evaluations of computer-aided drug design methodologies applied to a series of 2-aminothiophene derivatives with antifungal activities. Molecules 17, 2298–2315.

Seeliger, H.P.R., 1958. Pilzemmende Wirkung eines neuen Benzimidazol Derivatives. Mykosen 1, 162–171.

Serrano-Wu, M.H., St. Laurent, D.R., Mazzucco, C.E., Stickle, T.M., Barrett, J.F., Vyas, D.M., Balasubramanian, B.N., 2002a. Oxime derivatives of sordaricin as potent antifungal agents. Bioorg. Med. Chem. Lett. 12, 943–946.

Serrano-Wu, M.H., St. Laurent, D.R., Chen, Y., Huang, S., Lam, K.R., Matson, J.A., Mazzucco, C.E., Stickle, T.M., Tully, T.P., Wong, H.S., Vyas, D.M., Balasubramanian, B.N., 2002b. Sordarin oxazepine derivatives as potent antifungal agents. Bioorg. Med. Chem. Lett. 12, 2757–2760.

Shapiro, A.L., Rothman, S., 1945. Undecylenic acid in the treatment of dermatomycosis. Arch. Dermatol. Syphilol. 52, 166–171.

Sharkey, P.K., Graybill, J.R., Rinaldi, M.G., Stevens, D.A., Tucker, R.M., Peterie, J.D., Hoeprich, P.D., Greer, D.L., Frenkel, L., Counts, G.W., et al., 1990. Itraconazole treatment of phaeohyphomycosis. J. Am. Acad. Dermatol. 23 (3 Part 2), 577–586.

Sharma, D., Narasimhan, B., Kumar, P., Judge, V., Narang, R., De Clercq, E., Balzarini, J., 2009. Synthesis, antimicrobial and antiviral evaluation of substituted imidazole derivatives. Eur. J. Med. Chem. 44 (6), 2347–2353.

Shastry, M., Nielsen, J., Ku, T., Hsu, M.J., Liberator, P., Anderson, J., Schmatz, D., Justice, M.C., 2001. Species-specific inhibition of fungal protein synthesis by sordarin: identification of a sordarin-specificity region in eukaryotic elongation factor 2. Microbiology 147, 383–390.

Sheehan, D.J., Hitchcock, C.A., Sibley, C.M., 1999. Current and emerging azole antifungal agents. Clin. Microbiol. Rev. 12, 40–79.

Shen, Y., Sun, Y., Sang, Z., Sun, C., Dai, Y., Deng, Y., 2012. Synthesis, characterization, antibacterial and antifungal evaluation of novel monosaccharide esters. Molecules 17, 8661–8673.

Shrestha, S.K., Fosso, M.Y., Green, K.D., Garneau-Tsodikova, S., 2015. Amphiphilic tobramycin analogues as antibacterial and antifungal agents. Antimicrob. Agents Chemother. 59, 4861–4869.

Shrivastava, S.P., Nareshvarma, S., Rai, R., 2012. Synthesis and antimicrobial activity of new thiazolidinone derivatives with the use of γ-Ferrite catalyst. E. J. Chem. 9 (2), 825–831.

Sigg, H.P., Stoll, C., 1969. Antibiotic SL 2266. U.S. Patent 3,432,598.

Singh, O.M., Singh, S.J., Devi, M.B., Devi, L.N., Singh, N.I., Lee, S.G., 2008. Synthesis and in vitro evaluation of the antifungal activities of dihydropyrimidinones. Bioorg. Med. Chem. Lett. 18 (24), 6462–6467.

Skala, P., Machacek, M., Vejsova, M., Kubicova, L., Kunes, J., Waisser, K., 2009. Synthesis and antifungal evaluation of hydroxy-3-phenyl-2H-1,3-benzoxazine-2,4(3H)-diones and their thioanalogs. J. Heterocyclic Chem. 46, 873–880.

Smiljkovic, M., Matsoukas, M.T., Kritsi, E., Zelenko, U., Grdadolnik, S.G., Calhelha, R.C., Ferreira, I.C.F.R., Sankovic-Babic, S., Glamoclija, J., Fotopoulou, T., Koufaki, M., Zoumpoulakis, P., Sokovic, M., 2018. Nitrate esters of heteroaromatic compounds as *Candida albicans* CYP51 enzyme inhibitors. ChemMedChem 13 (3), 251–258.

Souza, B.C.C., 2011. Síntese e Avaliação das atividades Antifúngicas e Antitumorais de novos derivados 2-[(arilideno)amino]-cicloalquil[*b*]tiofenos-3-carbonitrila. M.Sc. Thesis, Graduate Program in Pharmaceutical Sciences. Federal University of Pernambuco, Recife, Brazil.

Sreedevi, M., Prasad, A.R.G., Spoorthy, Y.N., Ravindranath, L.R.K.R., 2013. Synthesis and antimicrobial evaluation of certain novel thiazoles. Adv. Pharm. Bull. 3 (1), 227–230.

Srinivas, A., Nagaraj, A., Reddy, C.S., 2008. Synthesis of some novel methylene-bis-pyrimidinyl-spiro-4-thiazolidinones as biologically potent agents. J. Heterocycl. Chem. 45, 1121–1125.

Stana, A., Vodnar, D.C., Tamaian, R., Pirnau, A., Vlase, L., Ionut, I., Oniga, O., Tiperciuc, B., 2017. Design, synthesis and antifungal activity evaluation of new thiazolin-4-ones as potential lanosterol 14α-demethylase inhibitors. Int. J. Mol. Sci. 18, 177.

Sun, Q.N., Fothergill, A.W., McCarthy, D.I., Rinaldi, M.G., Graybill, J.R., 2002. In vitro activities of posaconazole, itraconazole, voriconazole, amphotericin B, and fluconazole against 37 clinical isolates of Zygomycetes. Antimicrob. Agents Chemother. 46, 1581–1582.

Tandon, V.K., Singh, R.V., Yadav, D.B., 2004a. Synthesis and evaluation of novel 1,4-naphthoquinone derivatives as antiviral, antifungal and anticancer agents. Bioorg. Med. Chem. Lett. 14, 1079–1083.

Tandon, V.K., Singh, R.V., Yadav, D.B., 2004b. Synthesis and evaluation of novel 1,4-naphthoquinone derivatives as antiviral, antifungal and anticancer agents. Bioorg. Med. Chem. Lett. 14, 2901–2904.

Tandon, V.K., Yadav, D.B., Singh, R.V., Chaturvedi, A.K., Shukla, P.K., 2005. Synthesis and biological evaluation of novel (L)-a-amino acid methyl ester, heteroalkyl, and aryl substituted 1,4-naphthoquinone derivatives as antifungal and antibacterial agents. Bioorg. Med. Chem. Lett. 15, 5324–5328.

Tassel, D., Madoff, M.A., 1968. Treatment of Candida sepsis and Cryptococcus meningitis with 5-fluorocytosine. A new antifungal agent. JAMA 206 (4), 830–832.

Tratrat, C., Haroun, M., Xenikakis, I., Liaras, K., Tsolaki, E., Eleftheriou, P., Petrou, A., Aldhubiab, B., Attimarad, M., Venugopala, K.N., Harsha, S., Elsewedy, H.S., Geronikaki, A., Sokovic, M., 2019. Design, synthesis, evaluation of antimicrobial activity and docking studies of new thiazole-based chalcones. Curr. Top. Med. Chem. 19 (5), 356–375.

Tse, B., Balkovec, J.M., Blazey, C.M., Hsu, M.J., Nielsen, J., Schmatz, D., 1998. Alkyl side-chain derivatives of sordaricin as potent antifungal agents against yeast. Bioorg. Med. Chem. Lett. 8, 2269–2272.

Uchida, K., Yokota, N., Yamaguchi, H., 2001. In vitro antifungal activity of posaconazole against various pathogenic fungi. Int. J. Antimicrob. Agents 18, 167–172.

Upadhyay, A., Srivastava, S.K., Srivastava, S.D., 2010. Conventional and microwave assisted synthesis of some new N-[(4-oxo-2-substituted aryl-1,3-thiazolidine)-acetamidyl]-5-nitroindazoles and its antimicrobial activity. Eur. J. Med. Chem. 45 (9), 3541–3548.

Van Burik, J., Ratanatharathorn, V., Lipton, J., Miller, C., Bunin, N., Walsh, T.J., 2002. Randomized, double-blind trial of micafungin versus fluconazole for prophylaxis of invasive fungal infections in patients undergoing hematopoietic stem cell transplant. In: Program and Abstracts of the 42nd Interscience Conference on Antimicrobial Agents and Chemotherapy 2002. San Diego, CA, USA. 401. (Abstract M-1238).

Villanueva, A., Arathoon, E.G., Gotuzzo, E., Berman, R.S., DiNubile, M.J., Sable, C.A., 2001. A randomized double-blind study of caspofungin versus amphotericin for the treatment of candidal esophagitis. Clin. Infect. Dis. 33, 1529–1535.

Wagner, C., Graninger, W., Presterl, E., Joukhadar, C., 2006. The echinocandins: comparison of their pharmacokinetics, pharmacodynamics and clinical applications. Pharmacology 78 (4), 161–177.

Waller, D.G., Sampson, A.P., 2018. Medical Pharmacology and Therapeutics, Fifth Edition.

Watanabe, S., Takahashi, H., Nishikawa, T., Takiuchi, I., Higashi, N., Nishimoto, K., Kagawa, S., Yamaguchi, H., Ogawa, H., 2006. A comparative clinical study between 2 weeks of luliconazole 1% cream treatment and 4 weeks of bifonazole 1% cream treatment for tinea pedis. Mycoses 49 (3), 236–241.

Wu, J.J., Pang, K.R., Huang, D.B., Tyring, S.K., 2004. Therapy of systemic fungal infections. Dermatol. Ther. 17, 532–538.

Zhao, S., Zhang, X., Wei, P., Su, X., Zhao, L., Wu, M., Hao, C., Liu, C., Zhao, D., Cheng, M., 2017. Design, synthesis and evaluation of aromatic heterocyclic derivatives as potent antifungal agents. Eur. J. Med. Chem. 137, 96–107.

Zhu, J., Lu, J., Zhou, Y., Li, Y., Cheng, J., Zheng, C., 2006. Design, synthesis, and antifungal activities in vitro of novel tetrahydroisoquinoline compounds based on the structure of lanosterol 14ademethylase (CYP51) of fungi. Bioorg. Med. Chem. Lett. 16, 5285–5289.

Zhu, X.Y., Mardenborough, L.G., Li, S., Khan, A., Zhang, W., Fan, P., Jacob, M., Khan, S., Walker, L., Ablordeppey, S.Y., 2007. Synthesis and evaluation of isosteres of *N*-methyl indolo[3,2-b]-quinoline (cryptolepine) as new antiinfective agents. Bioorg. Med. Chem. 15 (2), 686–695.

Zoumpoulakis, P., Camoutsis, C., Pairas, G., Sokovic, M., Glamoclija, J., Potamitis, C., Pitsas, A., 2012. Synthesis of novel sulfonamide-1,2,4-triazoles, 1,3,4-thiadiazoles and 1,3,4-oxadiazoles, as potential antibacterial and antifungal agents. Biological evaluation and conformational analysis studies. Bioorg. Med. Chem. 20 (4), 1569–1583.

Chapter 6

Future trends

Marina Soković[a] and Konstantinos Liaras[b]
[a]*Mycological Laboratory, Institute for Biological Research "Siniša Stanković", University of Belgrade, National Institute of Republic of Serbia, Belgrade, Serbia,* [b]*Department of Pharmaceutical Chemistry, School of Pharmacy, Aristotle University of Thessaloniki, Thessaloniki, Greece*

Chapter outline

6.1 Connecting traditional medicines with modern technologies

The progress of Western medicine over the past 400 years was mainly based on the development of essential compounds that can be found in traditionally used plant extracts. The first major accomplishments in this field took place around 200 years ago when a series of alkaloids were purified and proved to be very useful in medicine. This process focused the scientific interest on the synthesis of pure compounds and their analogues, in order to improve the therapeutic potential of the natural products.

In the end of 20th century the paradigm in Western medicine was "single target—single compound" and consequently, mixtures of plant extracts in many cases were not regarded as equally scientific. Nevertheless, the slow-paced drug development process and the example of successful treatment of HIV infected patients with a combination of different medicines, questions the "single target—single compound" model (Verpoorte et al., 2005). Moreover, the treatment of elderly people with several different single compounds, each with diverse mode of action, is established as common practice, not always taking into account all the scientific data concerning their efficacy and safety.

Pharmacogenomic insights concerning pharmaceutical industry include the following statement "The vast majority of drugs—more than 90%—work only in 30%–50% of the people" (Connor, 2003). This conclusion is also compatible

Antifungal Compounds Discovery. https://doi.org/10.1016/B978-0-12-815824-1.00006-8

with this rethinking and the apprehension that traditional methods can be helpful, but at the same time should be adapted as knowledge and technology progress.

It is established that more or less, synthetic approaches always need a model to copy or improve upon. Moreover, "learning from nature, learning from our ancestors" is the lesson derived from around 50 years of thinking that biotechnology, combinatorial chemistry and synthesis would turn to be self-sufficient as methods for developing drugs (Verpoorte, 2016). The strong point of Western medicine is the continuous effort to develop more effective and safer treatments, having as basis the extensive research in the field, employment of new perspectives, and utilization of modern technologies. Systems biology approaches are being also enabled, integrating results with biochemical and physiological data. These approaches bring new opportunities for better understanding and use of medical knowledge of the past (Briggs, 2014; Leung et al., 2014; Wang and Xu, 2014; Williamson et al., 2015; Witt et al., 2015). It has been reported that combinatory use of different compounds (including prodrugs) that are able to exhibit synergistic and polyvalent effects, are involved in the therapeutic potential of medicinal plants (Sarris et al., 2011; Yoshitake et al., 2004). Moreover, combining pure compounds with known synergistic and poly-pharmacological profiles could provide an innovative way for more personalized treatments (Gertsch, 2011). Therefore, holistic, systematic approaches could turn to be very useful for the study and better comprehension of natural products that could ultimately lead to evidence-based applications, or even to novel lead compounds and drug candidates. Although it is recognized that an important part of Western medicine has its basis in traditional therapeutics, the significance of traditional medicine was undoubtedly recognized with the 2016 Nobel Prize for Medicine that was awarded to Youyou Tu, for the development of the antimalaria compound **artemisinin** from a Chinese traditional medicine (Efferth et al., 2015; Tu, 2011) and for the production of **ivermectin** from microbes to treat roundworms by biodiversity screening (Van Voorhis et al., 2015) (Fig. 6.1).

Traditional European and Mediterranean materia medica and medical thinking brought the principles of modern pharmacology, prolonging life expectancy and affecting epidemiology positively in prosperous societies. There are numerous neglected drugs and medical recipes that have been documented during the past two millennia and need to be studied thoroughly in the spirit of adopting a diachronic perspective of continuity and discontinuity, taking into consideration the epidemiological restrictions of the past. The aim is to create an ethnopharmacological database that would connect past and present therapeutic applications. In this way, new scientific ideas and hypotheses would turn out, that could be studied by using modern technologies and methods that include systems biology and network pharmacology.

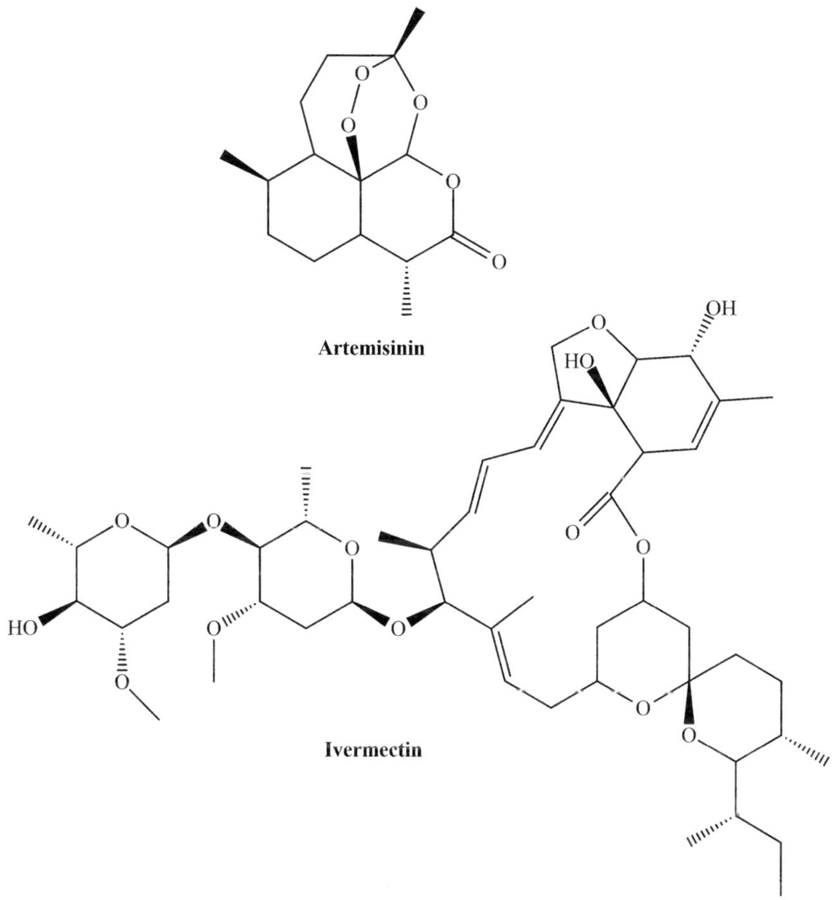

FIG. 6.1 Chemical structures of artemisinin and ivermectin.

6.2 Novel approaches to overcome antimicrobial resistance to existing drugs

The difficulties in antifungal therapy are associated with the limited amount of antifungal drugs, the high toxicity of existing antifungal agents, and finally the similarities between fungal and human cells. In addition, these pathogens have developed resistance mechanisms, increasing the demand for understanding these mechanisms.

Different approaches should be discussed and applied in research with the aim to develop new strategies, primarily for preventing resistance and ultimately to discover new and effective antifungal drugs. Researchers have to consider and consolidate their efforts in searching new approaches and drugs, to pay attention in the development of new formulations for antifungal agents, also take into account the synergistic effect of different substances, either natural or synthetic, and be able to combine different methodologies (in vitro and in vivo). They should also try to include nanoparticles as new types of carriers of antifungal drugs, possibly modify known chemical structures or traditional antifungals and also to consider improving the activity and pharmacokinetic parameters, or to activate the immune response and induce the production of antifungal molecules by the host. Another important factor is reproducibility and the correlation between in vitro and in vivo tests and results (Fig. 6.2).

One of the perspective future trends is to utilize the fact that many target proteins have orthologs in human cells. The differences in protein structure between host and pathogen give us the opportunity to identify molecules with selectivity for the fungal protein. The new approach should not replace the previous ones, but it could be a parallel and additive option. Furthermore, it could be used to expand the possibilities and pool of methods and tools available for application. After all, it is very important and necessary to bring preclinical compound testing to the real clinical trials and in the end into the market.

A series of approaches have been processed in order to address the problems that fungal pathogens cause. Some of the conventional strategies that were employed include the avoidance of infection sources, vector management, modification of cultural practices, improved protection and chemical control. While some of these strategies have been effective, there is always need for new ways for dealing with these issues. Moreover, a number of fungicides are being withdrawn

FIG. 6.2 Antifungal therapy in the future.

or they are not so often used due to their harmful effect in the environment and nature. Although this kind of data is useful for their economic aspects, such as the share of fungicides costs in the output or in the variable costs, the information is less useful as far as the environmental aspects are concerned, i.e., the amount of money spent on fungicides says little about the types and quantities used. Until now, synthetic antifungals are being used in most of the cases as the main weapon to confront this issue, making the need for innovative approaches to control diseases even more urgent. The introduction of new and innovative control methods for the diseases that are caused by fungi is of great importance especially for pathogens that can't get under control by employing conventional therapeutic ways. In the recent years, scientists have turned their interest on the use of natural products and their derivatives for a wide spectrum of applications such as therapeutic, environmental and industrial. With the emergence of resistance and constant need for safer and less expensive antifungal drugs, the screening of a series of other natural sources, i.e., micro-, macrofungi, and mosses that could prove to have bioactive secondary metabolites, turns into a necessity. Additionally, the discovery of new classes of compounds that act with different modes of action compared to existing drugs is always at the top of the scientific agenda. In order to serve these purposes, innovative techniques are constantly under development, testing already known molecules, compounds of natural origin, or combining the use of synthetic compounds with natural ones. It should be noted though, that to find successful therapeutic solutions, scientific efforts are focused also in elucidating, in vitro and in vivo, the underlying biology of fungi.

The majority of the results come from the investigation of antifungal activity of natural products and usage of conventional research methods for testing antifungal activity or introducing some alterations in the used assays for each group of compounds and microorganisms in order to receive more concrete data. As an example, it could be mentioned that for assaying compounds in small quantities, employment of microdilution and microatmosphere methods for testing volatile compounds could be the choice of preference. Moreover, bioautogaphic method on TLC plates is recommended for screening plant and fungi extracts as well as fractions. All these methodologies should be combined in order to obtain the best possible results.

Taking into account the fact that plants and fungi (micro- and macro-) can be a useful source of novel pharmacologically active compounds, special focus has been given on studying the biological potential of plant and fungal metabolites and their possible application in treating fungal diseases and protecting environment. They demonstrate great diversity and they possess excellent acquisition skills. This turns them into an ideal research material for the screening of natural products. As a result, it has been observed that as series of medicinal and therapeutic properties are attributed to active substances that derive from plants and fungi. A series of this kind of compounds are being studied because they trigger some procedures and mechanisms that are essential for plants and fungi, while other substances are assayed blindly for their antifungal potential.

Research effort should keep going on, in order to unveil as much important data as possible, including toxicity against animals or humans, modes of action, in vivo effects, positive and negative interactions with common antifungals. As a result, such scientific efforts would lead to production of compounds that could qualify for clinical trials.

Fungal infections have become very invasive and a global health problem. In spite of all the promising advances that are made in antifungal therapy, through the development of novel compounds and the new uses of existing antifungal agents, more efforts are needed that will expand the limited number of safe drugs.

We could summarize by saying that it is our strong belief that testing plants, fungi, and other natural sources of antifungal compounds is crucial, but the employment of appropriate methods of investigation is of great importance for the scientific groups working in this field.

References

Briggs, J.P., 2014. A global scientific challenge: learning the right lessons from ancient healing practices. The art and science of traditional medicine Part 1: TCM today—a case for integration. Science 346 (6216 Suppl), S7–S9.

Connor, S.G., 2003. Our Drugs Do Not Work on Most Patients. The Independent, Independent News and Media Limited, London, England. http://www.independent.co.uk/news/science/glaxo-chief-our-drugs-do-not-work-on-most-patients-575942.html.

Efferth, T., Zacchino, S., Georgiev, M.I., Liu, L., Wagner, H., Panossian, A., 2015. Nobel prize for artemisinin brings phytotherapy into the spotlight. Phytomedicine 22, A1–A3.

Gertsch, J., 2011. Botanical drugs, synergy, and network pharmacology: forth and back to intelligent mixtures. Planta Med. 77, 1086–1098.

Leung, E.L.-H., Wong, K.-W.V., Jang, Z.-H., Li, T., Liu, L., 2014. Integrated network based medicine: the role of traditional Chinese medicine in developing a new generation of medicine. The art and science of traditional medicine Part 1: TCM today—a case for integration. Science 346 (6216 Suppl), S16–S18.

Sarris, J., Panossian, A., Schweitzer, I., Stough, C., Scholey, A., 2011. Herbal medicine for depression, anxiety and insomnia: a review of psychopharmacology and clinical evidence. Eur. Neuropsychopharmacol. 21 (12), 841–860.

Tu, Y., 2011. The discovery of artemisinin (qinghaosu) and gifts from Chinese medicine. Nat. Med. 17, 1217–1220.

Van Voorhis, W.C., Hooft van Huijsduijnen, R., Wells, T.N., 2015. Profile of William C. Campbell, Satoshi Ōmura, and Youyou Tu, 2015 Nobel laureates in physiology or medicine. Proc. Natl. Acad. Sci. U. S. A. 112, 15773–15776.

Verpoorte, R., 2016. Ride the Wave of Success of Natural Products Research, The Easter & Qing-Ming Festival Edition. The March-April 2016 Newsletter of the GP-TCM Research Association.

Verpoorte, R., Choi, Y.H., Kim, H.K., 2005. Ethnopharmacology and systems biology: a perfect holistic match. J. Ethnopharmacol. 100 (1–2), 53–56.

Wang, Y., Xu, A., 2014. Zheng: a systems biology approach to diagnosis and treatments. The art and science of traditional medicine. Part 1: TCM today-a case for integration. Science 346 (6216), S13–S15.

Williamson, M., Chan, K., Xu, O., Nachtergael, A., Brunel, V., Zhang, L., Ouedraogo, M., Nortier, J., Qu, F., Shaw, D., Liu, X., Stévigny, C., Kahumba, J., Pelkonen, O., Duez, P., 2015. Evaluating the safety of herbal medicines: integrated toxicological approaches. The art and science of traditional medicine. Part 2: Multidisciplinary approaches for studying traditional medicine. Science 347 (6219 Suppl), S47–S49.

Witt, C.M., Liu, J., Robinson, N., 2015. Combining 'omics and comparative effectiveness research: evidence-based clinical research decision-making for Chinese medicine. The art and science of traditional medicine. Part 2: Multidisciplinary approaches for studying traditional medicine. Science 347 (6219 Suppl), S50–S51.

Yoshitake, T., Iizuka, R., Yoshitake, S., Weikop, P., Muller, W., Ogren, S., Kehr, J., 2004. *Hypericum perforatum* L. (St John's wort) preferentially increases extracellular dopamine levels in the rat prefrontal cortex. Br. J. Pharmacol. 142, 414–418.

Chapter 7

Concluding remarks

Marina Soković[a] and Konstantinos Liaras[b]
[a]Mycological Laboratory, Institute for Biological Research "Siniša Stanković", University of Belgrade, National Institute of Republic of Serbia, Belgrade, Serbia, [b]Department of Pharmaceutical Chemistry, School of Pharmacy, Aristotle University of Thessaloniki, Thessaloniki, Greece

The recent resurgence of interest in natural sources of bioactive compounds may, in part, be attributed to improved methods and instrumentation that has greatly reduced the time and effort required in natural product discovery programs. This interest is also associated with several other factors, including the realization that nature has already been selected as a source for compounds for very specific biological activities. Furthermore, the biological activities of relatively few of the known natural products have been fully characterized. All of the described and possible secondary metabolites have some kind of inherent activity, but in many cases these activities have not yet been discovered. Only the methods to detect their possible, perhaps until now unknown, type of activity remain to be developed.

The number of molecules in nature is unlimited, because of the innumerous possible combinations of the arrangement of the atoms. Starting from 1800 till 2015, it can be estimated that more than 101 million organic and inorganic molecules have been discovered and cataloged worldwide. The majority of these are produced by natural, synthetic, and bioengineering methods. It seems that this huge number is just a piece of the compounds' universe that still remains unexplored. From all of these molecules, hundreds of presently known bioactive metabolites originally were discovered as "inactive" and their activity was discovered only later, by investigation with new, more specific, and accurate methods. It should be mentioned that many of them were even reisolated (sometimes from different species). In our days, in fact, there would not be any reason to talk about "bioactive" or "inactive" secondary metabolites and treat them separately.

The study of natural products may lead to the discovery of novel target sites, and/or new classes of chemical compounds that can be developed for pathogen management. Structural diversity has been, and still remains, an invaluable source of lead compounds in developing novel products. A recent study on

Antifungal Compounds Discovery. https://doi.org/10.1016/B978-0-12-815824-1.00007-X

complementary synthetic and natural products confirmed that the latter generally have higher molecular weights than the former. Such diversity may be useful to synthetic and medicinal chemists in developing new classes of fungicides.

One indirect and important benefit of the chemical composition and structural characteristics of natural products (the absence of "unnatural" ring structures and the low content of heavy atoms) is that most of them are rapidly degraded in the natural environment to benign products. In addition to their structural features, natural products tend to have, in many cases, different target sites from conventional antifungals. Development of safe, effective, and inexpensive formulations should be in the focus of future studies of synergism, compatibility, and activity in different systems.

An essential part of obtaining natural fungicides is the development of bioassays. The development of fungal resistance to synthetic drugs poses a serious long-term trait to plant, animal, human health, and environmental requirements. Additionally, this could also have as a result significant financial issues. The advantage of natural products compared to synthetic is not only in their expected, in many cases, less or even nontoxic characteristics, but also their low cost. Growing of medicinal and aromatic plants and fungi is well established and, in general, economically justified. Identification and isolation of active components from plants and fungi is also well elaborated. Natural products with antifungal activity usually act in very small concentrations, especially essential oils, and for their application only small amounts are needed. The future of fungicide-management will probably be significantly influenced by research on natural products. Modern instrumentation has simplified the isolation and identification of lead compounds from which antifungals will be derived. The reviewed studies clearly demonstrate that natural products from plants and fungi present great potential for medical procedures and pharmaceutical industries. Nevertheless, finding suitable formulations for these products and their derivatives is an issue that needs to be addressed, especially for nontopical applications.

This book highlighted scientific validation of the results and findings reported by scientists all over the world and connection between the in vitro and partly in vivo bioactivity, as well as how to use that information to develop novel either natural and/or synthetic drugs that would be hopefully safe for humans. Despite the fact that around 49% of the new therapeutic drugs approved by the Drug Administration are naturally derived drugs, a great number of natural compounds do not meet the minimum of necessary requirements for human use, particularly for oral application. Apart from the biological efficacy of the molecule itself, in several cases this is a result of poorly designed, superficial basic research that does not provide consistent evidence to support the clinical testing of natural products and derived molecules. Microbial infection research is in general focused on basic aspects (microbial inhibitory effects), while testing of some pathological and physiological aspects of the infections (e.g., biofilm, antiquorum, and other mechanisms) has not been so extensively studied so far. In fact, in the majority of the studies the mechanisms of action of tested compounds remain completely unknown. The employment of only classical and traditional methodologies in testing of natural products' bioactivity instead of modern methods and new aspects, including identification of molecular targets in pathogens and human genome, bioinformatics, pharmacoengineering, high-throughput screening, appropriate statistical models, etc., could give only limited information. All these together as a complete strategy allow comprehensive characterization of the extract, fraction, or single compound prior to the clinical testing concerning bioactivity, local or systemic toxicity, pharmacokinetics, and pharmacodynamics.

Holistic approach will enable better understanding and recognition of the infection, disease, and the identification of potential candidates for drug development. It's equally important and required to recruit a multidisciplinary team of experts from biologists, chemical engineers, pharmacists, and physicians in the academic and industrial settings, not only to cover the technical aspects of sustainability, chemical synthesis, and pharmaceutical preparation, but also to comprehensively address the actual disease mechanisms and patient's therapeutic needs. The testing and improvement of existing antifungal agents, as well as the development of novel natural and synthetic agents, but also testing of their synergistic effects toward pathogenic fungi leads to better understanding of pathogens and the corresponding induced diseases. Using and combining knowledge from different fields of science could give as the desired results and new options for the treatment of fungal infections. In this way, we could possibly overcome the limitations of current treatments.

Until now, some progress has been achieved, but there is still a long and vague road in front of us in this struggle against pathogenic fungal species and we still need to make a solid and well-founded bridge, based on scientific research. The testing of already known as well as the development of novel natural and synthetic agents, alongside with their synergistic effects on certain types of fungal species, leads to better and deeper understanding of pathogens but also of the diseases that they induce. As it was stated previously, combining knowledge from different fields of science could, hopefully, give us the desired results and new options for the treatment of fungal infections, in our ongoing effort to overcome the limitations of conventional treatments.

Index

Note: Page numbers followed by *f* indicate figures and *t* indicate tables.

D

E

F

G

H

I

J

K

L

M

T

U

V

9780128158241